**Fragment-based Approaches
in Drug Discovery**

*Edited by
Wolfgang Jahnke and
Daniel A. Erlanson*

Methods and Principles in Medicinal Chemistry

Edited by R. Mannhold, H. Kubinyi, G. Folkers

Editorial Board
H.-D. Höltje, H. Timmerman, J. Vacca, H. van de Waterbeemd, T. Wieland

Previous Volumes of this Series:

R. Seifert, T. Wieland (eds.)

G-Protein Coupled Receptors as Drug Targets
Vol. 24

2005, ISBN 3-527-30819-9

O. Kappe, A. Stadler

Microwaves in Organic and Medicinal Chemistry
Vol. 25

2005, ISBN 3-527-31210-2

W. Bannwarth, B. Hinzen (eds.)

Combinatorial Chemistry
Vol. 26, 2nd Ed.

2005, ISBN 3-527-30693-5

G. Cruciani (ed.)

Molecular Interaction Fields
Vol. 27

2005, ISBN 3-527-31087-8

M. Hamacher, K. Marcus, K. Stühler, A. van Hall, B. Warscheid, H. E. Meyer (eds.)

Proteomics in Drug Design
Vol. 28

2005, ISBN 3-527-31226-9

D. Triggle, M. Gopalakrishnan, D. Rampe, W. Zheng (eds.)

Voltage-Gated Ion Channels as Drug Targets
Vol. 29

2006, ISBN 3-527-31258-7

D. Rognan (ed.)

Ligand Design for G Protein-coupled Receptors
Vol. 30

2006, ISBN 3-527-31284-6

D. A. Smith, H. van de Waterbeemd, D. K. Walker

Pharmacokinetics and Metabolism in Drug Research
Vol. 31, 2nd Ed.

2006, ISBN 3-527-31368-0

T. Langer, R. D. Hofmann (eds.)

Pharmacophores and Pharmacophore Searches
Vol. 32

2006, ISBN 3-527-31250-1

E.R. Francotte, W. Lindner (eds.)

Chirality in Drug Research
Vol. 33

2006, ISBN 3-527-31076-2

Fragment-based Approaches in Drug Discovery

Edited by
Wolfgang Jahnke and Daniel A. Erlanson

WILEY-VCH

WILEY-VCH Verlag GmbH & Co. KGaA

Series Editors

Prof. Dr. Raimund Mannhold
Molecular Drug Research Group
Heinrich-Heine-Universität
Universitätsstrasse 1
40225 Düsseldorf
Germany
Raimund.mannhold@uni-duesseldorf.de

Prof. Dr. Hugo Kubinyi
Donnersbergstrasse 9
67256 Weisenheim am Sand
Germany
kubinyi@t-online.de

Prof. Dr. Gerd Folkers
Collegium Helveticum
STW/ETH Zürich
8092 Zürich
Switzerland
folkers@collegium.ethz.ch

Volume Editors

Dr. Wolfgang Jahnke
Novartis Institutes for Biomedical Research
Novartis Pharma AG
Lichtstrasse
4002 Basel
Switzerland
wolfgang.jahnke@novartis.com

Dr. Daniel A. Erlanson
Sunesis Pharmaceuticals, Inc.
341 Oyster Point Boulevard
South San Francisco, CA 94080
USA
erlanson@sunesis.com

1st Reprint 2008

■ All books published by Wiley-VCH are carefully produced. Nevertheless, authors, editors, and publisher do not warrant the information contained in these books, including this book, to be free of errors. Readers are advised to keep in mind that statements, data, illustrations, procedural details or other items may inadvertently be inaccurate.

Library of Congress Card No.: applied for

British Library Cataloguing-in-Publication Data:
A catalogue record for this book is available from the British Library

Bibliographic information published by Die Deutsche Bibliothek
Die Deutsche Bibliothek lists this publication in the Deutsche Nationalbibliografie; detailed bibliographic data is available in the Internet at http://dnb.ddb.de.

© 2006 WILEY-VCH Verlag GmbH & Co. KGaA, Weinheim, Germany

All rights reserved (including those of translation into other languages). No part of this book may be reproduced in any form – by photoprinting, microfilm, or any other means – nor transmitted or translated into a machine language without written permission from the publishers. Registered names, trademarks, etc. used in this book, even when not specifically marked as such, are not to be considered unprotected by law.

Cover   SCHULZ Grafik Design, Fußgönheim
Composition   ProSatz Unger, Weinheim
Printing   Strauss GmbH, Mörlenbach
Bookbinding   Litges & Dopf GmbH, Heppenheim

Printed in the Federal Republic of Germany
Printed on acid-free paper

ISBN-13: 978-3-527-31291-7
ISBN-10: 3-527-31291-9

Contents

Preface *XV*

A Personal Foreword *XVII*

List of Contributors *XIX*

Part I: Concept and Theory

1 The Concept of Fragment-based Drug Discovery *3*
Daniel A. Erlanson and Wolfgang Jahnke
1.1 Introduction *3*
1.2 Starting Small: Key Features of Fragment-based Ligand Design *4*
1.2.1 FBS Samples Higher Chemical Diversity *4*
1.2.2 FBS Leads to Higher Hit Rates *5*
1.2.3 FBS Leads to Higher Ligand Efficiency *6*
1.3 Historical Development *6*
1.4 Scope and Overview of this Book *7*
References *9*

2 Multivalency in Ligand Design *11*
Vijay M. Krishnamurthy, Lara A. Estroff, and George M. Whitesides
2.1 Introduction and Overview *11*
2.2 Definitions of Terms *12*
2.3 Selection of Key Experimental Studies *16*
2.3.1 Trivalency in a Structurally Simple System *17*
2.3.2 Cooperativity (and the Role of Enthalpy) in the "Chelate Effect" *18*
2.3.3 Oligovalency in the Design of Inhibitors to Toxins *18*
2.3.4 Bivalency at Well Defined Surfaces (Self-assembled Monolayers, SAMs) *18*
2.3.5 Polyvalency at Surfaces of Viruses, Bacteria, and SAMs *18*
2.4 Theoretical Considerations in Multivalency *19*
2.4.1 Survey of Thermodynamics *19*
2.4.2 Additivity and Multivalency *19*

Fragment-based Approaches in Drug Discovery. Edited by W. Jahnke and D. A. Erlanson
Copyright © 2006 WILEY-VCH Verlag GmbH & Co. KGaA, Weinheim
ISBN: 3-527-31291-9

2.4.3	Avidity and Effective Concentration (C_{eff})	22
2.4.4	Cooperativity is Distinct from Multivalency	24
2.4.5	Conformational Entropy of the Linker between Ligands	25
2.4.6	Enthalpy/Entropy Compensation Reduces the Benefit of Multivalency	26
2.5	Representative Experimental Studies	26
2.5.1	Experimental Techniques Used to Examine Multivalent Systems	26
2.5.1.1	Isothermal Titration Calorimetry	26
2.5.1.2	Surface Plasmon Resonance Spectroscopy	27
2.5.1.3	Surface Assays Using Purified Components (Cell-free Assays)	27
2.5.1.4	Cell-based Surface Assays	27
2.5.2	Examination of Experimental Studies in the Context of Theory	28
2.5.2.1	Trivalency in Structurally Simple Systems	28
2.5.2.2	Cooperativity (and the Role of Enthalpy) in the "Chelate Effect"	29
2.5.2.3	Oligovalency in the Design of Inhibitors of Toxins	29
2.5.2.4	Bivalency in Solution and at Well Defined Surfaces (SAMs)	30
2.5.2.5	Polyvalency at Surfaces (Viruses, Bacteria, and SAMs)	31
2.6	Design Rules for Multivalent Ligands	32
2.6.1	When Will Multivalency Be a Successful Strategy to Design Tight-binding Ligands?	32
2.6.2	Choice of Scaffold for Multivalent Ligands	33
2.6.2.1	Scaffolds for Oligovalent Ligands	33
2.6.2.2	Scaffolds for Polyvalent Ligands	35
2.6.3	Choice of Linker for Multivalent Ligands	36
2.6.3.1	Rigid Linkers Represent a Simple Approach to Optimize Affinity	36
2.6.3.2	Flexible Linkers Represent an Alternative Approach to Rigid Linkers to Optimize Affinity	37
2.6.4	Strategy for the Synthesis of Multivalent Ligands	37
2.6.4.1	Polyvalent Ligands: Polymerization of Ligand Monomers	38
2.6.4.2	Polyvalent Ligands: Functionalization with Ligands after Polymerization	38
2.7	Extensions of Multivalency to Lead Discovery	39
2.7.1	Hetero-oligovalency Is a Broadly Applicable Concept in Ligand Design	39
2.7.2	Dendrimers Present Opportunities for Multivalent Presentation of Ligands	40
2.7.3	Bivalency in the Immune System	40
2.7.4	Polymers Could Be the Most Broadly Applicable Multivalent Ligands	42
2.8	Challenges and Unsolved Problems in Multivalency	44
2.9	Conclusions	44
	Acknowledgments	45
	References	45

3	**Entropic Consequences of Linking Ligands** 55
	Christopher W. Murray and Marcel L. Verdonk
3.1	Introduction 55
3.2	Rigid Body Barrier to Binding 55
3.2.1	Decomposition of Free Energy of Binding 55
3.2.2	Theoretical Treatment of the Rigid Body Barrier to Binding 56
3.3	Theoretical Treatment of Fragment Linking 57
3.4	Experimental Examples of Fragment Linking Suitable for Analysis 59
3.5	Estimate of Rigid Body Barrier to Binding 61
3.6	Discussion 62
3.7	Conclusions 64
	References 65

4	**Location of Binding Sites on Proteins by the Multiple Solvent Crystal Structure Method** 67
	Dagmar Ringe and Carla Mattos
4.1	Introduction 67
4.2	Solvent Mapping 68
4.3	Characterization of Protein–Ligand Binding Sites 69
4.4	Functional Characterization of Proteins 71
4.5	Experimental Methods for Locating the Binding Sites of Organic Probe Molecules 71
4.6	Structures of Elastase in Nonaqueous Solvents 72
4.7	Organic Solvent Binding Sites 73
4.8	Other Solvent Mapping Experiments 75
4.9	Binding of Water Molecules to the Surface of a Protein 78
4.10	Internal Waters 79
4.11	Surface Waters 80
4.12	Conservation of Water Binding Sites 81
4.13	General Properties of Solvent and Water Molecules on the Protein 82
4.14	Computational Methods 83
4.15	Conclusion 85
	Acknowledgments 85
	References 85

Part 2: Fragment Library Design and Computational Approaches

5	**Cheminformatics Approaches to Fragment-based Lead Discovery** 91
	Tudor I. Oprea and Jeffrey M. Blaney
5.1	Introduction 91
5.2	The Chemical Space of Small Molecules (Under 300 a.m.u.) 92
5.3	The Concept of Lead-likeness 94
5.4	The Fragment-based Approach in Lead Discovery 96
5.5	Literature-based Identification of Fragments: A Practical Example 99

5.6	Conclusions *107*
	Acknowledgments *109*
	References *109*

6	**Structural Fragments in Marketed Oral Drugs** *113*
	Michal Vieth and Miles Siegel
6.1	Introduction *113*
6.2	Historical Look at the Analysis of Structural Fragments of Drugs *113*
6.3	Methodology Used in this Analysis *115*
6.4	Analysis of Similarities of Different Drug Data Sets Based on the Fragment Frequencies *118*
6.5	Conclusions *123*
	Acknowledgments *124*
	References *124*

7	**Fragment Docking to Proteins with the Multi-copy Simultaneous Search Methodology** *125*
	Collin M. Stultz and Martin Karplus
7.1	Introduction *125*
7.2	The MCSS Method *125*
7.2.1	MCSS Minimizations *126*
7.2.2	Choice of Functional Groups *126*
7.2.3	Evaluating MCSS Minima *127*
7.3	MCSS in Practice: Functionality Maps of Endothiapepsin *132*
7.4	Comparison with GRID *135*
7.5	Comparison with Experiment *137*
7.6	Ligand Design with MCSS *138*
7.6.1	Designing Peptide-based Ligands to Ras *138*
7.6.2	Designing Non-peptide Based Ligands to Cytochrome P450 *140*
7.6.3	Designing Targeted Libraries with MCSS *140*
7.7	Protein Flexibility and MCSS *141*
7.8	Conclusion *143*
	Acknowledgments *144*
	References *144*

Part 3: Experimental Techniques and Applications

8	**NMR-guided Fragment Assembly** *149*
	Daniel S. Sem
8.1	Historical Developments Leading to NMR-based Fragment Assembly *149*
8.2	Theoretical Foundation for the Linking Effect *150*
8.3	NMR-based Identification of Fragments that Bind Proteins *152*
8.3.1	Fragment Library Design Considerations *152*

8.3.2	The "SHAPES" NMR Fragment Library	154
8.3.3	The "SAR by NMR" Fragment Library	156
8.3.4	Fragment-based Classification of protein Targets	160
8.4	NMR-based Screening for Fragment Binding	163
8.4.1	Ligand-based Methods	163
8.4.2	Protein-based Methods	165
8.4.3	High-throughput Screening: Traditional and TINS	167
8.5	NMR-guided Fragment Assembly	167
8.5.1	SAR by NMR	167
8.5.2	SHAPES	169
8.5.3	Second-site Binding Using Paramagnetic Probes	169
8.5.4	NMR-based Docking	170
8.6	Combinatorial NMR-based Fragment Assembly	171
8.6.1	NMR SOLVE	171
8.6.2	NMR ACE	173
8.7	Summary and Future Prospects	176
	References	177

9	**SAR by NMR: An Analysis of Potency Gains Realized Through Fragment-linking and Fragment-elaboration Strategies for Lead Generation**	**181**
	Philip J. Hajduk, Jeffrey R. Huth, and Chaohong Sun	
9.1	Introduction	181
9.2	SAR by NMR	182
9.3	Energetic Analysis of Fragment Linking Strategies	183
9.4	Fragment Elaboration	187
9.5	Energetic Analysis of Fragment Elaboration Strategies	188
9.6	Summary	190
	References	191

10	**Pyramid: An Integrated Platform for Fragment-based Drug Discovery**	**193**
	Thomas G. Davies, Rob L. M. van Montfort, Glyn Williams, and Harren Jhoti	
10.1	Introduction	193
10.2	The Pyramid Process	194
10.2.1	Introduction	194
10.2.2	Fragment Libraries	195
10.2.2.1	Overview	195
10.2.2.2	Physico-chemical Properties of Library Members	196
10.2.2.3	Drug Fragment Library	197
10.2.2.4	Privileged Fragment Library	197
10.2.2.5	Targeted Libraries and Virtual Screening	197
10.2.2.6	Quality Control of Libraries	201
10.2.3	Fragment Screening	201
10.2.4	X-ray Data Collection	202

10.2.5	Automation of Data Processing 203
10.2.6	Hits and Diversity of Interactions 205
10.2.6.1	Example 1: Compound 1 Binding to CDK2 205
10.2.6.2	Example 2: Compound 2 Binding to p38α 207
10.2.6.3	Example 3: Compound 3 Binding to Thrombin 207
10.3	Pyramid Evolution – Integration of Crystallography and NMR 207
10.3.1	NMR Screening Using Water-LOGSY 208
10.3.2	Complementarity of X-ray and NMR Screening 210
10.4	Conclusions 211
	Acknowledgments 211
	References 212

11 Fragment-based Lead Discovery and Optimization Using X-Ray Crystallography, Computational Chemistry, and High-throughput Organic Synthesis 215
Jeff Blaney, Vicki Nienaber, and Stephen K. Burley

11.1	Introduction 215
11.2	Overview of the SGX Structure-driven Fragment-based Lead Discovery Process 217
11.3	Fragment Library Design for Crystallographic Screening 218
11.3.1	Considerations for Selecting Fragments 218
11.3.2	SGX Fragment Screening Library Selection Criteria 219
11.3.3	SGX Fragment Screening Library Properties 220
11.3.4	SGX Fragment Screening Library Diversity: Theoretical and Experimental Analyses 220
11.4	Crystallographic Screening of the SGX Fragment Library 221
11.4.1	Overview of Crystallographic Screening 222
11.4.2	Obtaining the Initial Target Protein Structure 224
11.4.3	Enabling Targets for Crystallographic Screening 225
11.4.4	Fragment Library Screening at SGX-CAT 225
11.4.5	Analysis of Fragment Screening Results 226
11.4.6	Factor VIIa Case Study of SGX Fragment Library Screening 228
11.5	Complementary Biochemical Screening of the SGX Fragment Library 230
11.6	Importance of Combining Crystallographic and Biochemical Fragment Screening 232
11.7	Selecting Fragments Hits for Chemical Elaboration 233
11.8	Fragment Optimization 234
11.8.1	Spleen Tyrosine Kinase Case Study 234
11.8.2	Fragment Optimization Overview 240
11.8.3	Linear Library Optimization 241
11.8.4	Combinatorial Library Optimization 242
11.9	Discussion and Conclusions 243
11.10	Postscript: SGX Oncology Lead Generation Program 245
	References 245

12	**Synergistic Use of Protein Crystallography and Solution-phase NMR Spectroscopy in Structure-based Drug Design: Strategies and Tactics** *249*	
	Cele Abad-Zapatero, Geoffrey F. Stamper, and Vincent S. Stoll	
12.1	Introduction *249*	
12.2	Case 1: Human Protein Tyrosine Phosphatase *252*	
12.2.1	Designing and Synthesizing Dual-site Inhibitors *252*	
12.2.1.1	The Target *252*	
12.2.1.2	Initial Leads *252*	
12.2.1.3	Extension of the Initial Fragment *254*	
12.2.1.4	Discovery and Incorporation of the Second Fragment *256*	
12.2.1.5	The Search for Potency and Selectivity *257*	
12.2.2	Finding More "Drug-like" Molecules *258*	
12.2.2.1	Decreasing Polar Surface Area on Site 2 *258*	
12.2.2.2	Monoacid Replacements on Site 1 *258*	
12.2.2.3	Core Replacement *259*	
12.3	Case 2: MurF *261*	
12.3.1	Pre-filtering by Solution-phase NMR for Rapid Co-crystal Structure Determinations *261*	
12.3.1.1	The Target *261*	
12.3.1.2	Triage of Initial Leads *261*	
12.3.1.3	Solution-phase NMR as a Pre-filter for Co-crystallization Trials *262*	
12.4	Conclusion *263*	
	Acknowledgments *264*	
	References *264*	
13	**Ligand SAR Using Electrospray Ionization Mass Spectrometry** *267*	
	Richard H. Griffey and Eric E. Swayze	
13.1	Introduction *267*	
13.2	ESI-MS of Protein and RNA Targets *268*	
13.2.1	ESI-MS Data *268*	
13.2.2	Signal Abundances *268*	
13.3	Ligands Selected Using Affinity Chromatography *271*	
13.3.1	Antibiotics Binding Bacterial Cell Wall Peptides *272*	
13.3.2	Kinases and GPCRs *272*	
13.3.3	Src Homology 2 Domain Screening *273*	
13.3.4	Other Systems *274*	
13.4	Direct Observation of Ligand–Target Complexes *275*	
13.4.1	Observation of Enzyme–Ligand Transition State Complexes *276*	
13.4.2	Ligands Bound to Structured RNA *276*	
13.4.3	ESI-MS for Linking Low-affinity Ligands *277*	
13.5	Unique Features of ESI-MS Information for Designing Ligands *282*	
	References *282*	

14	**Tethering** *285*
	Daniel A. Erlanson, Marcus D. Ballinger, and James A. Wells
14.1	Introduction *285*
14.2	Energetics of Fragment Selection in Tethering *286*
14.3	Practical Considerations *289*
14.4	Finding Fragments *289*
14.4.1	Thymidylate Synthase: Proof of Principle *289*
14.4.2	Protein Tyrosine Phosphatase 1B: Finding Fragments in a Fragile, Narrow Site *292*
14.5	Linking Fragments *293*
14.5.1	Interleukin-2: Use of Tethering to Discover Small Molecules that Bind to a Protein–Protein Interface *293*
14.5.2	Caspase-3: Finding and Combining Fragments in One Step *296*
14.5.3	Caspase-1 *299*
14.6	Beyond Traditional Fragment Discovery *300*
14.6.1	Caspase-3: Use of Tethering to Identify and Probe an Allosteric Site *300*
14.6.2	GPCRs: Use of Tethering to Localize Hits and Confirm Proposed Binding Models *303*
14.7	Related Approaches *306*
14.7.1	Disulfide Formation *306*
14.7.2	Imine Formation *307*
14.7.3	Metal-mediated *307*
14.8	Conclusions *308*
	Acknowledgments *308*
	References *308*

Part 4: Emerging Technologies in Chemistry

15	**Click Chemistry for Drug Discovery** *313*
	Stefanie Röper and Hartmuth C. Kolb
15.1	Introduction *313*
15.2	Click Chemistry Reactions *314*
15.3	Click Chemistry in Drug Discovery *316*
15.3.1	Lead Discovery Libraries *316*
15.3.2	Natural Products Derivatives and the Search for New Antibiotics *317*
15.3.3	Synthesis of Neoglycoconjugates *320*
15.3.4	HIV Protease Inhibitors *321*
15.3.5	Synthesis of Fucosyltranferase Inhibitor *323*
15.3.6	Glycoarrays *324*
15.4	*In Situ* Click Chemistry *325*
15.4.1	Discovery of Highly Potent AChE by *In Situ* Click Chemistry *325*
15.5	Bioconjugation Through Click Chemistry *328*
15.5.1	Tagging of Live Organisms and Proteins *328*

15.5.2	Activity-based Protein Profiling	*330*
15.5.3	Labeling of DNA	*332*
15.5.4	Artificial Receptors	*333*
15.6	Conclusion	*334*
	References	*335*

16 Dynamic Combinatorial Diversity in Drug Discovery *341*
Matthias Hochgürtel and Jean-Marie Lehn

16.1	Introduction	*341*
16.2	Dynamic Combinatorial Chemistry – The Principle	*342*
16.3	Generation of Diversity: DCC Reactions and Building Blocks	*343*
16.4	DCC Methodologies	*346*
16.5	Application of DCC to Biological Systems	*347*
16.5.1	Enzymes as Targets	*349*
16.5.2	Receptor Proteins as Targets	*355*
16.5.3	Nucleotides as Targets	*357*
16.6	Summary and Outlook	*359*
	References	*361*

Index *365*

Preface

"The whole is more than the sum of its parts" is a phrase that is attributed to the Greek philosopher Aristotle (384–322 BC), who in his book Metaphysica compared a syllable with its individual letters. Applied to medicinal chemistry, it means that an active molecule is more than its parts and pieces. In this respect, we need not step down to the level of individual atoms, it is just enough to consider larger fragments of a protein ligand. More than 30 years ago, Green dissected the avidin ligand biotin into a methyl-substituted imidazolinone, hexanoic acid, and a sulfur atom. The binding affinities of the two organic fragments were several decades lower than the affinity of the original ligand or of desthio-biotin. This was a clear indication that the proper combination of fragments may lead to high-affinity ligands. However, the result corresponded to expectation and seemingly nobody concluded to go the other way, i.e. to combine fragments to a high-affinity ligand. Later, Page and Jencks formulated the "anchor principle": if two molecules A and B, both interacting with different pockets of the binding site of a protein, are combined to A–B, one molecule may be considered as a substituent of the other one. The entropy loss from freezing translational and rotational degrees of freedom can be attributed to one of the molecules; the other molecule contributes to affinity with its "intrinsic" free energy of binding, without an unfavorable entropy term. In this manner, a higher affinity of A-B is observed than expected from the affinities of the original molecules A and B. Of course, both fragments have to be combined in a relaxed manner and the final molecule has to fit the binding site without steric or other constraints.

In the following years, several authors confirmed the observation that the affinity of a ligand is more than the "sum of its fragments". Surprisingly, only ten years ago this principle was used for a systematic design of protein ligands from fragments by the "SAR by NMR" method, developed by Fesik and his group at Abbott Laboratories. Several other techniques followed, using protein crystallography, NMR, MS, cysteine tethering, or the dynamic assembly of ligands, to mention only some approaches. Within a short time, fragment-based design became a hot topic in drug discovery and in experimental techniques, as well as in cheminformatics and virtual screening. In addition to drug- and lead-likeness, desirable properties of fragment libraries were also defined and libraries for screening and

Fragment-based Approaches in Drug Discovery. Edited by W. Jahnke and D. A. Erlanson
Copyright © 2006 WILEY-VCH Verlag GmbH & Co. KGaA, Weinheim
ISBN: 3-527-31291-9

docking were generated, using these property definitions. It is clear that the combination of a limited number of fragments generates a multitude of different combinations, making this approach as attractive as combinatorial chemistry - without the need for producing millions of molecules.

This book is the very first to provide a comprehensive overview on this fascinating area, which opens a new perspective for the rational design of potential drugs. It is hoped that its content stimulates further research and strengthens the role of structure-based design in drug discovery.

We would like to express our gratitude to the editors Wolfgang Jahnke and Dan Erlanson, who assembled this book in short time, despite their hard work and responsibilities in their companies. We are also very grateful to all chapter authors, who accepted the invitation to contribute and to deliver their manuscripts in time. Of course, we appreciate the ongoing support of Renate Dötzer and Frank Weinreich, WILEY-VCH, for this book series and their valuable collaboration in this project.

May 2006

Raimund Mannhold, Düsseldorf
Hugo Kubinyi, Weisenheim am Sand
Gerd Folkers, Zürich

A Personal Foreword

The dilemma of rapidly emerging fields is that reviews are often outdated before they are printed. To make a contribution that would endure, we knew we had to go beyond a snapshot of the current state of fragment-based drug discovery and instead provide a framework for upcoming advances. To achieve this goal, we needed to convince leading scientists to take time from their busy schedules to write chapters. Fortunately, nearly all those we approached agreed; and what you hold in your hands is a virtual, although not comprehensive, "Who's Who" in fragment-based drug discovery. We are extremely grateful to all of our contributors for the quality of their chapters.

One striking feature of this book is that more than half of the chapters come from industry-based researchers, and even many of the academic contributors have close ties to industry. It has been alleged that the best science is done in academia; this book proves that this is not necessarily the case. Indeed, industrial researchers have largely pioneered fragment-based drug discovery strategies. Part of the reason may be that many of the techniques involved require expensive equipment and infrastructure as well as large collaborations between scientists from disparate disciplines - collaborations that would be difficult to set up outside industry. The multidisciplinary nature of fragment-based approaches shows in this volume: contributors include computational chemists, NMR spectroscopists, X-ray crystallographers, mass-spectrometrists, as well as organic and medicinal chemists.

Although fragment-based strategies for drug discovery have now pervaded laboratories across the world, the ultimate success of any drug discovery technology is measured in the quantity and quality of drugs that it produces. Fragment-based drug discovery has only been practical for the past decade, too soon to expect it to produce marketed drugs, but we believe these will come in time. Moreover, many of the techniques and concepts described in this book will alter drug discovery endeavors in subtle, tangential ways. Ideally, readers will be inspired to improve the methods described here, or even to develop fundamentally new methods for fragment-based drug discovery. But even if this book only changes the way medicinal chemists approach lead optimization, or persuades them to look more closely at weak but validated hits, it will have served its purpose.

March 2006

Wolfgang Jahnke, Basel
Daniel A. Erlanson, San Francisco

Fragment-based Approaches in Drug Discovery. Edited by W. Jahnke and D. A. Erlanson
Copyright © 2006 WILEY-VCH Verlag GmbH & Co. KGaA, Weinheim
ISBN: 3-527-31291-9

List of Contributors

Cele Abad-Zapatero
Abbott Laboratories
Department of Structural Biology
R46Y, AP-10
100 Abbott Park Road
Abbott Park, IL 60064–6098
USA

Marcus D. Ballinger
Sunesis Pharmaceuticals, Inc.
341 Oyster Point Boulevard
South San Francisco, CA 94080
USA

Jeffrey M. Blaney
SGX Pharmaceuticals, Inc.
10505 Roselle Street
San Diego, CA 92121
USA

Stephen K. Burley
SGX Pharmaceuticals, Inc.
10505 Roselle Street
San Diego, CA 92121
USA

Thomas G. Davies
Astex Therapeutics Ltd
436 Cambridge Science Park
Milton Road
Cambridge, CB4 0QA
UK

Daniel A. Erlanson
Sunesis Pharmaceuticals, Inc.
341 Oyster Point Boulevard
South San Francisco, CA 94080
USA

Lara A. Estroff
Department of Materials Science and Engineering
Cornell University
214 Bard Hall
Ithaca, NY 14853
USA

Richard H. Griffey
SAIC, San Diego
10260 Campus Point Drive
San Diego, CA 92121
USA

Philip J. Hajduk
Abbott Laboratories
R46Y, AP 10
100 Abbott Park Road
Abbott Park, IL 60064–6098
USA

Matthias Hochgürtel
Alantos Pharmaceuticals AG
Im Neuenheimer Feld 584
69120 Heidelberg
Germany

Fragment-based Approaches in Drug Discovery. Edited by W. Jahnke and D. A. Erlanson
Copyright © 2006 WILEY-VCH Verlag GmbH & Co. KGaA, Weinheim
ISBN: 3-527-31291-9

Jeffrey R. Huth
Abbott Laboratories
R46Y AP 10
Abbott Park, IL 60064
USA

Wolfgang Jahnke
Novartis Institutes for Biomedical
Research
Novartis Pharma AG
Lichtstrasse
4002 Basel
Switzerland

Harren Jhoti
Astex Therapeutics Ltd
436 Cambridge Science Park
Milton Road
Cambridge, CB4 0QA
UK

Martin Karplus
Department of Chemistry and
Chemical Biology
Harvard University
12 Oxford Street
Cambridge, MA 02138
USA

Hartmuth C. Kolb
Department of Molecular and
Medical Pharmacology
University of California
Los Angeles, CA 90095
USA

Vijay M. Krishnamurthy
Department of Chemistry and
Chemical Biology
Harvard University
12 Oxford Street
Cambridge, MA 02138
USA

Jean-Marie Lehn
Laboratoire de Chimie
Supramoléculaire
ISIS/ULP
8, allée Gaspard Monge
67083 Strasbourg Cedex
France

Carla Mattos
Department of Molecular and Structural
Biochemistry
North Carolina State University
Campus Box 7622
128 Polk Hall
Raleigh, NC 27695
USA

Rob L. M. van Montfort
Astex Therapeutics Ltd
436 Cambridge Science Park
Milton Road
Cambridge, CB4 0QA
UK

Christopher W. Murray
Astex Therapeutics Ltd
436 Cambridge Science Park
Milton Road
Cambridge, CB4 0QA
UK

Vicki Nienaber
SGX Pharmaceuticals, Inc.
10505 Roselle Street
San Diego, CA 92121
USA

Tudor I. Oprea
Division of Biocomputing
University of New Mexico
School of Medicine
MSC11 6145
Albuquerque, NM 87131
USA

Dagmar Ringe
Departments of Biochemistry and
Chemistry, and
Rosenstiel Basic Medical Sciences
Research Center
Brandeis University, MS 029
415 South Street
Waltham, MA 02454-9110
USA

Stefanie Röper
Department of Chemistry
The Scripps Research Institute
10550 North Torrey Pines Road
La Jolla, CA 92037
USA

Daniel S. Sem
Chemical Proteomics Facility at
Marquette
Department of Chemistry
Marquette University
535 North 14th Street
Milwaukee, WI 53233
USA

Miles Siegel
Discovery Chemistry Research,
DC 1920
Lilly Research Laboratories
Indianapolis, IN 46285
USA

Geoffrey F. Stamper
Abbott Laboratories
Department of Structural Biology
R46Y, AP-10
100 Abbott Park Road
Abbott Park, IL 60064–6098
USA

Vincent S. Stoll
Abbott Laboratories
Department of Structural Biology
R46Y, AP-10
100 Abbott Park Road
Abbott Park, IL 60064–6098
USA

Collin M. Stultz
Division of Health Sciences and
Technology, and
Department of Electrical Engineering
and Computer Science
Massachusetts Institute of Technology,
32-310
77 Massachusetts Ave.
Cambridge, MA 02139
USA

Chaohong Sun
Abbott Laboratories
R46Y AP 10
Abbott Park, IL 60064
USA

Eric E. Swayze
Isis Pharmaceuticals
1896 Rutherford Rd.
Carlsbad, CA 92008
USA

Marcel L. Verdonk
Astex Therapeutics Ltd
436 Cambridge Science Park
Milton Road
Cambridge, CB4 0QA
UK

Michal Vieth
Discovery Chemistry Research,
DC 1930
Lilly Research Laboratories
Indianapolis, IN 46285
USA

James A. Wells
Department of Pharmaceutical
Chemistry & Molecular and Cellular
Pharmacology
University of California, San Francisco
Box 2552, QB3
1700 4th Street
San Francisco, CA 94143-2552
USA

George M. Whitesides
Department of Chemistry and
Chemical Biology
Harvard University
12 Oxford Street
Cambridge, MA 02138
USA

Glyn Williams
Astex Therapeutics Ltd
436 Cambridge Science Park
Milton Road
Cambridge, CB4 0QA
UK

Part I: Concept and Theory

1
The Concept of Fragment-based Drug Discovery

Daniel A. Erlanson and Wolfgang Jahnke

1.1
Introduction

Fragment-based drug discovery builds drugs from small molecular pieces. It combines the empiricism of random screening with the rationality of structure-based design. Though the concept was articulated decades ago, the approach has become practical only recently.

Historically, most drugs have been discovered by one of two methods. The first of these was famously summarized by Nobel Laureate Sir James Black, who noted that the best way to find a new drug is to start with an existing one. Indeed, any successful drug spawns a surge of similar molecules, as illustrated by the number of chemically similar COX-2 inhibitors or HIV protease inhibitors on the market and in development. Though often disparaged as "me-too" or "patent-busting", such efforts are productive. The first drug to market is rarely the best; one need only consider the state of HIV medication now compared to a decade ago to appreciate this fact. Even the search for new drugs often begins with known starting points in the form of natural ligands such as substrates, co-factors or inhibitors.

For diseases and targets where no drug or other starting point exists, the second major route of drug discovery, random screening, is essential. This approach to drug discovery is perhaps the oldest and most venerable but requires serendipity. Indeed, it was a serendipitous observation of bacterial killing by fungus that led Alexander Fleming to the discovery of the natural product penicillin. Many highly successful drugs, from cyclosporine to paclitaxel, have been discovered by screening collections of compounds. With each medicinal chemistry program, more chemical compounds and their analogs are added to corporate screening libraries.

The invention of combinatorial chemistry in the late 1980s and early 1990s vastly expanded the number of compounds in chemical collections, just as the development of sophisticated automation equipment and miniaturization of biological assays led to the advent of high-throughput screening, or HTS. Today, most major pharmaceutical companies and many biotechnology companies have in-house collections of hundreds of thousands or even millions of molecules.

Fragment-based Approaches in Drug Discovery. Edited by W. Jahnke and D. A. Erlanson
Copyright © 2006 WILEY-VCH Verlag GmbH & Co. KGaA, Weinheim
ISBN: 3-527-31291-9

In parallel to HTS, more rational routes for drug discovery have been sought. Structure-based drug design attempts to design inhibitors *in silico* on the basis of the three-dimensional structure of the target protein.

Among the latest developments in drug discovery is a concept called fragment-based drug design, or fragment-based screening (FBS). In contrast to conventional HTS, where fully built, "drug-sized" chemical compounds are screened for activity, FBS identifies very small chemical structures ("fragments") that may only exhibit weak binding affinity. Follow-up strategies are then applied to increase affinity by elaborating these minimal binding elements. Fragment-based drug design thus attempts to build a ligand piece-by-piece, in a modular fashion. Structural information plays a central role in most follow-up strategies. Therefore, fragment-based drug design can be viewed as the synthesis of random screening and structure-based design.

1.2
Starting Small: Key Features of Fragment-based Ligand Design

Fragment-based screening promises to have a great impact on drug discovery because of several advantages, which are summarized in the following sections.

1.2.1
FBS Samples Higher Chemical Diversity

Typical chemical libraries used for HTS contain 10^5 to 10^6 individual compounds. Though a million-compound library sounds vast, it covers only a very small portion of "drug space", the theoretical set of possible small, drug-like molecules. In fact, a widely quoted estimate (actually a back-of-the-envelope calculation in a footnote in a review of structure-based drug design) places this number at 10^{63} molecules [1], a number beyond the comprehension of anyone except perhaps astrophysicists. A recent estimate of the total number of molecules available for screening in all the commercial and academic institutions on the Earth is around 100 million, or 10^8, so even a planet-wide screening effort would not even scratch the surface of diversity space [2]. This will never change in any meaningful way. To understand why, imagine assembling a library of 10^{63} molecules. Even if miniaturization advances to the point where we need only 1 pmol of each molecule (about 0.5 ng for a 500-Da molecule), this would still require gathering 5×10^{47} tons of material, roughly 26 orders of magnitude larger than the mass of our planet. Clearly, libraries screened in HTS will always explore only a tiny fraction of drug space.

The explored fraction of diversity space swells when working with smaller molecules ("fragments"), because there are fewer possible small molecules than possible large molecules. If we screen small molecular fragments, rather than drug-sized molecules, we can cover exponentially larger swaths of diversity space with much smaller collections of molecules. To illustrate, imagine two sets of compounds, each consisting of 1000 fragments. If we were to exhaustively make all binary combinations with a single asymmetric linker, this would yield (1000 molecules) × (1000 molecules) = 1 000 000 molecules to synthesize and screen, a daunting task. In contrast, if we could identify the five best fragments in each set and only combine and screen those, we would only need to synthesize and test [(1000 molecules) + (1000 molecules)] + [(5 molecules) × (5 molecules)] = 2025 molecules. This number is clearly much more manageable, and still covers the same chemical diversity space.

A first-principles computational analysis suggests that there are roughly 13.9×10^6 stable, synthetically feasible small molecules with a molecular weight less than or equal to 160 Da (44×10^6 once stereoisomers are considered, although the approach excludes compounds containing three- and four-membered rings and elements other than carbon, hydrogen, oxygen, nitrogen, and halogens) [3]. This is still a large number, but it is at least a comprehensible number, especially compared with 10^{63}. It shows that, with fragment-based screening, a higher (although still very small) proportion of diverse drug space can be covered. From a technical standpoint as well, focusing on these smaller fragments could simplify many aspects of the drug discovery process, from compound acquisition and synthesis through data management.

1.2.2
FBS Leads to Higher Hit Rates

Imagine a small fragment with high but imperfect complementarity to a target protein. Now imagine adding a methyl group at exactly the right spot to increase complementarity even further: rendering the fragment more complex in the right manner leads to slightly increased affinity to the target protein. But imagine adding the methyl group at any other spot, so that it protrudes from this fragment towards the receptor such that the modified fragment can no longer bind to the target: rendering the fragment more complex in the "wrong" manner ablates affinity for the receptor. Notably, there are many more ways to increase complexity in the "wrong" manner, and doing so often leads to a decrease of binding affinity by several orders of magnitude, whereas in the lucky case of increasing complexity in the "right" manner, binding is generally only enhanced by one or two orders of magnitude. This simple example makes sense intuitively, and a more rigorous theoretical analysis comes to the same conclusion: as molecules become more complex, additional chemical groups are much more likely to ablate binding than to enhance it [4]. The probability of binding (the "hit rate" in screening) thus decreases with increasing ligand complexity. Libraries containing smaller compounds ("fragments") are expected to exhibit higher hit rates, although the resulting affinities are generally weak and so require sensitive detection methods.

1.2.3
FBS Leads to Higher Ligand Efficiency

Screening drug-sized molecules is thought to favor ligands with several sub-optimal binding interactions, rather than those with a few optimal interactions. This is schematically shown in Fig. 1.1: the drug-sized molecule on the left side is identified by HTS since it binds to the receptor. However, none of the binding interactions are optimal, since establishing one optimal interaction would disrupt another interaction. All binding interactions are thus compromised and do not retain the full strength they would have without the molecular strain.

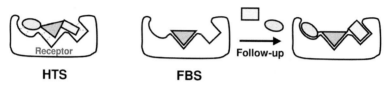

Fig. 1.1
Potential drawback of HTS (left), and principle and advantages of FBS (right): In HTS, fully assembled, "drug-sized" ligands are identified, but with multiple compromised, non-optimal binding interactions. In FBS, ligands for individual subpockets are identified separately, and show few but good binding interactions. Follow-up strategies such as fragment elaboration or linking are used to increase ligand affinity.

Relative to their molecular size, fragments can thus show more favorable binding energies than drug-sized molecules. The binding energy, normalized by the number of heavy atoms in the ligand, is referred to by the term ligand efficiency [5]. Smaller fragments can have higher ligand efficiency, leading to smaller drugs with better chances for favorable pharmacokinetics [6, 7]. This concept is also being applied to conventional HTS with the advent of "lead-like", instead of "drug-like," compound libraries [8].

1.3
Historical Development

The basic concept of fragment-based drug discovery was developed about 25 years ago by William Jencks, who wrote in 1981 that the affinities of whole molecules could be understood as a function of the affinities of separate parts:

"It can be useful to describe the Gibbs free energy changes for the binding to a protein of a molecule, A–B, and of its component parts, A and B, in terms of the "intrinsic binding energies" of A and B (ΔG_A^i and ΔG_B^i) and a "connection Gibbs energy" (ΔG^s) that is derived largely from changes in translational and rotational entropy [9]."

This paper received considerable attention and spawned academic interest in ligand–receptor interactions. Nakamura and Abeles studied the inhibition of HMG-CoA reductase by small molecule inhibitors, and found that those inhibitors could be understood as a linkage of two fragments, each binding to distinct sites on the enzyme [10]. The paper also provided a theoretical framework for understanding a very early study in which biotin was deconstructed into component fragments, which were found to bind weakly to streptavidin [11].

Despite these developments, Jencks' formulation did not immediately have an impact on drug discovery. The practical implementation of the theoretical promise required overcoming two difficult barriers: finding fragments and linking them.

Finding weakly binding fragments is inherently difficult because the binding interactions are easily disrupted. Moreover, there are hidden hazards in looking for weak binders: apparent hits could be "false positives". For example, compounds forming aggregates at low to mid-micromolar concentrations can inhibit biochemical functional assays without specifically interacting with the target [12–14].

But even if a true weak hit was identified, what could be done with it? Jencks provided an elegant theoretical framework for combining two weakly binding fragments into a single molecule, but enormous practical difficulties remained: First, one had to find a suitable fragment. One then had to find a second fragment; and that second fragment had to bind in close proximity to the first. Finally, one had to figure out how to link these two fragments while not distorting the binding mode of either. It was no wonder that the field remained largely theoretical and computational for well over a decade.

All this changed in 1996, when researchers at Abbott published the first practical demonstration of fragment-based drug discovery, called SAR-by-NMR [15]. In this approach, Shuker, Hajduk, Meadows, and Fesik used nuclear magnetic resonance (NMR) as a robust binding assay with sensitivity for weak interactions to identify fragments and to determine their binding sites, which revealed how to link the fragments. A flood of papers followed, initially from Abbott but soon from other research groups as well. Today, well over a dozen companies, from small biotechs to multinational pharmaceutical companies, as well as a large number of academic laboratories, are pursuing some form of fragment-based drug discovery. Many of the laboratories that have been leading the conceptual, theoretical and experimental development of FBS are represented in the following chapters. Below, we provide an overview of what to expect.

1.4
Scope and Overview of This Book

The text can be roughly divided into three sections: background and computational approaches are covered in Chapters 1–7, experimental methods and applications are covered in Chapters 8–14, and the last two Chapters, 15 and 16, describe related and emerging fields in chemistry that have the potential to inform and transform fragment-based drug discovery.

An intellectual sister to fragment-based drug discovery is the concept of multivalency, which is covered in a comprehensive review by Krishnamurthy, Estroff, and Whitesides in Chapter 2. This chapter focuses on the application of multivalency to the design of high-avidity ligands, and thus sets the conceptual framework for fragment-based drug design.

A special case of multivalency, hetero-oligovalency, is often encountered in fragment-based drug design when two ligands that bind to adjacent binding pockets are linked to form a high-affinity ligand. Murray and Verdonk discuss the entropic effects associated with this process in theory and experiment in Chapter 3. Basic concepts such as ligand efficiency and ligand hot spots are introduced in this chapter as well.

The identification and characterization of a protein binding site is key for ligand design. One way to achieve this is by mapping organic solvent binding sites in a protein, as in the multiple solvent crystal structure method, described in chapter 4 by Ringe & Mattos.

The quality of the fragment library is a crucial success factor for fragment-based drug design. The design of fragment libraries is the topic of Chapters 5 and 6. Oprea and Blaney outline the concepts of chemical space, lead-likeness, and fragment-like leads, using both *de novo* calculations and data mining in Chapter 5. This chapter illustrates the difficulty in trying to sample even a relatively small section of "fragment space", and provides a number of specific examples to direct researchers toward the most fruitful regions.

In Chapter 6, Vieth and Siegel "dissect" existing drugs into their component fragments and demonstrate that there are considerable differences between oral and injectable drugs. This clearly has implications for the choice of fragments in a screening collection.

In the years after Jencks' formulation of fragment-based drug discovery, before experimental methods became sufficiently sensitive to discover fragments, computational approaches were the dominant activity. In Chapter 7, Stultz and Karplus discuss the multi-copy simultaneous search (MCSS) program, one of the earliest and most powerful approaches to *in silico* fragment-based drug discovery, and its use for ligand design.

Chapter 8, by Sem, begins the "applied" section of the book and covers NMR-based approaches to fragment assembly. This chapter is a comprehensive review of the subject and covers the theory, various approaches, and specific examples.

The following chapter, by Hajduk, Huth, and Sun, discusses the original "SAR-by-NMR" approach and summarizes successes with this method. It also considers the success and requirements of fragment linking versus fragment elaboration from both a theoretical and an experimental vantage point. The authors draw important conclusions regarding the limits of these approaches as well as the sizes of libraries that should be assembled to maximize the likelihood of success in fragment elaboration approaches.

Chapters 10 and 11 focus on X-ray crystallography applications for fragment-based drug design. Davies, van Montfort, Williams, and Jhoti describe the process established at Astex by using NMR and X-ray for fragment screening, and X-ray crystallography as the basis for fragment optimization. Blaney, Nienaber, and Burley outline the crystallography-driven fragment-based ligand design at SGX and illustrate it with case studies.

Chapters 8–11 represent predominantly either NMR or X-ray crystallography approaches, while Chapter 12, by Abad-Zapatero, Stamper, and Stoll, describes the synergies that can result by marrying these techniques. The authors describe in two case studies how the combined use of these powerful biophysical techniques can rapidly advance medicinal chemistry programs.

Chapter 13 covers two somewhat unusual topics: use of mass spectrometry (MS) to identify fragments, and fragment-based discovery methods applied to an RNA target. Although the structural resolution of MS is necessarily less than that of either NMR or X-ray, Griffey and Swayze demonstrate that the technique can be powerfully applied to a challenging drug target.

A further use of MS, Tethering, is discussed in Chapter 14. This technique differs from other approaches in that it uses a transient covalent bond between the fragment and the target protein. The technology can be used to both identify fragments as well as to link two fragments. Erlanson, Ballinger, and Wells review the theory and practice of this method of fragment-based drug discovery.

Finally, the last two chapters touch on two fields that are themselves areas of vibrant research and that also overlap with fragment-based drug discovery. Chapter 15, by Röper and Kolb, introduces the powerful technique of Click chemistry, while Chapter 16, by Hochgürtel and Lehn, discusses dynamic combinatorial chemistry. Both of these approaches have been successfully applied to fragment-based drug discovery, albeit in a few limited studies. It is therefore fitting that we should end this volume here, at the intersection of emerging fields, where the opportunities are great, if only dimly perceived.

References

1 Bohacek, R.S., McMartin, C., Guida, W.C. **1996**, The art and practice of structure-based drug design: a molecular modeling perspective. *Med. Res. Rev.* 16, 3–50.

2 Hann, M.M., Oprea, T.I. **2004**, Pursuing the leadlikeness concept in pharmaceutical research. *Curr. Opin. Chem. Biol.* 8, 255–263.

3 Fink, T., Bruggesser, H., Reymond, J.L. **2005**, Virtual exploration of the small-molecule chemical universe below 160 Daltons. *Angew. Chem. Int. Ed. Engl.* 44, 1504–1508.

4 Hann, M.M., Leach, A.R., Harper, G. **2001**, Molecular complexity and its impact on the probability of finding leads for drug discovery. *J. Chem. Inf. Comput. Sci.* 41, 856–864.

5 Hopkins, A.L., Groom, C.R., Alex, A. **2004**, Ligand efficiency: a useful metric for lead selection. *Drug. Discov. Today* 9, 430–431.

6 Kuntz, I.D., Chen, K., Sharp, K.A., Kollman, P.A. **1999**, The maximal affinity of ligands. *Proc. Natl Acad. Sci. USA* 96, 9997–10002.

7 Lipinski, C.A., Lombardo, F., Dominy, B.W., Feeney, P.J. **1997**, Experimental and computational approaches to estimate solubility and permeability in drug discovery and development settings. *Adv. Drug Deliv. Rev.* 23, 3–25.

8 Teague, S.J., Davis, A.M., Leeson, P.D., Oprea, T. **1999**, The design of leadlike combinatorial libraries. *Angew. Chem. Int. Ed. Engl.* 38, 3743–3748.

9 Jencks, W.P. **1981**, On the attribution and additivity of binding energies. *Proc. Natl Acad. Sci. USA* 78, 4046–4050.
10 Nakamura, C.E., Abeles, R.H. **1985**, Mode of interaction of beta-hydroxy-beta-methylglutaryl coenzyme A reductase with strong binding inhibitors: compactin and related compounds. *Biochemistry* 24, 1364–1376.
11 Green, N.M. **1975**, Avidin. *Adv. Protein Chem.* 29, 85–133.
12 McGovern, S.L., Caselli, E., Grigorieff, N., Shoichet, B.K. **2002**, A common mechanism underlying promiscuous inhibitors from virtual and high-throughput screening. *J. Med. Chem.* 45, 1712–1722.
13 McGovern, S.L., Helfand, B.T., Feng, B., Shoichet, B.K. **2003**, A specific mechanism of nonspecific inhibition. *J. Med. Chem.* 46, 4265–4272.
14 Ryan, A.J., Gray, N.M., Lowe, P.N., Chung, C.W. **2003**, Effect of detergent on "promiscuous" inhibitors. *J. Med. Chem.* 46, 3448–3451.
15 Shuker, S.B., Hajduk, P.J., Meadows, R.P., Fesik, S.W. **1996**, Discovering high-affinity ligands for proteins: SAR by NMR. *Science* 274, 1531–1534.

2
Multivalency in Ligand Design

Vijay M. Krishnamurthy, Lara A. Estroff, and George M. Whitesides

2.1
Introduction and Overview

We define *multivalency* to be the operation of multiple molecular recognition events of the same kind occurring simultaneously between two entities (molecules, molecular aggregates, viruses, cells, surfaces; Fig. 2.1). We include in this definition hetero-multivalency (i.e., interactions in which two or more *different* types of molecular recognition events occur between the two entities), but do not discuss this type of system in detail (representative examples are sketched in Section 2.7.1; Fig. 2.1 b). Hetero-multivalency is probably a more broadly applicable concept than homo-multivalency, but one whose underlying principles are the same. Homo-multivalency is, however, simpler to understand than hetero-multivalency. We elaborate on our definition of multivalency in Section 2.2.1.

Multivalency is a design principle that can convert inhibitors with low affinity ($K_d^{\text{affinity}} \sim$ mM – µM) to ones with high avidity ($K_d^{\text{avidity}} \sim$ nM) and/or biological "activity" (gauged by some relevant parameter: for example, values of IC$_{50}$, the concentration of free ligand, often approximated as the total ligand, that reduces the experimental signal to 50% of its initial value) [1]. We discuss the distinction between "affinity" and "avidity" in Section 2.2.1, but emphasize here that high

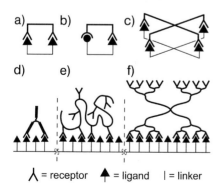

Fig. 2.1
Types of multivalent systems: (**a**) a bivalent ligand binding to a bivalent receptor ($N = M = i = 2$), (**b**) a heterobivalent ligand binding to a protein with two different kinds of binding sites ($N = M = i = 2$), (**c**) a tetravalent ligand binding to a tetravalent receptor (oligovalent) ($N = M = i = 4$), (**d**) a bivalent antibody binding to a surface ($N =$ unknown, $M = i = 2$), (**e**) a polymer binding to a surface ($N =$ unknown, $M = 8, i = 4$), and (**f**) a dendrimer binding to a surface ($N =$ unknown, $M = 16, i = 8$).

Fragment-based Approaches in Drug Discovery. Edited by W. Jahnke and D. A. Erlanson
Copyright © 2006 WILEY-VCH Verlag GmbH & Co. KGaA, Weinheim
ISBN: 3-527-31291-9

avidity does not necessarily require high affinity. Multivalency provides a strategy for designing ligands against defined oligovalent systems containing multiple, identical binding sites (e.g., antibodies, complement, multi-subunit toxins, such as those with an AB_5 structure). In addition, multivalent approaches can be effective in generating high-avidity ligands for proteins with multiple binding sites from low-affinity ligands. (One or more of these binding sites can, in some instances, even have low structural specificity and still be effective in increasing avidity: a hydrophobic patch adjacent to the active site of a monovalent enzyme can, for example, serve as the second binding site in a hetero-bivalent interaction.)

Multivalent ligands (primarily polyvalent ones, see Section 2.2.1) are especially well suited for inhibiting or augmenting interactions at biological surfaces (e.g., surfaces of bacteria, viruses, cells; Fig. 2.1d–f): they can prevent adhesion of these surfaces to other surfaces [2, 3] (e.g., by grafting polymers to the surfaces of viruses to prevent adhesion to cells) [4–10], or cluster cell-surface receptors to induce downstream effects [11–13]. By using polymers that display multiple kinds of ligands as side-chains, multivalency can convert a surface having one set of properties into one with different properties [14, 15].

This chapter sketches a theoretical analysis of multivalent systems that is intended to guide the application of multivalency to the *design* of high-avidity ligands for appropriate biological targets. The chapter has seven parts: (i) introduce the nomenclature of multivalency and the qualitative concepts that characterize it, (ii) present key experimental studies to provide examples of particularly tight-binding (high-avidity) multivalent ligands, (iii) present theoretical models that describe multivalent systems, (iv) explore representative multivalent ligands in the context of these models, (v) provide design rules for multivalent ligands, (vi) discuss extensions of multivalency to lead discovery, and (vii) mention challenges and unsolved problems for the application of multivalency in ligand design.

2.2
Definitions of Terms

In this chapter, we focus on representative systems that exhibit the principles of multivalency, and that suggest approaches to new types of drug leads α ligands (i.e., the design of ligands to interact tightly with multivalent receptors – primarily proteins). There are several thorough reviews on both experimental [16–18] and theoretical [1, 19, 20] aspects of multivalency. The book by Choi is an excellent compilation of experimental results; it also discusses potential targets (e.g., receptors on pathogens, multivalent proteins, etc.) for multivalent ligands [16].

We use the terms "ligand" and "receptor" to identify the individual components of multivalent species: the *receptor* is the component that accepts the *ligand*, using a declivity or pocket on its surface (Fig. 2.1). We refer to the entire molecule or cluster of molecules that present the receptors as "oligovalent receptors" (with an analogous relationship between ligand and "oligovalent ligands"). The *linker* is the tether between ligands in the oligovalent ligand. While there is also a linker between receptors, we do not discuss this linker in this chapter because it is usually not subject to manipulation by the investigator (it is part of a naturally occurring structure).

We make a distinction between the total number of ligands (and receptors) in an oligovalent species and the number of interactions between an oligovalent ligand and receptor. We denote the total number of ligands (the valency of the oligovalent ligand) with N and the total number of receptors with M. The number of receptor–ligand interactions between the two oligovalent species (in a particular state: Section 2.4.2) is i (Fig. 2.1).

We divide multivalency, on the basis of the number of interactions (i) between the multivalent receptor and ligand, into three categories: (a) bivalency (Fig. 2.1a, d), with two interactions between the different species ($i = 2$; for example, IgG and IgE binding to a cell surface; Fig. 2.1d), (b) oligovalency (Fig. 2.1c), with a discrete number (which we arbitrarily define as $i \leq 10$, a number that includes the pentameric immunoglobulin, IgM [21], and some interactions involving pentameric toxins) of interactions, and (c) polyvalency (Fig. 2.1e, f; usually associated with polymers), with a large number ($i > 10$) of interactions between the two species (the exact number of which is often unknown). While bivalency and oligovalency are distinct in mechanism, the thermodynamics of the two are sufficiently similar that we discuss them together in the remainder of the chapter and refer to them, collectively, as oligovalency. Polyvalency differs fundamentally from oligovalency both in terms of thermodynamics and mechanism; and we discuss it separately.

There are several different thermodynamic terms in the literature that have been used to describe the binding strength of multivalent ligands to multivalent receptors. The *affinity* of a monovalent interaction is defined by its dissociation constant (K_d^{affinity}); and this constant usually has units of concentration (typically, molarity; Fig. 2.2a). We define the *avidity* (K_d^{avidity}) of a multivalent interaction to be the dissociation constant ($K_{d,N}$) of the completely associated receptor–ligand complex with N receptor ligand–interactions ($i = N$) relative to the completely dissociated ($i = 0$) forms of the multivalent receptor and ligand, Eq. (1):

$$K_d^{\text{avidity}} \equiv K_{d,N} \tag{1}$$

An example with a bivalent ligand and receptor ($N = 2$) is shown in Fig. 2.2b. Kitov and Bundle [22] proposed an alternative definition of avidity ($K_d^{\text{avidity,KB}}$), given in Eq. (2):

$$K_d^{\text{avidity,KB}} \equiv (1/K_{d,N} + 1/K_{d,N-1} + \ldots + 1/K_{d,1})^{-1} \tag{2}$$

where $K_{d,N}$, $K_{d,N-1}$, and $K_{d,1}$ are the dissociation constants of receptor-ligand species with N receptor–ligand interactions ($i = N$), $N–1$ interactions ($i = N–1$), and one interaction ($i = 1$), respectively, to the completely dissociated multivalent receptor and multivalent ligand ($i = 0$; Fig. 2.2).

This definition of avidity [Eq. (2)] is more general than that in Eq. (1) because it takes into account all of the receptor–ligand species in solution and will prove par-

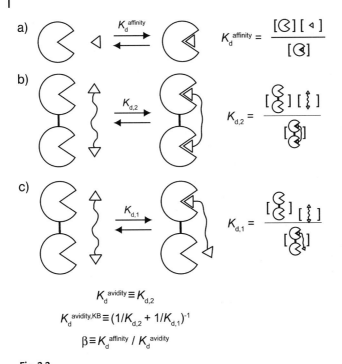

Fig. 2.2
Thermodynamic equilibria used for the definitions of affinity, avidity, and enhancement.
(a) A monovalent ligand binds a monovalent receptor with a dissociation constant of $K_d^{affinity}$.
(b) The oligovalent ligand (here, bivalent) binds a receptor of the same valency with a dissociation constant of $K_{d,2}$ for the equilibrium between the fully complexed receptor and free receptor and ligand. (c) The bivalent receptor can also bind the bivalent ligand with only one receptor–ligand interaction; the complex has a dissociation constant of $K_{d,1}$. We define the avidity ($K_d^{avidity}$) as $K_{d,2}$ (the equilibrium in b). Kitov and Bundle [22] defined the avidity ($K_d^{avidity,KB}$) to take into account all receptor–ligand complexes in solution. The enhancement (β) is the ratio of the affinity to the avidity. In this case, the enhancement contains a contribution from a statistical factor of 2.

ticularly useful when the binding of even one receptor of a multivalent receptor by a multivalent ligand is enough to achieve the desired response (e.g., in certain cases of inhibition of binding to a surface). Application of Eq. (2) is less convenient than of Eq. (1), because the distribution of receptor–ligand complexes must often be modeled and often cannot be measured directly. This difficulty prevents a simple analysis of the separated thermodynamic parameters of enthalpy and entropy (see Section 2.4.2) that is the focus of the approach that we present here. Further, the definition that we propose [Eq. (1)] will be more relevant than Eq. (2) for multivalent systems in which all of the receptors of the multivalent receptor must be bound to achieve the desired biological response (e.g., inhibiting multiple catalytic sites of a multimeric enzyme). Finally, in the design of ligands that are to serve as the starting points for possible drugs to bind oligovalent receptors with fewer than three receptors, the assumption that the fully associated complex

($i = N$) is the predominant receptor–ligand complex is justifiable because, in many of these cases, the multiple receptors are in close proximity to one another. We, thus, only discuss the avidity as defined by Eq. (2) in depth, but provide a general model for the free energy of binding of all species in solution in Section 2.4.2 before restricting our analysis to the fully associated state ($i = N$).

We previously defined the enhancement (β) due to multivalency as the ratio of $K_d^{avidity}$ to $K_d^{affinity}$ (Fig. 2.2); this parameter gives a measure of the benefit of having several ligands linked together [1]. This definition of enhancement does not take statistical factors (or influences of the topology of the ligand; Section 2.4.2) into account; and so the enhancement will increase with valency on a statistical basis alone (Fig. 2.2). This definition, however, enables the use of a general, empirical parameter for describing multivalency in many systems (and in those systems with an unknown number of interactions between species).

For polyvalent ligands (e.g., polymers), values of avidity and enhancement can be defined as above: $K_d^{avidity}$ would be the concentration of free polyvalent ligand that results in a measured signal of one-half of the maximum signal [22]. An enhancement calculated on this basis would have a significant contribution from statistical effects due to the large number of ligands (even without any "real" benefit of linking the ligands together). A theoretical understanding of the enhancement (Section 2.4.2) is complicated, because the number of ligands directly binding to receptors is sometimes unknown, and often known only approximately [4–8, 10]. The attachment of a polyvalent ligand to a surface, for example, may be very effective in changing the properties of that surface, even if only a few of the receptors on the surface are occupied [1, 4, 15] Another approach to defining the avidity of a polyvalent ligand is to calculate $K_d^{avidity}$ on a "per ligand" basis. An enhancement calculated using the "corrected" $K_d^{avidity}$ gives a qualitative correction for statistical factors, although again the unknown number of interacting ligands complicates rigorous analysis. Kitov and Bundle discussed the shortcomings of this approach based on "corrected" values of $K_d^{avidity}$ (Section 2.4.2), but there is no obviously better, general approach [22].

Frequently, estimates of avidity ($K_d^{avidity}$) are difficult to obtain experimentally and to interpret theoretically: polymers provide a prime example. Another term, related to avidity, but sometimes experimentally more tractable, is "biological activity" (sometimes measured, in familiar terms, as an IC_{50}; Section 2.5.1). Values of biological activity have the advantage that they are often measured under conditions that better simulate biological environments (e.g., at surfaces) than do the conditions usually used in the laboratory to measure values of $K_d^{avidity}$ (e.g., in solution). The two kinds of values (biological activities and $K_d^{avidity}$) are qualitatively similar (e.g., the qualitative ordering of ligands will usually be the same for both), but the exact values of biological activities are often assay-specific, while values of $K_d^{avidity}$ are less sensitive to the details of the assay. Kitov and Bundle have recently provided a theoretical framework for interpreting values of biological activities in certain contexts [22].

2.3
Selection of Key Experimental Studies

We have selected five representative examples that highlight key theoretical aspects of multivalent systems (Fig. 2.3). All of these systems are studied in aqueous solutions. We have not considered examples in organic solvents [23–28]. These examples may illustrate general principles of multivalency, but have no direct relevance to biochemistry or biology occurring in aqueous solutions. We do not mean the examples discussed here to be comprehensive, but merely representative of

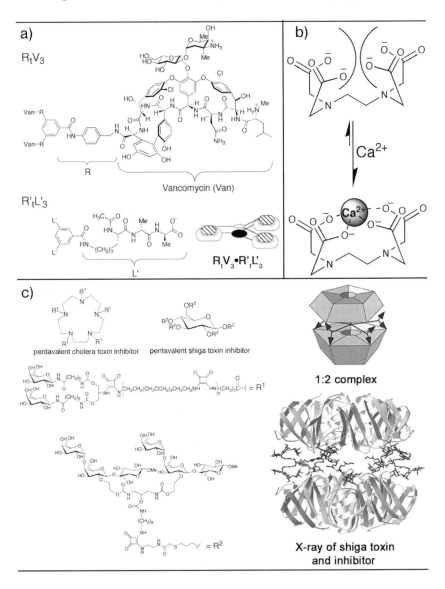

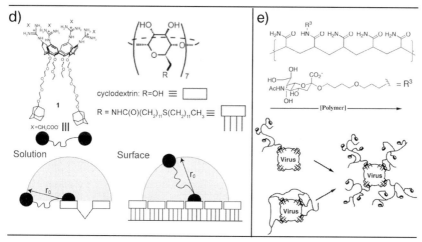

Fig. 2.3
Key experimental systems discussed in the text. (a) Trivalent Vancomycin-D-Ala-D-Ala [29, 30] (adapted with permission from [30]; copyright 2000, American Chemical Society). (b) EDTA/Ca^{2+} [31]. (c) AB$_5$ toxin inhibitors: Shiga-like [32] and cholera [33] toxin inhibitors (adapted with permission from [33]; copyright 2002, American Chemical Society). The ribbon diagram was generated using Swiss PDB Viewer and atomic coordinates: PDB 1QNU [32]. (d) Bivalent adamantane binding to bivalent cyclodextrin in solution and to SAMs displaying cyclodextrins [34] (adapted with permission from [34]; copyright 2004, American Chemical Society). (e) Polymers (polyacrylamide) displaying sialic acid for binding flu [8] (adapted with permission from [8]; copyright 1996, American Chemical Society).

studies that illustrate elements of multivalency. In Section 2.7 we present further examples and applications of multivalency to lead discovery. Choi reviewed experimental examples of multivalency in a more comprehensive fashion than we do here [16].

2.3.1
Trivalency in a Structurally Simple System

The trivalent vancomycin · D-Ala-D-Ala system is an example of the use of oligovalency to convert an interaction that is moderately strong for the monovalent species ($K_d^{\text{affinity}} \sim \mu M$) into one that is very strong – in fact the highest affinity known ($K_d^{\text{avidity}} \sim 10^{-17}$ M) for a ligand–receptor interaction involving species of low molecular weight (relative to those of proteins; Fig. 2.3a) [29, 30]. This system also illustrates the difference in mechanism (manifested in the kinetics) between tight-binding oligovalent systems and tight-binding monovalent ones: the trivalent vancomycin · D-Ala-D-Ala complex dissociates rapidly in the presence of competing monovalent ligand (equilibration of 3 μM of the trivalent complex vancomycin · D-Ala-D-Ala with 86 mM (~17 000 K_d^{affinity}) of diacetyl-L-Lys-D-Ala-D-Ala was complete in <45 min), while monovalent complexes of comparable affinity (e.g., biotin-avidin, $K_d^{\text{affinity}} \sim 10^{-15}$ M) dissociated slowly (half-life for dissociation

~ 200 days) [35]. This kinetic observation illustrates that the mechanism for dissociation (and association) of oligovalent species is qualitatively different than that for tight-binding monovalent species such as biotin–avidin.

2.3.2
Cooperativity (and the Role of Enthalpy) in the "Chelate Effect"

Toone and co-workers have determined the enthalpy and entropy (using isothermal titration calorimetry) for the chelation of calcium(II) ion by the tetravalent carboxylate ligand, ethylenediaminetetraacetic acid (EDTA; Fig. 2.3b) [31]. Their results indicate that enthalpy drives the binding (in contrast to the classic explanation of entropy as the origin of the "chelate effect") and that the unbound state (unassociated metal ion and ligand) is coulombically destabilized relative to the fully bound complex of metal and ligand. This work emphasizes the necessity of considering the thermodynamics of both bound *and* dissociated states.

2.3.3
Oligovalency in the Design of Inhibitors to Toxins

Toxins in the AB_5 family are ideal protein targets for multivalent ligands: the toxins have five-fold symmetry and bind to monovalent sugars with low affinity ($K_d^{\text{affinity}} \sim$ mM; Fig. 2.3c) [36, 37]. Kitov, Bundle, and co-workers designed decavalent ligands, with a glucose scaffold, that bound to Shiga-like toxin with an enhancement of 10^7 ($K_d^{\text{avidity}} \sim$ nM) [22, 32]. In related work, Hol, Fan, and co-workers targeted cholera toxin and heat-labile *E. coli* enterotoxin with both pentavalent and decavalent inhibitors with a pentacyclen scaffold; they also observed large enhancements in binding (10^5–10^6) [38–40].

2.3.4
Bivalency at Well Defined Surfaces (Self-assembled Monolayers, SAMs)

Reinhoudt, Huskens, and co-workers have studied the differences between bivalent binding in solution and at a structurally well defined surface [34, 41]. They examined the binding of a bivalent adamantane to a bivalent, soluble cyclodextrin, and to a self-assembled monolayer of cyclodextrin (Fig. 2.3d). The enhancement (β) in binding (relative to the monovalent interaction) was 10^3 greater at a surface than in solution. They rationalized this difference in enhancement by postulating that the effective concentration (Section 2.4.3) of cyclodextrin in the vicinity of an unbound adamantane was much larger at the surface than in solution.

2.3.5
Polyvalency at Surfaces of Viruses, Bacteria, and SAMs

Polymeric ligands are effective at binding to the surfaces of viruses, bacteria, and SAMs [2, 4–10, 13–15, 42–44]. For example, we have examined the ability of polymers presenting sialic acid to block the adhesion of influenza virus particles to erythrocytes (Fig. 2.3e) [4–10]. We observed large enhancements on a per sialic acid basis (~10^9; $IC_{50}^{avidity}$ ~ pM relative to $IC_{50}^{affinity}$ ~ mM for monovalent sialic acid).

2.4 Theoretical Considerations in Multivalency

2.4.1 Survey of Thermodynamics

To understand the origins of the large enhancements of multivalent systems introduced in Sections 2.3.1–2.3.6, we need a theoretical model for multivalency. We introduce one here with the primary goal of demonstrating how the thermodynamics of multivalent systems differ from those of their monovalent components. A theoretical understanding will facilitate the design of tight-binding multivalent ligands to a variety of multivalent receptor targets.

The key thermodynamic principle in multivalency is that, ideally, the enthalpy of binding of a multivalent system is more favorable than that of the monovalent species, with little or no corresponding increase in the unfavorable translational and rotational entropy of binding (Fig. 2.2) [1, 45]. The enthalpy of interaction of a multivalent ligand with a multivalent receptor is, in principle, additive (the enthalpy of interaction of three ligands with a receptor is three times the enthalpy of interaction of a single ligand), while the entropy of interaction is not (since the three ligands are connected, association of one ligand with one receptor increases the local concentration of the other ligands and receptors, and decreases the unfavorable entropic penalty "paid" to bring ligands and receptors together).

In practice, several factors substantially complicate this simple picture: (i) strain in the oligovalent ligand and/or receptor, if the geometry of the ligands and receptors do not match, (ii) loss in entropy caused by constraining the linker in the receptor–ligand complex, and (iii) energetically favorable or unfavorable interactions between the linker and the receptor (e.g., the surface of the protein). We proceed to address these three factors.

2.4.2 Additivity and Multivalency

We take a simple theoretical approach (Fig. 2.4): we treat the individual receptor–ligand interactions between the two multivalent entities as additive and then include terms from multivalency that affect the free energy of binding favorably (e.g., "chelate effect" [45, 46], favorable contacts of the linker, positive cooperativity, statistical effects) or unfavorably (e.g., unfavorable contacts of the linker, loss in conformational entropy of the linker, negative cooperativity). The model accounts for complexes of the multivalent ligand and receptor with different numbers of receptor–ligand interactions; for example, a trivalent ligand can be bound

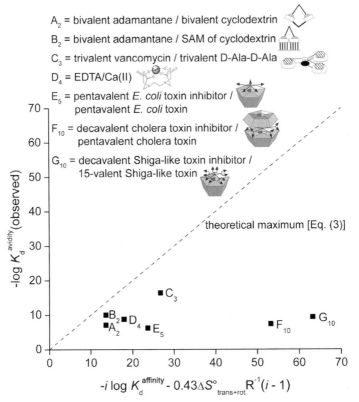

Fig. 2.4
Variation of $K_d^{avidity}$ with $K_d^{affinity}$ corrected for the valency of the interaction and for the translational and rotational entropic benefit of multivalency, for a number of experimental multivalent systems. The subscripts for the labeled points correspond to the number of interactions (i) between the two multivalent species. The abscissa results from the right-hand side of Eq. (3), assuming that the last four terms are negligible: interaction of the linker with the receptor, loss in conformational entropy of the linker, cooperativity between binding sites, and avidity entropy. The translational and rotational entropy was taken as -20 cal mol^{-1} K^{-1} ($-T\Delta S°_{trans+rot} \sim +6$ kcal mol^{-1}) for the loss in the modes of motion of the bound ligand relative to the receptor; this value is in the middle of those reported in the literature [17, 45]. The dashed line shows the theoretical maximum from multivalency [Eq. (3)]. All of the oligovalent ligands exhibit values of $K_d^{avidity}$ well below those expected from theory, perhaps due to large losses in the conformational entropy of the linker between ligands ($T\Delta S°_{conf}$). Further, values of the logarithm of $K_d^{avidity}$ cluster in a relatively narrow range and do not scale with the abscissa.

to a trivalent receptor by one ligand (two unbound ligands), by two ligands (one unbound ligand), or by three ligands (no unbound ligands; the fully-bound complex; Fig. 2.5). The model does not take into account aggregation of the multivalent receptor by the multivalent ligand (it assumes that the receptor is too dilute for this process to occur). We discuss aggregation of receptors in the context of multivalent toxins (Section 2.5.2.3) and antibodies (Section 2.7.3).

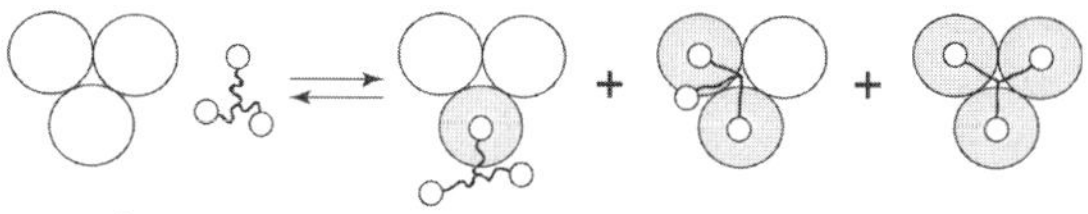

i	0	1	2	3
Ω_i	1	9	18	6
$-RT \ln(\Omega_i/\Omega_0)$ (kcal mol^{-1})	0	-1.3	-1.7	-1.1

Fig. 2.5
Graphical representation of Ω_i ("avidity entropy") as a function of the ligand–receptor interactions (i). The trivalent receptor is assumed to be a completely rigid assembly of three subunits; while the ligand has radial topology. The receptors are shaded gray when they are occupied by a ligand. For such a system, the number of degenerate states can be calculated from $\Omega_i = \dfrac{N!\,M!}{(N-i)!(M-i)!\,i!}$ where N and M are the valencies of the oligovalent ligand and receptor, respectively (here, $N = M = 3$); and i is the number of ligand–receptor interactions [22]. The number of states reaches a maximum value at an occupancy (i) of less than the valency of the ligand (N). The free energy of binding for this term (difference in energy between the fully unassociated state and state with i interactions) is virtually the same for all of the different occupancies.

Equation (3) gives the theoretical free energy of binding ($\Delta G°_N(i)$; Fig. 2.4) for a multivalent ligand with N ligands, as a function of the number (i) of ligands that are bound to receptors, where $i = 1 \ldots N$:

$$\Delta G°_N(i) = i\Delta H°_{\text{affinity}} - iT\Delta S°_{\text{affinity}} + (i-1)T\Delta S°_{\text{trans+rot}} + (i-1)\Delta H°_{\text{linker}}$$
$$- (i-1)T\Delta S°_{\text{conf}} + (i-1)\Delta G°_{\text{coop}} - RT\ln(\Omega_i/\Omega_0) \qquad (3)$$

The first term ($i\Delta H°_{\text{affinity}}$) is the product of the number of bound ligands (i) for the particular receptor–ligand state and the monovalent enthalpy of binding ($\Delta H°_{\text{affinity}}$). The second term ($-iT\Delta S°_{\text{affinity}}$) is analogous to the first but deals with the monovalent entropy of binding ($-T\Delta S°_{\text{affinity}}$). The sum of these first two terms ($i\Delta H°_{\text{affinity}} - iT\Delta S°_{\text{affinity}}$) is the free energy of binding that would be observed if the i receptor–ligand interactions occurred independently (i.e., if there were no effect of oligovalency). The third term [$(i-1)T\Delta S°_{\text{trans+rot}}$] deals with the classic "chelate effect" [45, 46] and conceptually is the center of multivalency: the unfavorable translational and rotational entropy of binding is approximately the same for the multivalent interaction as for the monovalent one. We note that the translational and rotational entropies show weak (logarithmic) dependences on the mass and dimensions of particles [1, 47]. This weak dependence justifies the assumption that the monovalent and multivalent interactions have equal translational and rotational entropies. The correlation between enthalpy of binding and translational and rotational entropy of binding (enthalpy/entropy compensation) [48–50] will, however, complicate this analysis. We discuss this effect in Section 2.4.6. The effect of the term $(i-1)T\Delta S°_{\text{trans+rot}}$ is to "add back" the unfavorable entropy of complexation for all receptor–ligand interactions but one [i.e., for $(i-1)$ interactions]. The fourth term [$(i-1)\Delta H°_{\text{linker}}$] deals with any enthalpic contacts (favorable or unfavorable) between the linker(s) and the oligovalent receptor. This simple analysis, summarized by Eq. (3), assumes that each receptor–linker interaction occurs with the same enthalpy (ΔH°_{linker}) [51, 52]. The fifth term ($-(i-1)T\Delta S^\circ_{conf}$) accounts for the loss in conformational entropy of the linker(s) between the ligands (and between protein subunits, if applicable; Section 2.4.5) [17, 46, 53]. The sixth term [$(i-1)\Delta G^\circ_{coop}$] addresses effects of cooperativity between binding sites (either between protein subunits or individual ligands). Such effects originate from the influence of one binding event on subsequent ones (Section 2.4.4) [54]. This term is usually near zero (i.e., the individual binding sites behave independently). The final term is a statistical factor dealing with the degeneracy (Ω_i) for each receptor–ligand complex (each with a different number of receptor–ligand interactions, i; Fig. 2.5). Kitov and Bundle discussed the importance of this term (which they define as the "avidity entropy") in multivalent systems [22].

Our assumption for *bivalent* and *trivalent* systems, in which the number of ligands is equal to the number of receptors ($N = M$), is that all ligands (of the oligovalent ligand) are bound to receptors (on the same oligovalent receptor) in the only bound state of the complex [i.e., $\Delta G^\circ_N(N) \ll \Delta G^\circ_N(N-1), \Delta G^\circ_N(N-2),..., \Delta G^\circ_N(1)$]. This assumption is not valid for higher *oligovalent* or *polyvalent* systems. Kitov and Bundle proposed a more general but less convenient definition than ours (see Section 2.2.1) [22]. Under the assumption above, Eq.(3) simplifies to Eq. (4) for the only populated state of a bivalent system ($i = 2$):

$$\Delta G^\circ_2(2) = 2\Delta H^\circ_{affinity} - 2T\Delta S^\circ_{affinity} + T\Delta S^\circ_{trans+rot} + \Delta H^\circ_{linker} - T\Delta S^\circ_{conf} + \Delta G^\circ_{coop} - RT\ln 2 = \Delta G^\circ_{avidity} \quad (4)$$

The free energy of binding for this only occupied state for the receptor-ligand complex is equal to the avidity free energy ($\Delta G^\circ_{avidity}$; Fig. 2.2).

Figure 2.4 shows a comparison of observed and theoretical $K_d^{avidity}$ as a function of the number of receptor–ligand interactions (i) and $K_d^{affinity}$ for a number of multivalent systems. For clarity in the plot, only the first three terms in Eq. (3) are included in the abscissa (Fig. 2.4). All of the experimental systems show avidities much lower than that expected from the theory (Section 2.8).

2.4.3
Avidity and Effective Concentration (C_{eff})

Several investigators [34, 41, 55, 56] have discussed avidity in terms of the effective concentration (C_{eff}) of an unbound ligand near an unbound receptor when the oligovalent receptor and ligand are bound at another site (Fig. 2.6). Lees and co-workers [56] estimated C_{eff} for the binding of an oligovalent ligand to a rigid oligovalent receptor from polymer theory [57, 58]. They assumed that the linker between the ligands was subject to random Gaussian chain statistics and that C_{eff} was proportional to the probability that the distance between the unbound ligand and the bound ligand (the "ends" of the polymer) was equal to the distance between receptors (Fig. 2.6c). The application of random-walk statistics requires

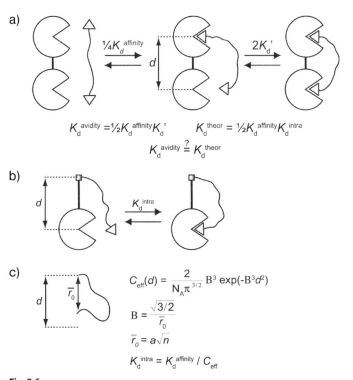

Fig. 2.6
Graphical representation illustrating a method to assess cooperativity in multivalent systems. (a) The binding of an oligovalent ligand to an oligovalent receptor (with two subunits which are a distance d apart) can be conceptualized as occurring in two steps: an intermolecular one and an intramolecular one. $K_d^{avidity}$ is obtained experimentally, while K_d^{theor} is obtained using Eq. (5) and requires independent measurements of $K_d^{affinity}$ and K_d^{intra}. A comparison of $K_d^{avidity}$ and K_d^{theor} allows the assessment of cooperativity between the two receptors.
(b) A hypothetical reaction to measure the dissociation constant for intramolecular binding (K_d^{intra}). The long, "unphysical" pole is shown to maintain the fixed distance (d) used for the separation between protein subunits in (a).
(c) A method to estimate K_d^{intra} using effective concentration (C_{eff}): the probability that the ends of a polymer (at variable distance of r_0) will be the appropriate distance (d) apart. The equations assume a random-coil model and show the dependence of the effective concentration on the distance between binding sites (d) and the number of subunits (n) in the linker. N_A is Avogadro's number. The value a is a constant that characterizes the stiffness of different chains and varies between 1.5 and 5.5 for most linkers [57]. Flexible linkers, such as oligo(ethylene glycol) are characterized by higher values of a [59].

that the linker behaves as a random coil polymer [57]. If the linker were too short, this condition would not be met, and the two "ends" would not rotate freely relative to one another [57]. Steric effects along the backbone between non-adjacent units of the linker (e.g., transannular strain in the linker in the fully bound receptor–ligand complex) would also invalidate the assumption of Gaussian chain statistics [58]. In addition, contacts between the linker and the oligovalent receptor could effectively shorten the linker and result in a higher value for C_{eff} than that predicted by the theory.

2.4.4
Cooperativity is Distinct from Multivalency

Cooperativity occurs when the binding of one ligand to a receptor affects the binding (that is, the dissociation constant) of additional ligands to the same receptor [54]. Cooperativity has been defined rigorously for the binding of multiple, *monovalent* ligands to a multivalent receptor (usually a protein). In biochemistry, a cooperative interaction occurs when the binding of one monovalent ligand to one site of a multivalent protein results in a change in the conformation of the protein (or stabilizes that alternative conformation) that extends to other binding sites; this conformational change affects the binding of subsequent ligands to the protein [60, 61]. In positive cooperativity, $K_{d,n}$, the dissociation constant for the binding of a ligand to a receptor already bound to n ligands, is smaller than $K_{d,n-1}$, the dissociation constant for the binding of a ligand to the receptor bound to $n-1$ ligands (after correction of both values for statistical factors). In negative cooperativity, $K_{d,n}$ is greater than $K_{d,n-1}$. In a system with no cooperativity (i.e., independent binding sites), $K_{d,n}$ is equal to $K_{d,n-1}$.

Ercolani discussed a method of assessing cooperativity in multivalent systems (Fig. 2.6a) by comparing $K_d^{avidity}$ with K_d^{theor} [54]. K_d^{theor} is defined in Eq. (5):

$$K_d^{theor} = c K_d^{affinity}(K_d^{intra})^{n-1} \tag{5}$$

Here, c is a statistical factor and K_d^{intra} describes a hypothetical intramolecular binding process (Fig. 2.6b) [54]. If $K_d^{avidity}$ is less than K_d^{theor}, positive cooperativity is at work. If $K_d^{avidity}$ is greater than K_d^{theor}, negative cooperativity is occurring in the system. If the two values ($K_d^{avidity}$ and K_d^{theor}) are equal, the system is non-cooperative. Several investigators proposed using a theoretical estimate of C_{eff} to estimate K_d^{intra} (Fig. 2.6c) [34, 41, 55, 56]; an assessment of cooperativity could then follow the method of Ercolani [Eq. (5)].

The design of a suitable intramolecular reference reaction to measure K_d^{intra} directly can be difficult in biological systems. Estimating C_{eff} for an oligovalent ligand faces the difficulties described in Section 2.4.3. Given these complications, we propose that a comparison of the experimentally observed enthalpy terms from Eq. (3) [ΔH_{theor}^{o}, defined in Eq. (6)] with the observed enthalpy of the multivalent interaction ($\Delta H_{avidity}^{o}$) can be used to assess cooperativity in multivalent systems:

$$\Delta H_{theor}^{o} = i\Delta H_{affinity}^{o} + (i-1)\Delta H_{linker}^{o} \tag{6}$$

where the terms are as defined for Eq. (3). If $\Delta H_{avidity}^{o}$ is more favorable than ΔH_{theor}^{o}, the system is positively cooperative. If $\Delta H_{avidity}^{o}$ is less favorable than ΔH_{theor}^{o}, the system is negatively cooperative (perhaps due to strain from a sub-optimal linker; Section 2.6.3). If $\Delta H^\circ_{\text{avidity}}$ is equal to $\Delta H^\circ_{\text{theor}}$, the system is non-cooperative (and the two binding events occur independently). This approach avoids the difficulties with the approach based on K_d^{intra} [Eq. (5); which uses only free energies of binding], because it removes the entropy of binding [the primary contributor to multivalency; Eq. (3)] from the examination of the multivalent system. The application of Eq. (6) requires accurate measurements of the enthalpy and entropy of binding for complexation; and the best technique for such measurements is isothermal titration calorimetry (Section 2.5.1). To date, there have been no reported *multivalent biological* examples (where both the ligand and receptor are multivalent) in which cooperativity contributes to avidity.

2.4.5
Conformational Entropy of the Linker between Ligands

We [1, 53] and others [17, 32, 38, 40, 46] suggested that the linker for the oligovalent ligand should be rigid in order to minimize the loss in conformational entropy [$T\Delta S^\circ_{\text{conf}}$ in Eqs. (3), (4)] that occurs upon binding the receptor (Fig. 2.7). The intuitive argument is that flexible linkers have more entropy to lose (more accessible conformational states) upon association than rigid ones, and thus should be avoided to achieve tight binding. A rough estimate of the conformational entropy that is lost upon complexation for a flexible linker is $RT \ln 3 \sim 0.7$ kcal mol^{-1} per three-fold rotor (freely rotating bond) of the linker (Fig. 2.7). The situation is, in fact, substantially more complex than this simple analysis might suggest: for instance, residual mobility of the linker in the bound complex, and a pre-organized ligand (and linker) when unassociated with receptor, would both reduce the loss in conformational entropy. We discuss design principles of ligands more extensively in Section 2.6.3, together with alternative models for linker flexibility. One of these models gives a much lower loss in conformational entropy of a flexible linker than that predicted here.

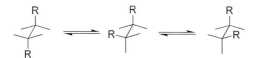

Fig. 2.7
Conformational entropy about rotors. We assume that the three conformational states above are degenerate. When one conformation is populated exclusively on binding of a protein to a ligand with this linker, the loss in conformational entropy (at $T = 298$ K) will be ~ 0.7 kcal mol^{-1} per rotor (freely rotating bond) of the linker ($-T\Delta S^\circ_{\text{conf}} \sim mRT \ln 3$, where m is the number of rotors in the linker).

2.4.6
Enthalpy/Entropy Compensation Reduces the Benefit of Multivalency

Our model assumes that the primary benefit of multivalency is the entropic enhancement of linking the ligands together [third term in Eq. (3)], so that the rotational and translational entropy of binding for a multivalent ligand is the same as for the component monovalent ligand ($T\Delta S^{\circ}_{trans+rot}$). The complication of using Eq. (3) quantitatively to estimate free energies of binding of multivalent systems is that more exothermic interactions are correlated with greater rotational and translational entropic costs of association, a phenomenon known as enthalpy/entropy compensation (EEC) [48–50]. This greater entropic cost is attributed to residual mobility of the ligand in the receptor–ligand complex [48, 49, 62]. A multivalent interaction (with no cooperativity between binding sites) has an enthalpy of binding that is the sum of the monovalent interactions [Eq. (6)]. EEC predicts that this multivalent interaction will have a greater entropic cost than the monovalent interaction ($-T\Delta S^{\circ\,avidity}_{trans+rot} > -T\Delta S^{\circ}_{trans+rot}$) and not the same entropic cost as the monovalent interaction [predicted by Eq. (3)]. This compensation will decrease the magnitude of the free energy of binding (increase $K_d^{avidity}$) relative to that expected from Eq. (3). This greater entropic cost of association does not completely compensate for the more favorable enthalpy of binding in most cases; the free energy of binding does, generally, become more favorable with more favorable enthalpy [49, 62]. Dunitz presented a simple theoretical model for EEC [48]. Williams and co-workers discussed EEC in the context of residual mobility of the *receptor* as well as the ligand in the receptor–ligand complex [49].

2.5
Representative Experimental Studies

2.5.1
Experimental Techniques Used to Examine Multivalent Systems

A number of experimental techniques have been used to study multivalent systems in solution (ITC) and at surfaces (SPR, ELISA, hemagglutination). We summarize these techniques briefly in this section.

2.5.1.1 Isothermal Titration Calorimetry
Isothermal titration calorimetry (ITC) measures heats of association for receptor–ligand complexation, as one component is titrated into the other. Analysis of these heats as a function of the concentration of the ligand relative to the receptor yields values for the enthalpy of binding (ΔH°) and the dissociation constant (K_d; when this value is the range of µM–nM) [63, 64]. Turnbull and Daranas suggested that ITC can also determine K_d, but perhaps not ΔH°, for lower affinity (~ mM) systems [65]. ITC can thus determine both the free energy and the enthalpy of bind-

ing; and the entropy of binding can be determined from these two values ($T\Delta S° = \Delta H° - \Delta G°$). Because ITC is carried out at one temperature (and so without concerns for changes in protein structure with temperature, and without assuming the change in heat capacity upon complexation is independent of temperature), it generates much more reliable thermodynamic information for biological systems than does van't Hoff analysis, which requires measurements at different temperatures [66–69]. For thermodynamic analysis of multivalent systems, the accuracy provided by ITC is essential in understanding both the entropic and enthalpic contributions to binding, in determining the thermodynamic advantage gained from multivalency (relative to monovalency), and in assessing the cooperativity of the system (Section 2.4.4).

2.5.1.2 Surface Plasmon Resonance Spectroscopy

Surface plasmon resonance (SPR) [70–72] spectroscopy provides both kinetic and thermodynamic (K_d) data for the binding of small molecules and proteins to surfaces. One component of the binding interaction is covalently attached to the surface; and the refractive index near the surface (proportional to the amount of material adsorbed to the surface) is monitored as the other component is flowed across the surface. Surfaces that have been used include self-assembled monolayers (SAMs) [34, 41, 44, 72, 73], which permit rigorous physical–organic characterization of protein–ligand interactions, because the structure of the monolayer at the surface is well defined [73], and the matrices of gels (e.g., dextran) [74–76] which allow for high levels of binding (and thus a stronger signal than with SAMs) but with ambiguities concerning the partitioning of the soluble component into the gel.

2.5.1.3 Surface Assays Using Purified Components (Cell-free Assays)

There are several other techniques (e.g., ELISA) for measuring binding of multivalent ligands or receptors to surfaces using *in vitro* assays [8, 32, 77]. These assays provide parameters (e.g., IC_{50}) for characterizing binding that are more empirical than K_d^{avidity}, because they may measure more complex processes. For instance, measurement of the binding of multivalent ligands to multivalent toxins is often based on the initial non-specific adsorption of the toxin to a microwell plate [32, 78]. The surface-bound toxin is then allowed to equilibrate with the multivalent ligand and a monovalent competitor linked to a reporter molecule; the reported signal is proportional to the concentration of free toxin (i.e., toxin not bound to the multivalent ligand).

2.5.1.4 Cell-based Surface Assays

Some surface assays (e.g., hemagglutination [79], optical tweezers [10, 80], fluorescent cells attached to surfaces [81]) are based on whole cells and thus have the benefit that they are measuring biological activity at authentic biological surfaces.

Hemagglutination assays involve observing the agglutination of erythrocytes by bacteria, viruses, or microspheres into a gel, and determining the concentration of multivalent ligand required to inhibit this process (i.e., to allow the erythrocytes to sediment into a compact pellet). Additional examples of cell-based assays are shear-flow experiments. The assay of Kiessling and co-workers involves incubating selectin-transfected cells (models for leukocytes) with different concentrations of a multivalent ligand and then microscopically monitoring the rolling of the cells on substrates coated with ligands to selectin [2, 3]. This assay system is more physiologically relevant for the inhibition of attachment of leukocytes to endothelial cells than static cell-free assays, which do not involve shear.

2.5.2
Examination of Experimental Studies in the Context of Theory

We now examine the examples presented earlier (Section 2.3) in the context of theoretical aspects of multivalency (introduced in Section 2.4).

2.5.2.1 Trivalency in Structurally Simple Systems

The thermodynamic stability of the trivalent vancomycin·D-Ala-D-Ala system (Fig. 2.3a) seems to derive, in large part, from the geometric match between the two components [29, 30]. The relatively rigid scaffolds (aromatic rings) and linkers [p-substituted aromatic rings for the vancomycin derivative and short alkyl chains (butyl) for D-Ala-D-Ala] make this match possible. ITC measurements revealed that the enthalpy of binding ($\Delta H^\circ_{\text{avidity}} = -40$ kcal mol^{-1}) was approximately three times that of the monovalent interaction [$\Delta H^\circ_{\text{affinity}} = -12$ kcal mol^{-1}, as predicted by theory; Eqs. (3), (6)] [29]. The unfavorable entropy of binding ($-T\Delta S^\circ_{\text{avidity}} = 18$ kcal mol^{-1}), however, was approximately 4.5 times that of the monovalent interaction ($-T\Delta S^\circ_{\text{affinity}} = 4.1$ kcal mol^{-1}); and the theoretical maximum for avidity would have the entropy of binding the same for both processes [Eq. (3)]. Using the semi-quantitative analysis of the binding of monovalent D-Ala-D-Ala to vancomycin by Williams et al. [82], we estimated a conformational entropy loss ($-T\Delta S^\circ_{\text{conf}}$) of ~ 41 kcal mol^{-1} upon complexation [30]. Averaging this entropy over the 27 total rotors of the trivalent molecules that were frozen upon complexation, gave $-T\Delta S^\circ_{\text{conf}} \sim 1.5$ kcal mol^{-1} per rotor frozen upon complexation (in good agreement with the estimate of Page and Jencks [46], but higher than a simple theoretical estimate of 0.7 kcal mol^{-1}; Section 2.4.5). The remarkable result was that the loss in conformational entropy was almost exactly offset by the gain from translational and rotational entropy of linking the ligands (and receptors) together ($-T\Delta S^\circ_{\text{conf}} \sim -2T\Delta S^\circ_{\text{trans,rot}}$) in the most likely model. Even though the observed enhancement in binding for this system fell far short (by a factor of ~ 10^{11}) of the maximum expected theoretically [Eq. (3), Fig. 2.4], it remains one of the tightest-binding examples of a low molecular weight ligand–receptor interaction in water. This system also demonstrates that large enhancements ($\beta = 10^{11}$) are possible, even if the theoretical, maximum benefit of multivalency is not obtained.

2.5.2.2 Cooperativity (and the Role of Enthalpy) in the "Chelate Effect"

Toone and co-workers have interpreted their thermodynamic studies of the chelation of Ca(II) ion by the tetravalent chelating agent EDTA to indicate that the high avidity binding is primarily due to a favorable enthalpy, not a favorable entropy [31]. This result is in contrast to the theoretical treatment [Eq. (3)] that predicts an entropic origin for the increased avidity of multivalent systems. The monovalent interaction (acetate binding to Ca(II)) has an unfavorable enthalpy ($\Delta H^{\circ}_{\text{affinity}}$ = +1.3 kcal mol^{-1}; $-T\Delta S^{\circ}_{\text{affinity}}$ = –3.0 kcal mol^{-1}) compared to the favorable value for tetravalent EDTA ($\Delta H^{\circ}_{\text{avidity}}$ = –5.3 kcal mol^{-1}; $-T\Delta S^{\circ}_{\text{avidity}}$ = –6.7 kcal mol^{-1}). According to our definition of cooperativity (Section 2.4.4), this interaction is therefore an example of cooperativity of the ligand sites [$\Delta H^{\circ}_{\text{avidity}} < 4\Delta H^{\circ}_{\text{affinity}}$; Eq. (6)]. The investigators hypothesized that the origin of this cooperative behavior is a relief of charge–charge repulsion [83–85] in the unbound state of EDTA upon binding to Ca(II). This charge–charge repulsion may also make the unbound state of the ligand rigid; and this restricted motion would lead to an insignificant loss of conformational entropy (Section 2.4.5) upon binding to Ca(II) [$-T\Delta S^{\circ}_{\text{conf}} \sim$ 0 kcal mol^{-1}; Eq. (3)], since the ligand is also rigid when fully associated with the calcium ion. Indeed, the investigators observed similar entropies of binding for a series of homologous ligands with equivalent valency but with different numbers of rotors between the ligands [31].

2.5.2.3 Oligovalency in the Design of Inhibitors of Toxins

Kitov, Bundle, Hol, Fan, and co-workers designed penta- and decavalent glycoside ligands for AB$_5$ family toxins, including Shiga-like toxin, cholera toxin, and heat labile *E. coli* enterotoxin [22, 32, 38–40]. The ligands were designed with approximate five-fold symmetry, to match that of the toxins. The linkers in these designs are long (~12 ethylene glycol units) and flexible; these characteristics ensure that the pendant glycosides can bind to the receptors around the periphery of the toxin (Fig. 2.3c). X-ray crystal structures of both Shiga-like [32] and cholera [39, 40] toxins bound to the penta- and decavalent ligands were unable to resolve the central core (scaffold) and linkers of the ligands; these results suggest a highly disordered, flexible region. While theory predicts that long, flexible linkers would be subject to large losses of conformational entropy ($-T\Delta S^{\circ}_{\text{conf}}$) upon binding (Section 2.4.5), the toxin ligands still bind with high avidities (values of β up to 10^7). We discuss this apparent contradiction in Section 2.6.3. The observed avidities are much lower than the maxima expected from theory, however (Fig. 2.4); and so large losses in conformational entropy could be making unfavorable contributions to the avidities. There are no calorimetric data for the binding of these penta- and decavalent ligands to their toxin receptors, so we are not able to comment on the relative enthalpic and entropic (e.g., $-T\Delta S^{\circ}_{\text{conf}}$) contributions to the avidity of binding.

The work with decavalent ligands introduces another, important design principle: using one multivalent receptor (protein) to pre-organize and present a multivalent ligand to another multivalent receptor (Fig. 2.3c). Crystal structures and dynamic light-scattering experiments demonstrate that the decavalent inhibitors of both Shiga-like [32] and cholera [39] toxin dimerize the pentameric toxins. Fan, Hol, and co-workers extended this strategy to form heterodimers of two different pentavalent proteins (cholera toxin and human serum amyloid P component) using a hetero-bivalent ligand [86]. The hetero-bivalent molecule and one protein were pre-incubated, forming a new pentavalent ligand that was then introduced to the other protein. They observed enhancements (β) of up to three orders of magnitude in the value of IC_{50} (Section 2.5.1) for the binding of the complex of hetero-bivalent ligand and cholera toxin to surface-bound human serum amyloid.

In ELISA-like assays, where only 1:1 binding is possible, the decavalent inhibitors bound to the toxins with a higher avidity (by a factor of 10–100) than do the corresponding pentavalent ligands [22, 39]. A possible explanation for this observation is that the avidity entropy [Eq. (3), Fig. 2.5] becomes important when the decavalent ligands bind to a pentavalent receptor [22].

2.5.2.4 Bivalency in Solution and at Well Defined Surfaces (SAMs)

Reinhoudt, Huskens, and co-workers studied the binding of a bivalent adamantane derivative (bivalent ligand), both to a bivalent cyclodextrin derivative (by ITC), and to a SAM displaying a saturating coverage of cyclodextrins (by SPR; Fig. 2.3 d) [34].

In solution, calorimetric data demonstrated that the enthalpy of binding of the bivalent adamantane to bivalent cyclodextrin (CD) was approximately two-fold that for bivalent adamantane to monovalent CD ($\Delta H°_{avidity}$ = –14.8 kcal mol^{-1}, $\Delta H°_{affinity}$ = –7.0 kcal mol^{-1}) [34]; a result that indicated that the binding sites were non-cooperative [Eq. (6)]. The entropy of binding to the bivalent CD was significantly more unfavorable than that for the monovalent CD ($-T\Delta S°_{avidity}$ = +5.1 kcal mol^{-1}, $-T\Delta S°_{affinity}$ = +0.6 kcal mol^{-1}), and much more unfavorable than expected from theory [Eq. (3) predicts the same entropic cost for the bivalent and monovalent associations for the theoretical maximum of avidity]; and this unfavorable term reduced the enhancement (β) to ~300 from the maximum enhancement from theory of ~10^9 (Fig. 2.4). Both the bivalent adamantane and bivalent CD contained linkers of oligo(ethylene glycol), which is expected to be very flexible and result in a large loss in conformational entropy ($-T\Delta S°_{conf}$ > 0) upon association (Section 2.4.5, Fig. 2.7).

The investigators observed that the bivalent adamantane bound more tightly (by ~10^3) to a SAM displaying a saturating coverage of CD than to the bivalent CD in solution [34]. They calculated the effective concentration (C_{eff}; Section 2.4.3) for binding in each (in solution, $C_{eff} \geq 1.8$ mM; at the surface, $C_{eff} \geq 200$ mM); this C_{eff} ~10^2 greater at the surface than in solution is consistent with the observation that binding at a surface is tighter than in solution. Figure 2.3 d shows the difference in C_{eff} graphically: there are more cyclodextrin receptors available on the surface of the SAM than in solution, within the probing volume of the uncomplexed adamantane (defined by the average end-to-end distance between the complexed and uncomplexed adamantanes). The investigators extended this work to the binding of ligands with different valencies (dendrimers with pendant adamantane or ferrocene moieties, and polymers displaying adamantane) to cyclodextrin SAM surfaces [41, 87–89] They did not, however, examine the binding of oligovalent adamantanes to surfaces with different mole fractions of cyclodextrin.

2.5.2.5 Polyvalency at Surfaces (Viruses, Bacteria, and SAMs)

Polymers are useful as polyvalent scaffolds in three circumstances: (i) when it is technically difficult, impractical, or undesirable to design a multivalent ligand that matches a multivalent receptor geometrically, (ii) when the objective of the association is not just to fill the active site of the receptors with ligand, but to serve some other function (e.g., putting a layer of polymer on a surface to prevent adhesion, presenting a second type of ligand at a surface, attaching a hapten or a fluorophore to the surface, or delivering a drug compound to the cell) [4–10, 14, 15, 43, 90, 91] (Fig. 2.8), and (iii) when one wishes to present multiple types of ligands, in a polyvalent fashion, to one multivalent species (e.g., a cell surface with multiple types of receptors [5, 7]) [1, 16].

In one example, we designed and synthesized polyacrylamides with side-chains of sialic acid (Fig. 2.3e); these polymers inhibited the binding of influenza virus to erythrocytes (an interaction that occurs via the binding of hemagglutinin, a capsid coat protein on influenza virus, to sialic acid groups at the termini of the oligosaccharides on the surface of erythrocytes) [4–10, 90]. In cell-based assays (hemagglutination), the best of these polymers displayed enhancements (β) of 10^9 (based

a) Steric Stabilization

b) Bifunctional Polymer

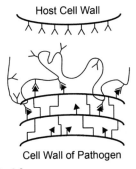
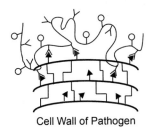

Fig. 2.8
Schematic representations of two different applications of polymers in binding to the cell walls of pathogens. (a) The polymer binds to the pathogen and prevents its association with a host cell through steric repulsion (steric stabilization). (b) A polymer presenting two different types of functional elements (bifunctional polymer) binds to the surface of a pathogen (mediated by one functionality, "Y") and presents the other functionality (o) at the surface of the pathogenic cell.

on an $IC_{50}^{avidity}$ calculated per sialic acid on the polymer) [5, 7]. Bi- and trifunctional polymers (that is, polymers that combine sialic acid with hydrophobic substituents and neuraminidase inhibitors) were better inhibitors of adhesion than were polymers that presented only sialic acid. This observation suggests that the polymer plays two roles in inhibition: polyvalent binding of the ligands to receptors, and steric shielding of the surface of the virus that prevents its approach to the surface of the erythrocyte (Fig. 2.8a) [4, 5, 7, 8].

Avidity entropy [Eq. (3), Fig. 2.5] could play a role in the binding of these polymers to influenza virus particles. Species of the virus-bound polymer where some of the ligands of the polymer are not interacting with viral receptors would have higher avidity entropies than the species where all ligands of the polymer were bound to viral receptors (Fig. 2.5) [6, 7]. The contribution of this term is challenging to estimate because it is not possible to measure or estimate the number of bound ligands. Regardless of the mechanism of enhancement, this work demonstrates that polyvalency is able to convert low-affinity interactions into high-avidity interactions.

2.6
Design Rules for Multivalent Ligands

2.6.1
When Will Multivalency Be a Successful Strategy to Design Tight-binding Ligands?

We believe that multivalent approaches will prove successful in the design of ligands for receptors that are themselves multivalent (for example, certain multimeric proteins and proteins displayed at high density on cell surfaces). Multivalency can also be applied to receptors that have multiple binding sites (and that are not themselves multimeric) in the form of hetero-oligovalency (Section 2.7.1). The primary benefit of multivalency is the conversion of ligands with low affinities ($K_d^{affinity} \sim$ mM–µM) to those with high avidities ($K_d^{avidity} \sim$ nM), by connecting the ligands together in a way that is straightforward experimentally. This approach is easier than the *de novo* design (rational design) of tight-binding inhibitors using computational approaches. De novo design has proven challenging due to the conformational flexibility of many protein targets and ligands, to the difficulty in estimating entropy accurately, and to poorly understood effects such as enthalpy/entropy compensation (Section 2.4.6) [48–50, 92].

While it is apparent from Eq. (3) that higher affinity ($K_d^{affinity}$) monovalent ligands will, in favorable circumstances, result in higher avidity ($K_d^{avidity}$) multivalent ligands, we can also predict the ideal partitioning of $K_d^{affinity}$ into enthalpy ($\Delta H_{affinity}^{\circ}$) and entropy ($\Delta S_{affinity}^{\circ}$) to optimize $K_d^{avidity}$. We predict that the highest-avidity ($K_d^{avidity}$) multivalent ligands will be generated from monovalent ligands that bind with the most favorable enthalpy ($\Delta H_{affinity}^{\circ}$), for a given $K_d^{affinity}$ (a constant $\Delta G_{affinity}^{\circ}$). This prediction is based on Eq. (3) and is easiest to understand if we re-arrange Eq. (3) to Eq. (7):

$$\Delta G^\circ_{\text{avidity}}(i) \approx i\Delta G^\circ_{\text{affinity}} + (i-1)T\Delta S^\circ_{\text{trans+rot}}$$

$$\approx i\Delta G^\circ_{\text{affinity}} + c(i-1)\Delta H^\circ_{\text{affinity}} \tag{7}$$

For clarity, Eq. (7) omits contributions of conformational entropy ($-T\Delta S^\circ_{\text{conf}}$), cooperativity between binding sites ($\Delta G^\circ_{\text{coop}}$), and the avidity entropy [$-RT \ln(\Omega_i/\Omega_0)$] to the avidity. From the concept of enthalpy/entropy compensation (Section 2.4.6), binding events with more favorable enthalpies of binding are associated with more unfavorable translational and rotational entropies of association; this idea is shown by relating $T\Delta S^\circ_{\text{trans+rot}}$ to $\Delta H^\circ_{\text{affinity}}$ by the constant c in Eq. (7). The more unfavorable entropy does not completely compensate for the more favorable enthalpy, and so $0 < c < 1$. Because this entropic term is "added back" in Eq. (7), multivalent interactions that are based on monovalent interactions with more favorable enthalpies of binding gain more in terms of free energy (more favorable) than those based on monovalent interactions with less favorable enthalpies.

2.6.2
Choice of Scaffold for Multivalent Ligands

The spacing of receptors in an oligovalent protein is defined by the system (as it occurs naturally). For this reason, our design rules focus on the multivalent ligand; specifically, we discuss the scaffold for tethering the ligands together (Section 2.6.2), and the linker to connect the ligands to the scaffold (Section 2.6.3; Fig. 2.9a). We do not attempt to discuss design of the ligands themselves and simply assume that the "best" choice of ligand is based on some combination of availability, ease of modification, affinity, and other biological and physical (e.g., solubility, stability) properties.

2.6.2.1 Scaffolds for Oligovalent Ligands

The design of oligovalent ligands often requires a scaffold that serves to present the ligands (attached by linkers) to the oligovalent receptor (Fig. 2.9a). Figure 2.9b shows some of the common scaffolds that have been used to connect ligands in the design of oligovalent ligands [16]. A common approach is to use a rigid central element to minimize the loss in conformational entropy upon complexation (Section 2.4.5). Most of the ligand scaffolds are planar, to facilitate the matching of distances and angles in the target oligovalent receptor. The scaffold should also match the symmetry of the target receptor (e.g., the design of a pentavalent ligand starting with a pentacyclen scaffold and five pendant ligands in studies of binding to pentavalent cholera toxin [36, 38, 40]). A mismatch in these design elements with the biological target will result in strain in the receptor–ligand interactions and could result in steric repulsion between scaffold and receptor; both of these effects will reduce the avidity.

In certain cases (primarily bivalent ligands), the ligands have been directly connected to one another via a linker without a scaffold (Fig. 2.9a) [100, 102, 104,

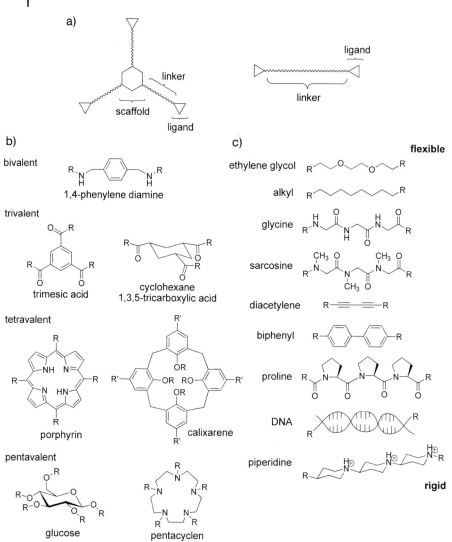

Fig. 2.9

Scaffolds and linkers for oligovalent ligands [16]. (a) Schematic showing the different parts of an oligovalent ligand: the scaffold, linkers, and ligands. The right-most structure shows a bivalent ligand that does not contain a scaffold. (b) Representative examples of scaffolds that have been used in oligovalent ligands of different valency: bivalent (1,4-phenylene diamine [93]), trivalent (trimesic acid [29, 30] and 1,3,5-cyclohexane tricarboxylic acid [94]), tetravalent (porphyrin [95] and calixarene [27]), and pentavalent (glucose [22, 32] and pentacyclen [38]). The only reported examples of oligovalent ligands containing scaffolds of 1,3,5-cyclohexane tricarboxylic acid or calixarene are studies in organic solvents. (c) Representative examples of linkers that have been used in oligovalent ligands: ethylene glycol [22, 32, 38], alkyl [29, 30], glycine [96], sarcosine [97], diacetylene [98], biphenyl [99], proline [100, 101], DNA [102], and piperidine [103]. The bivalent ligand containing the diacetylene linker was studied in an organic solvent (no aqueous examples are reported). A more expansive collection can be found in the book by Choi [16].

105]. This design maximizes the avidity entropy (Section 2.4.2) because the number of states for the fully associated receptor–ligand complex is at a maximum (there are more ways to arrange the receptor–ligand complex than when the ligands are bound to a rigid scaffold). Kitov and Bundle discussed the influence of topology of the oligovalent ligand on avidity entropy [22]. There is a trade-off between minimizing the loss in conformational entropy (Section 2.4.5) and maximizing the avidity entropy in selection of a scaffold.

2.6.2.2 Scaffolds for Polyvalent Ligands

The two common types of scaffolds for polyvalent ligands are linear, random-coil (e.g., polymers) or approximately spherical (e.g., dendrimers; Fig. 2.10) [16]. Un-

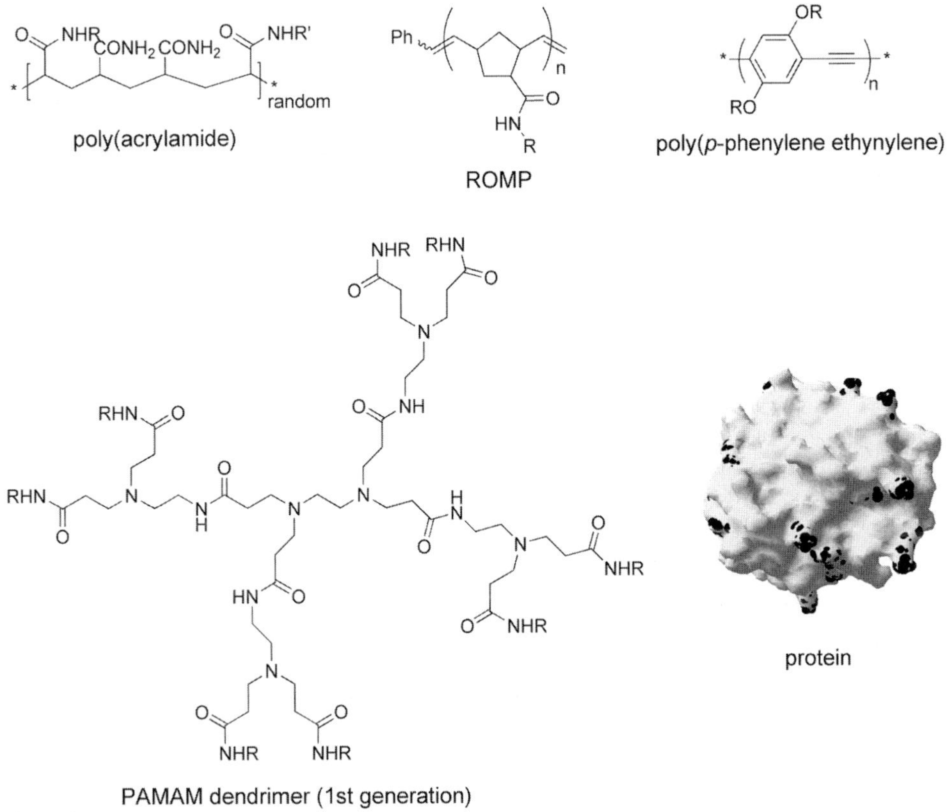

Fig. 2.10
Representative examples of scaffolds for polyvalent ligands: polyacrylamide [6, 7, 110], ROMP [106], poly(p-phenylene ethynylene), PAMAM dendrimer first generation [111], and protein [109, 112]. The protein was rendered as a molecular surface (shaded gray) with displayed ligands shown in black using Swiss PDB Viewer and deposited atomic coordinates (PDB 1V9I) [113]. A more expansive collection of scaffolds for oligovalent ligands can be found in the book by Choi [16].

branched polymers have the advantage that they may be better able to interact with a number of cell surface receptors than can dendrimers. Dendrimers have the advantage that they have a relatively small influence on the viscosity of a solution, while extended, linear polymers can make solutions intractably viscous. The flexibility of polymers can be varied by the selection of the backbone: flexible (e.g., polyacrylamide) [2, 4–10, 14, 15, 90, 106, 107] or rigid [e.g., poly(p-phenylene ethynylene)] [108]. Rigid polymers often have poor solubility in water, and are often not truly rigid [101]. We have reported the use of proteins as scaffolds for monodisperse polymers: lysine residues of proteins were perfunctionalized with ligands [109]. These scaffolds can be either roughly spherical (if the modified protein is allowed to re-fold into its native state) or extended (if analysis is conducted under denaturing conditions, in non-aqueous solvents, or if the native structure of the protein is so significantly perturbed that it cannot re-fold).

2.6.3
Choice of Linker for Multivalent Ligands

Figure 2.9c shows some common linkers that have been used in the literature to connect the ligands to the scaffold for multivalent ligands [16]; the linkers are arranged approximately in order of decreasing flexibility. We proceed to discuss the advantages and disadvantages of using rigid (Section 2.6.3.1) and flexible (Section 2.6.3.2) linkers in the design of multivalent ligands.

2.6.3.1 Rigid Linkers Represent a Simple Approach to Optimize Affinity

The simplest theory (Section 2.4.5) suggests using rigid linkers (those at the bottom of Fig. 2.9c) for multivalent ligands to optimize avidity for multivalent receptors. Such linkers (Fig. 2.11a) minimize the loss in conformational entropy upon complexation, but maximize the risk of unfavorable interactions between ligands, linkers, and receptors. The loss in conformational entropy reflects the restriction of modes of motion of the linker upon complexation (Fig. 2.7). The length of the linker should place the ligands at a distance that approximately matches the dis-

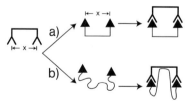

Fig. 2.11
Schematic representation of two different approaches to linker design for oligovalent systems: (a) a rigid linker whose length exactly matches the spacing of the receptor binding sites, (b) a flexible linker that interacts favorably with the surface of the receptor upon binding.

tance between binding sites of the multivalent receptor. This approach requires, of course, extensive experimental study into the system (e.g., spacing and geometry of binding sites, etc.); computational study may also be useful, although it has not so far been helpful. Otherwise, a poor fit (e.g., destabilizing, non-bonded interactions between the linker and the receptor, less-than-ideal interaction between each receptor and ligand due to sub-optimal positioning by linker) will result and yield a protein–ligand complex with low stability.

2.6.3.2 Flexible Linkers Represent an Alternative Approach to Rigid Linkers to Optimize Affinity

An alternative approach to using rigid linkers is to use flexible linkers in the design of multivalent ligands. A flexible linker can adopt a number of conformations without steric strain (unlike rigid linkers) and allow the multivalent ligand to sample conformational space to optimize the binding of the tethered ligands to multiple receptors. This sampling of conformational space reduces the possibility of a sterically obstructed fit, a circumstance that can occur readily with rigid linkers. The flexible linker should be longer than the spacing between receptors in the multivalent receptor target to allow this sampling of conformational space; but "how much longer is optimum?" is not a question that has, so far, been answered. This approach can, theoretically, generate a design that suffers from a significant loss in conformational entropy on binding (by requiring a defined conformational state for a number of rotors upon complexation; Fig. 2.7) and thus an increase in K_d^{avidity}. If the linker is able to interact favorably with the oligovalent receptor, favorable enthalpic contacts [$\Delta H_{\text{linker}}^\circ < 0$, Eq. (3)] might make association less unfavorable than expected based on this model (Fig. 2.11 b). Alternatively, a model based on effective concentration (Fig. 2.6) predicts a much smaller loss in conformational entropy upon complexation for long, flexible linkers than the model based on the assumption of bonds that are free rotors, which become completely restricted upon oligovalent receptor–ligand association. This C_{eff} model has been used to rationalize the tight-binding (low values of K_d^{avidity}) observed for oligovalent sugars tethered by flexible linkers binding to pentavalent toxins (Section 2.5.2.3) [38, 105].

2.6.4
Strategy for the Synthesis of Multivalent Ligands

The synthesis of oligovalent ligands (the connection of the individual ligands to the linkers and the scaffold) generally follows established methods of small-molecule organic synthesis and will not be discussed here. We also do not discuss the synthesis of polyvalent ligands such as dendrimers, because their synthesis has been discussed in detail elsewhere [114–116]. We focus instead on polymers (the most common types of polyvalent ligands). There are two general approaches to the synthesis of polyvalent polymers: polymerization of ligand monomers (Section 2.6.4.1), and reaction of ligands with a pre-formed activated polymer (Section 2.6.4.2).

2.6.4.1 Polyvalent Ligands: Polymerization of Ligand Monomers

Kiessling and co-workers polymerized ligand-functionalized monomers directly, using ring-opening metathesis polymerization (ROMP) [117, 118]; this process yields fully functionalized polymers of controllable valencies and lengths (Fig. 2.10) [13, 106, 112, 119–121]. Polymers with less than quantitative loading of ligand have been synthesized by co-polymerizing ligand-functionalized and unfunctionalized monomers [6, 110]. The loading of the polymer with ligand depends on the ratio of ligand-functionalized to unfunctionalized monomer and their relative reactivities [122]. This approach of polymerizing ligand-functionalized monomers offers the following advantages: (i) monomers can be fully characterized before polymerization, and (ii) controllable valencies are accessible using ROMP. There are several disadvantages: (i) the need for polymerization techniques that are compatible with the functional groups on the ligand, (ii) the difficulty in synthesizing ligand-functionalized monomers, (iii) the difficulty in predicting the loading density of ligands (when ligand-functionalized and unfunctionalized monomers are co-polymerized and have a difference in reactivity towards polymerization), and (iv) the difficulty in determining the distribution of ligands (when ligand-functionalized and unfunctionalized monomers are co-polymerized) along the polymer backbone (block co-polymers or random co-polymers often cannot be readily distinguished).

2.6.4.2 Polyvalent Ligands: Functionalization with Ligands after Polymerization

In the other common synthetic route to polyvalent polymers, monomers with reactive groups (e.g., activated carboxylic acids) have been synthesized and then polymerized [7, 107, 123]. In a subsequent step, the activated polymer is allowed to react with the desired ligands. The loading is controlled by the amount of ligand that is allowed to react with the polymer (reaction yields are often nearly quantitative [7]). This approach offers many advantages: (i) large amounts of activated polymer can be synthesized and then coupled with different amounts of ligand (allowing screening of the influence of ligand density of the polymer on biological activity), (ii) ligands with diverse functionality can be used because the ligands are introduced after polymerization, (iii) these ligands are often easier to synthesize than ligands activated for polymerization, (iv) synthesis of multifunctional polymers (containing two or more different functionalities) is straightforward, and (v) constant length and polydispersity of polymers are ensured if the same batch of activated polymer is coupled to different types or amounts of ligands. The constant polymer backbone removes a complicating variable from determining the effect of different types and densities of ligand(s) on biological activity. The approach has the following disadvantages: (i) it can be challenging to obtain fully functionalized polymer, (ii) the activated polymer (e.g., N-hydroxysuccinimidyl-containing polymer) is often susceptible to hydrolysis or other side reactions, and (iii) determining the composition of a polymer after functionalization with ligand can be challenging (^{1}H NMR can be used, but the resonances are usually quite broad. When the ligand contains chromophores, UV-Vis offers a more quantitative alternative to NMR) [15, 42].

2.7
Extensions of Multivalency to Lead Discovery

2.7.1
Hetero-oligovalency Is a Broadly Applicable Concept in Ligand Design

All of the theory regarding homo-bivalency applies to hetero-bivalent systems, with appropriate incorporation of multiple dissociation constants for the monovalent interactions and statistical factors. Further, hetero-bivalency is expected to be more broadly applicable than homo-bivalency, since it is applicable to problems requiring the ability to target monomeric proteins, which do not contain multiple, proximate, identical binding sites, but may have "sticky" sites adjacent to binding sites.

The examples presented here are merely meant to be representative of the application of hetero-bivalent ligands to target monomeric proteins (proteins which do not have two identical binding sites). The additional site can be either a hydrophobic patch on the protein or a binding site for a second substrate [124]. Using a model protein for ligand design – carbonic anhydrase II (CA) – we demonstrated this targeting of a secondary site (a hydrophobic patch) near the active site of an enzyme [96, 125, 126]. This patch was relatively non-discriminating towards the type of hydrophobic group in the ligand, depending only on the size of the group: for instance, ligands with benzyl, adamantyl, and octyl groups all bound with roughly the same (~nM) avidity [96]. Finn, Sharpless, and co-workers applied a Huisgen 1,3-dipolar cycloaddition templated by acetylcholinesterase to generate a tight-binding ligand for the enzyme ($K_d^{avidity}$ = 77–410 fM) from known monovalent ligands ($K_d^{affinity}$ ~ nM–μM) that bound two different sites on the enzyme: the active site and a "peripheral" site at the rim of the active site gorge [127]. Rosenberg, Fesik, and co-workers reported similar covalent tethering to synthesize a high-avidity ($K_d^{avidity}$ ≤ 1 nM) hetero-bivalent ligand for Bcl-2 family proteins from low-affinity monovalent ligands ($K_d^{affinity}$ ~ mM) to an active site and an adjacent hydrophobic patch (Chapter 9) [128]. Parang and Cole reviewed the application of hetero-bivalent ligands to protein kinases, enzymes that transfer a phosphate group from adenosine triphosphate (ATP) to protein targets [129]. These ligands usually consist of a nucleotide analog and a peptide to bind to the ATP-binding site and protein-binding site of the enzyme, respectively. Such bisubstrate inhibitors were shown to exhibit selectivity between the members of the family of protein kinases. Theravance (San Francisco, Calif.) [130] reported the synthesis [131, 132], *in vitro* antibiotic activity against conventional and antibiotic-resistant strains of bacteria [133–135], and phase II clinical trials of telavancin [136], a derivative of vancomycin containing a hydrophobic and a hydrophilic side-chain [131]. The anti-bacterial action of telavancin occurs by two mechanisms: disruption of cell wall biosynthesis (similar to vancomycin itself) in Gram-positive bacteria, and depolarization (i.e., disruption) of the bacterial cell membrane [135]. An injectable form of the compound is currently in Phase III clinical trials for the treatment of complicated skin and skin structure infections (cSSSI) and hospital-acquired pneumonia (HAP) [131]. Telavancin demonstrates the therapeutic potential of hetero-bivalent ligands.

2.7.2
Dendrimers Present Opportunities for Multivalent Presentation of Ligands

A number of sugar-displaying dendrimers have been evaluated for binding to lectins or the inhibition of hemagglutination induced by either bacteria or lectins. Lundquist and Toone reviewed these studies [17]. In particularly well controlled studies, Cloninger and co-workers examined the influence of PAMAM dendrimer size (generations 1–6) and loading density of mannose on Con A-induced hemagglutination of erythrocytes [111, 137, 138]. They found that, for fully mannose-functionalized dendrimers, only large dendrimers (generations 4–6 with >50 sugars per dendrimer) were able to inhibit hemagglutination with significant enhancements; they attributed this result to the large dendrimers being able to bind bivalently to Con A (the spacing between binding sites on Con A is ~6.5 nm) [137]. In a follow-up study, they examined the influence of loading density of mannose on the dendrimer on hemagglutination [111]. The enhancements (on a per *mannose* basis) for the different-sized dendrimers (generations 4–6) all peaked at ~50% of the maximal loading of mannose (maximum enhancements of 250–600, scaling with generation number, were observed). The investigators speculated that steric interactions between mannose residues decreased the enhancement at higher loadings. The enhancement (on a per *dendrimer* basis) increased monotonically with the loading density; this increase (with no peak) was attributed to statistical effects (avidity entropy).

Dendrimers are also promising as agents in human health care [115, 139]. Starpharma (Australia) [140] recently completed Phase I clinical trials on VivaGel, a topical vaginal microbicide that prevents infection by HIV *in vivo* (primate models) [141]. The active ingredient in VivaGel is a fourth generation dendrimer (SPL7013) decorated with naphthalene disulfonate groups; this polyanionic coating is believed to bind to the viral coat (presumably, via electrostatic interactions) and prevent the attachment of the virus to host T cells [142].

2.7.3
Bivalency in the Immune System

IgG and IgE antibodies, prime components of the immune system, are bivalent proteins containing two identical receptors (Fab sites; Fig. 2.12) [21]. When binding bivalently to a surface (Fig. 2.12a) or to a soluble bivalent ligand (Fig. 2.12b), we postulate that the enhancement (β) for a given antibody is inversely proportional to the monovalent dissociation constant (K_d^{affinity}) and directly proportional to the effective concentration (C_{eff}) of ligand near an available receptor (Fig. 2.12). If we assume C_{eff} to be constant for all antibodies (that is, that they have the same average distance between Fab sites), then greater enhancements will result from higher affinity (lower K_d^{affinity}) ligands. At cell surfaces, the enhancement for the binding of a polyclonal mixture of IgG with high monovalent affinity (average $K_d^{\text{affinity}} \sim 1$ nM) to the surface of *Bacillus* sp. was ~100 [143]. Cremer and co-workers examined the binding of a polyclonal mixture of IgG to phospholipid mem-

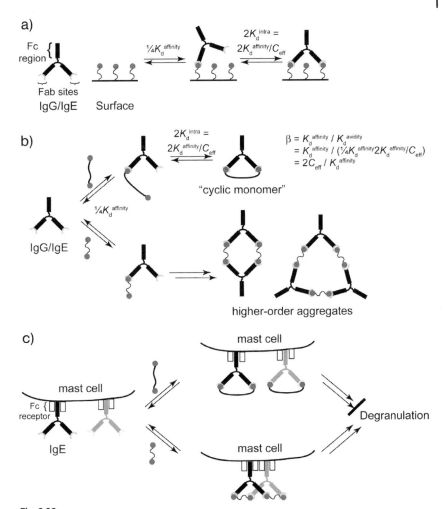

Fig. 2.12
Binding of bivalent antibodies (IgG or IgE) to bivalent ligands and surfaces. (a) The stepwise equilibria characterizing binding of an antibody bivalently to a surface displaying ligands are shown. The dissociation constant for the second step (K_d^{intra}) is taken to be the theoretical value assuming no cooperativity between Fab sites (Fig. 2.6). (b) Two different pathways are available to antibodies binding to bivalent ligands in solution depending on the length of the linker in the bivalent ligand [100, 145]. The dissociation constants for the top pathway (intramolecular ring closure) are analogous to those in (a). The enhancement (β) applies to both (a) and the top pathway in (b). (c) IgE bound to mast cells by their Fc regions can bind to bivalent ligands in two different pathways analogous to those in (b), again depending on the length of the linker in the bivalent ligand [102, 146]. These two pathways have different effects on degranulation of the mast cells: bivalent ligands with long linkers form "cyclic monomers" and inhibit degranulation, while bivalent ligands with short linkers cross-link the surface-bound IgEs and induce degranulation. The two IgEs are shaded differently to aid visualization of the aggregate.

branes containing ligand lipids; an enhancement of ~40 was observed for the weak-affinity system studied (K_d^{affinity} ~ 50 µM) [144].

Pecht, Licht, and co-workers used bivalent ligands with long, rigid poly(proline) linkers to examine the formation of soluble IgE "closed monomers": both receptor sites of the antibody were bound to the same bivalent ligand (Fig. 2.12 b, top pathway) [100]. They measured enhancements of roughly the same order of magnitude (β ~ 20) for medium-affinity ligands (K_d^{affinity} ~ 0.1 µM) containing sufficiently long linkers to bridge both Fab sites as for the cell-surface results. The relatively low enhancements observed in these studies indicate low values of C_{eff} of a ligand in proximity to an available (unbound) Fab site of the antibody. Consistent with this low value of C_{eff}, crystal structures revealed that the two receptors on an antibody are ~8 nm apart and are flexibly coupled to one another [147–149]. We expect this large distance and flexible coupling to increase the entropic cost of association (Section 2.4.5).

A number of investigators showed that bivalent ligands too short to form "closed monomers" can form discrete cyclic antibody aggregates (e.g., cyclic dimers where two antibodies are bridged by two bivalent ligands; Fig. 2.12 b, bottom pathway) [104, 150–156]. In solution, these aggregates were shown to be stable on relatively long time-scales (for analysis by HPLC and ultracentrifugation) [104, 155–157].

Holowka and Baird discussed the importance of aggregation of IgE (that are bound to mast cells by their Fc region) on the release of histamine, a process referred to as degranulation (Fig. 2.12 c) [158]. Degranulation of mast cells can be either inhibited or promoted by the binding of oligovalent ligands to IgE bound to mast cells: long oligovalent ligands that can form "closed monomers" (i.e., span both Fab sites of an IgE; bind *intra*molecularly to one IgE) can inhibit degranulation (Figure 2.12 c, top pathway), while short oligovalent ligands that aggregate IgEs *inter*molecularly result in degranulation (Fig. 2.12 c, bottom pathway). Baird and co-workers reported the inhibition of degranulation of mast cells by bivalent ligands containing long oligo(ethylene glycol) linkers (≥ 9 units) [146] and long DNA linkers (30-mer; bivalent ligands with shorter DNA linkers promoted degranulation) [102], and by large ligand-displaying dendrimers (smaller dendrimers promoted degranulation) [146]. They observed enhancements (β) of up to 100. Ligands that inhibit the degranulation of mast cells could be useful in the treatment of allergies.

2.7.4
Polymers Could Be the Most Broadly Applicable Multivalent Ligands

We demonstrated that polymers displaying two types of functionalities (bifunctional polymers) can bind to and display a functional element on the surface of synthetic surfaces (SAMs) and bacterial surfaces [15]. In this proof-of-principle demonstration, the polymer-labeled surface was efficiently bound by antibodies (directed against the functional element); this binding shows that a new functionality can be *introduced* onto a surface by using bifunctional polymers. This approach should be general: as long as a recognition element (ligand) with some affinity ($K_d^{\text{affinity}} \sim$ mM–μM) can be found, a tunable functional element (e.g., fluorophore, hapten) can be incorporated onto any surface.

There are a number of conventional methods to discover ligands of moderate affinity for target receptors (e.g., combinatorial chemistry [159, 160], phage display [161], *de novo* design [92]). Our work designing polymers that were effective against anthrax toxin demonstrated the discovery of a low-affinity ligand and the conversion of this ligand to a high-avidity inhibitor using polyvalency [42]. A peptide that inhibited the assembly of the components of anthrax toxin with low affinity ($IC_{50}^{\text{affinity}} = 150$ μM) was discovered by phage display and was converted by polyvalency into a high-avidity polymer ($IC_{50}^{\text{avidity}} = 20$ nM on a per-peptide basis). This polymer was effective in protecting an animal model against challenge with purified components of anthrax toxin [42].

Kiessling and co-workers demonstrated that the enhancement of sugar-displaying polymers in the inhibition of binding of model leukocytes to artificial surfaces is greater in dynamic, cell-rolling assays (Section 2.5.1; $\beta = 170$ on a per sugar basis) than in static, cell-free assays ($\beta = 5$ on a per sugar basis) [2]. Since the dynamic assays are more physiologically relevant than the static ones, these results suggest that this polymer could be more effective *in vivo* than predicted by static assays. Generalizing this single result to the potential *in vivo* efficacy of other polymers is, of course, not possible. Kiessling and co-workers also investigated the clustering of cell-surface receptors by multivalent polymers displaying galactose and the influence of this clustering on such downstream effects as bacterial chemotaxis [11–13, 162].

Copaxone (glatiramer acetate), marketed by Teva Pharmaceuticals (Israel), is an example of a highly successful polymer therapeutic [163]. Glatiramer acetate is a large (average MW = 5–11 kDa) synthetic polypeptide of L-Ala, L-Glu, L-Lys, and L-Tyr (a mimic of myelin basic protein) that is random in distribution but defined in composition [164]. It has been approved for the treatment of patients with relapsing-remitting multiple sclerosis (RR-MS; reducing the frequency of relapses and disease progression in clinical studies) [165]. While the exact mechanism of action is unknown, the polypeptide is known to attenuate patients' autoimmune response to myelin and to reduce both the inflammation and neurodegeneration associated with the disease [164]. An oral form of glatiramer acetate is effective in the treatment of models of MS *in vivo* (rodents and primates) [166, 167]. It has not yet, however, shown statistically significant results in human clinical trials (as of 2005, Phase II clinical trials were ongoing) [163]. The oral form of the drug has been shown to be non-toxic, providing hope that perhaps higher doses will be effective at treating the disease in humans [167, 168].

2.8
Challenges and Unsolved Problems in Multivalency

A number of unanswered questions about multivalency still remain: Why do we not achieve the theoretical maximum of enhancement in any multivalent system [even with seemingly well designed systems such as trivalent D-Ala-D-Ala/vancomycin (Section 2.5.2.1; Fig. 2.4)]? Why does the experimentally observed avidity not scale with valency (i; Fig. 2.4)? What principles of design will allow one to approach this theoretical maximum?

What is the best linker for a multivalent ligand (in terms of length, flexibility, chemical composition)? What is the mechanism of polyvalent binding to surfaces (e.g., polymers and dendrimers)? How do we design the most effective multivalent ligands? How can this effectiveness be demonstrated (e.g., what kinds of assays?) in a way that attracts the active interest of the pharmaceutical industry?

How should multivalent ligands (primarily polyvalent ones) be manufactured for therapeutic use? Will the innate polydispersity of polymers pose exceptional problems in regulatory clearance [169]?

2.9
Conclusions

Multivalency can convert weak monovalent ligands into oligovalent ligands effective at low concentrations. The discovery of low-affinity monovalent ligands ($K_d^{affinity} \sim$ mM–μM) is less challenging than that of high-affinity monovalent ligands ($K_d^{affinity} \sim$ nM). These weak monovalent ligands can be discovered through rational design or through screening efforts (e.g., combinatorial chemistry [159, 160], phage display [42, 161]). For a hetero-bivalent inhibitor, the second binding site can be as non-specific as a hydrophobic patch on a monomeric protein, or can involve an additional substrate binding site. Polyvalent ligands present new mechanisms of action that are not available to monovalent ligands. Examples include steric inhibition of binding to a surface [4, 5, 7, 8], shielding one set of receptors on a surface by binding to another set, incorporating a new functional element onto a cell surface (e.g., "painting" a bacterial surface with antigens) [14, 15, 170], "cloaking" the antigenicity of a surface, and forming structured aggregates of antibodies and other species.

The design principles and mechanism of action of multivalency are still not entirely clear. For instance, Kitov and Bundle designed a decavalent ligand to bind 1:1 with the pentavalent Shiga-like toxin (to take advantage of two of the three receptor sites per monomer) [32]. They discovered, surprisingly, that a 2:1 complex of toxin to inhibitor was favored (Section 2.5.2.3). This example illustrates some of the limitations of our understanding of multivalent ligand design.

Multivalent ligands are generally larger (higher molecular weight) than monovalent ones. This greater size can decrease bioavailability (especially oral bioavailability), decrease rates of excretion, and limit tissue permeation. Multivalent ligands are typically more difficult to synthesize and characterize than monovalent ones. Polymers (the most common polyvalent ligands) are not single chemical entities and require special techniques for reproducible synthesis and manufacture. Multivalent species (especially polymeric polyvalent species) are unfamiliar to the FDA, and from the point of regulation will require new rules and regulations before approval [169].

There are a number of potential therapeutic applications of multivalent ligands. Hetero-bivalent ligands can, in principle, be imagined for most proteins (exploiting a secondary site). The immune system relies fundamentally upon bivalency; to interact effectively with components of it (e.g., antibodies) requires a multivalent approach (Section 2.7.3). Bifunctional polymers are able to introduce a new functional element to cell surfaces [170]. Theoretically, any class of cells (e.g., pathogens, cancer cells) can be targeted by this approach, the only requirement being a ligand–receptor interaction of modest affinity ($K_d^{\text{affinity}} \sim \text{mM}-\mu\text{M}$) and a reasonable surface density of receptors. We believe that polymers will be particularly effective in places where their large size is an advantage rather than a disadvantage. Examples could include administration to the digestive tract, respiratory system, eye, superficial wounds, and vagina, where retaining the polyvalent ligand in that organ or structure is useful, and where release into the systemic circulation is undesirable.

Acknowledgments

Research on multivalency in our laboratory was supported by the National Institutes of Health (GM30367), DARPA, and the Space and Naval Warfare Systems Center, San Diego. V.M.K. acknowledges a pre-doctoral fellowship from the NDSEG. L.A.E. acknowledges a postdoctoral fellowship from the NIH (EB003361).

References

1 Mammen, M., S.-K. Choi, G. M. Whitesides. **1998**, Polyvalent Interactions in Biological Systems: Implications for Design and Use of Multivalent Ligands and Inhibitors. *Angew. Chem. Int. Ed. Engl.* 37, 2755–2794.

2 Sanders, W. J., E. J. Gordon, O. Dwir, P. J. Beck, R. Alon, L. L. Kiessling. **1999**, Inhibition of L-Selectin-Mediated Leukocyte Rolling by Synthetic Glycoprotein Mimics. *J. Biol. Chem.* 274, 5271–5278.

3 Mowery, P., Z. Q. Yang, E. J. Gordon, O. Dwir, A. G. Spencer, R. Alon, L. L. Kiessling. **2004**, Synthetic Glycoprotein Mimics Inhibit L-Selectin-Mediated Rolling and Promote L-Selectin Shedding. *Chem. Biol.* 11, 725–732.

4 Choi, S.-K., M. Mammen, G. M. Whitesides. **1996**, Monomeric Inhibitors of Influenza Neuraminidase Enhance the Hemagglutination Inhibition Activities of Polyacrylamides Presenting Multiple C-Sialoside Groups. *Chem. Biol.* 3, 97–104.

5 Choi, S.-K., M. Mammen, G. M. Whitesides. **1997**, Generation and in Situ Evaluation of Libraries of Poly(Acrylic Acid) Presenting Sialosides as Side Chains as Polyvalent Inhibitors of Influenza-Mediated Hemagglutination. *J. Am. Chem. Soc.* 119, 4103–4111.

6 Lees, W. J., A. Spaltenstein, J. E. Kingerywood, G. M. Whitesides. **1994**, Polyacrylamides Bearing Pendant α-Sialoside Groups Strongly Inhibit Agglutination of Erythrocytes by Influenza a Virus: Multivalency and Steric Stabilization of Particulate Biological Systems. *J. Med. Chem.* 37, 3419–3433.

7 Mammen, M., G. Dahmann, G. M. Whitesides. **1995**, Effective Inhibitors of Hemagglutination by Influenza-Virus Synthesized from Polymers Having Active Ester Groups – Insight into Mechanism of Inhibition. *J. Med. Chem.* 38, 4179–4190.

8 Sigal, G. B., M. Mammen, G. Dahmann, G. M. Whitesides. **1996**, Polyacrylamides Bearing Pendant α-Sialoside Groups Strongly Inhibit Agglutination of Erythrocytes by Influenza Virus: The Strong Inhibition Reflects Enhanced Binding through Cooperative Polyvalent Interactions. *J. Am. Chem. Soc.* 118, 3789–3800.

9 Sparks, M. A., K. W. Williams, G. M. Whitesides. **1993**, Neuraminidase-Resistant Hemagglutination Inhibitors – Acrylamide Copolymers Containing a C-Glycoside of N-Acetylneuraminic Acid. *J. Med. Chem.* 36, 778–783.

10 Mammen, M., K. Helmerson, R. Kishore, S. K. Choi, W. D. Phillips, G. M. Whitesides. **1996**, Optically Controlled Collisions of Biological Objects to Evaluate Potent Polyvalent Inhibitors of Virus-Cell Adhesion. *Chem. Biol.* 3, 757–763.

11 Gestwicki, J. E., L. E. Strong, L. L. Kiessling. Tuning Chemotactic Responses with Synthetic Multivalent Ligands. *Chem. Biol.* 7, 583–591.

12 Gestwicki, J. E., L. E. Strong, S. L. Borchardt, C. W. Cairo, A. M. Schnoes, L. L. Kiessling. **2001**, Designed Potent Multivalent Chemoattractants for *Escherichia coli*. *Bioorg. Med. Chem.* 9, 2387–2393.

13 Gestwicki, J. E., L. L. Kiessling. **2002**, Inter-Receptor Communication through Arrays of Bacterial Chemoreceptors. *Nature* 415, 81–84.

14 Li, J., S. Zacharek, X. Chen, J. Wang, W. Zhang, A. Janczuk, P. G. Wang. **1999**, Bacteria Targeted by Human Natural Antibodies Using α-Gal Conjugated Receptor-Specific Glycopolymers. *Bioorg. Med. Chem.* 7, 1549–1558.

15 Metallo, S. J., R. S. Kane, R. E. Holmlin, G. M. Whitesides. **2003**, Using Bifunctional Polymers Presenting Vancomycin and Fluorescein Groups to Direct Anti-Fluorescein Antibodies to Self-Assembled Monolayers Presenting D-Alanine-D-Alanine Groups. *J. Am. Chem. Soc.* 125, 4534–4540.

16 Choi, S.-K. **2004**, *Synthetic Multivalent Molecules: Concepts and Biomedical Applications*, John Wiley & Sons, Hoboken.

17 Lundquist, J. J., E. J. Toone. **2002**, The Cluster Glycoside Effect. *Chem. Rev.* 102, 555–578.

18 Kiessling, L. L., J. E. Gestwicki, L. E. Strong. **2000**, Synthetic Multivalent Ligands in the Exploration of Cell-Surface Interactions. *Curr. Opin. Chem. Biol.* 4, 696–703.

19 Mulder, A., J. Huskens, D. N. Reinhoudt. **2004**, Multivalency in Supramolecular Chemistry and Nanofabrication. *Org. Biomol. Chem.* 2, 3409–3424.

20 Kiessling, L. L., L. E. Strong, J. E. Gestwicki. **2000**, Principles for Multivalent Ligand Design. in *Annual Reports in Medicinal Chemistry, Vol 35*, ed. Hagmann, W., Doherty, A., Academic Press, New York.

21 Janeway, C. A. Jr., P. Travers, M. Walport, M. J. Shlomchik **2001**, *Immunobiology: The Immune System in Health and Disease*, Garland Publishing, New York.

22 Kitov, P. I., D. R. Bundle. **2003**, On the Nature of the Multivalency Effect: A Thermodynamic Model. *J. Am. Chem. Soc.* 125, 16271–16284.

23 Badjic, J. D., A. Nelson, S. J. Cantrill, W. B. Turnbull, J. F. Stoddart. **2005**, Multivalency and Cooperativity in Supramolecular Chemistry. *Acc. Chem. Res.* 38, 723–732.

24 Hennrich, G., W. M. David, Y. J. Bomble, E. V. Anslyn, J. S. Brodbelt, J. F. Stanton. **2004**, Self-Assembling Dimeric and Trimeric Aggregates Based on Solvophobic and Charge-Pairing Interactions. *Supramol. Chem.* 16, 521–528.

25 Wiskur, S. L., J. L. Lavigne, A. Metzger, S. L. Tobey, V. Lynch, E. V. Anslyn. **2004**, Thermodynamic Analysis of Receptors Based on Guanidinium/Boronic Acid

Groups for the Complexation of Carboxylates, α-Hydroxycarboxylates, and Diols: Driving Force for Binding and Cooperativity. *Chem. Eur. J.* 10, 3792–3804.

26 Fantuzzi, G., P. Pengo, R. Gomila, P. Ballester, C. A. Hunter, L. Pasquato, P. Scrimin. **2003**, Multivalent Recognition of Bis- and Tris-Zn-Porphyrins by *N*-Methylimidazole Functionalized Gold Nanoparticles. *Chem. Commun.* 2003, 1004–1005.

27 Baldini, L., P. Ballester, A. Casnati, R. M. Gomila, C. A. Hunter, F. Sansone, R. Ungaro. **2003**, Molecular Acrobatics: Self-Assembly of Calixarene-Porphyrin Cages. *J. Am. Chem. Soc.* 125, 14181–14189.

28 Anderson, H. L., S. Anderson, J. K. M. Sanders. **1995**, Ligand-Binding by Butadiyne-Linked Porphyrin Dimers, Trimers and Tetramers. *J. Chem. Soc. Perkin Trans. 1*, 2231–2245.

29 Rao, J. H., J. Lahiri, L. Isaacs, R. M. Weis, G. M. Whitesides. **1998**, A Trivalent System from Vancomycin-D-Ala-D-Ala with Higher Affinity Than Avidin-Biotin. *Science* 280, 708–711.

30 Rao, J., J. Lahiri, R. M. Weis, G. M. Whitesides. **2000**, Design, Synthesis, and Characterization of a High-Affinity Trivalent System Derived from Vancomycin and L-Lys-D-Ala-D-Ala. *J. Am. Chem. Soc.* 122, 2698–2710.

31 Christensen, T., D. M. Gooden, J. E. Kung, E. J. Toone. **2003**, Additivity and the Physical Basis of Multivalency Effects: A Thermodynamic Investigation of the Calcium EDTA Interaction. *J. Am. Chem. Soc.* 125, 7357–7366.

32 Kitov, P. I., J. M. Sadowska, G. Mulvey, G. D. Armstrong, H. Ling, N. S. Pannu, R. J. Read, D. R. Bundle. **2000**, Shiga-Like Toxins Are Neutralized by Tailored Multivalent Carbohydrate Ligands. *Nature* 403, 669–672.

33 Zhang, Z. S., J. Y. Liu, C. Verlinde, W. G. J. Hol, E. K. Fan. **2004**, Large Cyclic Peptides as Cores of Multivalent Ligands: Application to Inhibitors of Receptor Binding by Cholera Toxin. *J. Org. Chem.* 69, 7737–7740.

34 Mulder, A., T. Auletta, A. Sartori, S. Del Ciotto, A. Casnati, R. Ungaro, J. Huskens, D. N. Reinhoudt. **2004**, Divalent Binding of a Bis(Adamantyl)-Functionalized Calix [4] Arene to β-Cyclodextrin-Based Hosts: An Experimental and Theoretical Study on Multivalent Binding in Solution and at Self-Assembled Monolayers. *J. Am. Chem. Soc.* 126, 6627–6636.

35 Green, N. M. **1990**, Avidin and Streptavidin. *Methods Enzymol.* 184, 51–67.

36 Fan, E. K., E. A. Merritt, C. Verlinde, W. G. J. Hol. **2000**, AB_5 Toxins: Structures and Inhibitor Design. *Curr. Opin. Struct. Biol.* 10, 680–686.

37 Merritt, E. A., W. G. J. Hol. **1995**, Ab(5) Toxins. *Curr. Opin. Struct. Biol.* 5, 165–171.

38 Fan, E. K., Z. S. Zhang, W. E. Minke, Z. Hou, C. Verlinde, W. G. J. Hol. **2000**, High-Affinity Pentavalent Ligands of *Escherichia Coli* Heat-Labile Enterotoxin by Modular Structure-Based Design. *J. Am. Chem. Soc.* 122, 2663–2664.

39 Zhang, Z. S., E. A. Merritt, M. Ahn, C. Roach, Z. Hou, C. Verlinde, W. G. J. Hol, E. Fan. **2002**, Solution and Crystallographic Studies of Branched Multivalent Ligands That Inhibit the Receptor-Binding of Cholera Toxin. *J. Am. Chem. Soc.* 124, 12991–12998.

40 Merritt, E. A., Z. S. Zhang, J. C. Pickens, M. Ahn, W. G. J. Hol, E. K. Fan. **2002**, Characterization and Crystal Structure of a High-Affinity Pentavalent Receptor-Binding Inhibitor for Cholera Toxin and *E-Coli* Heat-Labile Enterotoxin. *J. Am. Chem. Soc.* 124, 8818–8824.

41 Huskens, J., A. Mulder, T. Auletta, C. A. Nijhuis, M. J. W. Ludden, D. N. Reinhoudt. **2004**, A Model for Describing the Thermodynamics of Multivalent Host-Guest Interactions at Interfaces. *J. Am. Chem. Soc.* 126, 6784–6797.

42 Mourez, M., R. S. Kane, J. Mogridge, S. Metallo, P. Deschatelets, B. R. Sellman, G. M. Whitesides, R. J. Collier. **2001**, Designing a Polyvalent Inhibitor of Anthrax Toxin. *Nat. Biotechnol.* 19, 958–961.

43 Disney, M. D., P. H. Seeberger. **2004**, The Use of Carbohydrate Microarrays to Study Carbohydrate-Cell Interactions and to Detect Pathogens. *Chem. Biol.* 11, 1701–1707.

44 Mann, D. A., M. Kanai, D. J. Maly, L. L. Kiessling. **1998**, Probing Low Affinity

and Multivalent Interactions with Surface Plasmon Resonance: Ligands for Concanavalin A. *J. Am. Chem. Soc.* 120, 10575–10582.

45 Jencks, W. P. **1981**, On the Attribution and Additivity of Binding Energies. *Proc. Natl Acad. Sci. USA* 78, 4046–4050.

46 Page, M. I., W. P. Jencks. **1971**, Entropic Contributions to Rate Accelerations in Enzymic and Intramolecular Reactions and Chelate Effect. *Proc. Natl Acad. Sci. USA* 68, 1678–1683.

47 Atkins, P. **1994**, *Physical Chemistry*, W. H. Freeman and Co., New York.

48 Dunitz, J. D. **1995**, Win Some, Lose Some: Enthalpy–Entropy Compensation in Weak Intermolecular Interactions. *Chem. Biol.* 2, 709–712.

49 Williams, D. H., E. Stephens, D. P. O'Brien, M. Zhou. **2004**, Understanding Noncovalent Interactions: Ligand Binding Energy and Catalytic Efficiency from Ligand-Induced Reductions in Motion within Receptors and Enzymes. *Angew. Chem. Int. Ed. Eng.* 43, 6596–6616.

50 Gilli, P., V. Gerretti, G. Gilli, P. A. Borea. **1994**, Enthalpy–Entropy Compensation in Drug–Receptor Binding. *J. Phys. Chem.* 98, 1515–1518.

51 Cappalonga Bunn, A. M., R. S. Alexander, D. W. Christianson. **1994**, Mapping Protein-Peptide Affinity: Binding of Peptidylsulfonamide Inhibitors to Human Carbonic Anhydrase II. *J. Am. Chem. Soc.* 116, 5063–5068.

52 Jain, A., S. G. Huang, G. M. Whitesides. **1994**, Lack of Effect of the Length of Oligoglycine- and Oligo(Ethylene Glycol)-Derived Para-Substituents on the Affinity of Benzenesulfonamides for Carbonic Anhydrase II in Solution. *J. Am. Chem. Soc.* 116, 5057–5062.

53 Mammen, M., E. I. Shakhnovich, G. M. Whitesides. **1998**, Using a Convenient, Quantitative Model for Torsional Entropy to Establish Qualitative Trends for Molecular Processes That Restrict Conformational Freedom. *J. Org. Chem.* 63, 3168–3175.

54 Ercolani, G. **2003**, Assessment of Cooperativity in Self-Assembly. *J. Am. Chem. Soc.* 125, 16097–16103.

55 Kramer, R. H., J. W. Karpen. **1998**, Spanning Binding Sites on Allosteric Proteins with Polymer-Linked Ligand Dimers. *Nature* 395, 710–713.

56 Gargano, J. M., T. Ngo, J. Y. Kim, D. W. K. Acheson, W. J. Lees. **2001**, Multivalent Inhibition of AB_5 Toxins. *J. Am. Chem. Soc.* 123, 12909–12910.

57 Jacobson, H., W. H. Stockmayer. **1950**, Intramolecular Reaction in Polycondensations 1. The Theory of Linear Systems. *J. Chem. Phys.* 18, 1600–1606.

58 Winnik, M. A. **1981**, Cyclization and the Conformation of Hydrocarbon Chains. *Chem. Rev.* 81, 491–524.

59 Knoll, D., J. Hermans. **1983**, Polymer–Protein Interactions – Comparison of Experiment and Excluded Volume Theory. *J. Biol. Chem.* 258, 5710–5715.

60 Monod, J., J. Wyman, J. P. Changeux. **1965**, On Nature of Allosteric Transitions – A Plausible Model. *J. Mol. Biol.* 12, 88–98.

61 Changeux, J. P., S. J. Edelstein. **2005**, Allosteric Mechanisms of Signal Transduction. *Science* 308, 1424–1428.

62 Calderone, C. T., D. H. Williams. **2001**, An Enthalpic Component in Cooperativity: The Relationship between Enthalpy, Entropy, and Noncovalent Structure in Weak Associations. *J. Am. Chem. Soc.* 123, 6262–6267.

63 Wiseman, T., S. Williston, J. F. Brandts, L.-N. Lin. **1989**, Rapid Measurement of Binding Constants and Heats of Binding Using a New Titration Calorimeter. *Anal. Biochem.* 179, 131–137.

64 Pierce, M. M., C. S. Raman, B. T. Nall. **1999**, Isothermal Titration Calorimetry of Protein–Protein Interactions. *Methods* 19, 213–221.

65 Turnbull, W. B., A. H. Daranas. **2003**, On the Value of C: Can Low Affinity Systems Be Studied by Isothermal Titration Calorimetry? *J. Am. Chem. Soc.* 125, 14859–14866.

66 Naghibi, H., A. Tamura, J. M. Sturtevant. **1995**, Significant Discrepancies between Van't Hoff and Calorimetric Enthalpies. *Proc. Natl Acad. Sci. USA* 92, 5597–5599.

67 Liu, Y. F., J. M. Sturtevant. **1995**, Significant Discrepancies between Van't Hoff

and Calorimetric Enthalpies 2. *Protein Sci.* 4, 2559–2561.

68 Liu, Y. F., J. M. Sturtevant. **1997**, Significant Discrepancies between Van't Hoff and Calorimetric Enthalpies 3. *Biophys. Chem.* 64, 121–126.

69 Horn, J. R., D. Russell, E. A. Lewis, K. P. Murphy. **2001**, Van't Hoff and Calorimetric Enthalpies from Isothermal Titration Calorimetry: Are There Significant Discrepancies? *Biochemistry* 40, 1774–1778.

70 Raether, H. **1977**, Surface Plasmon Oscillations and Their Applications. in *Physics of Thin Films*, ed. Haas, G., Francombe, M. H., Hoffman, R. W., Academic Press, New York.

71 Homola, J., S. S. Yee, G. Gauglitz. **1999**, Surface Plasmon Resonance Sensors: Review. *Sensors Actuators Chem.* 54, 3–15.

72 Brockman, J. M., B. P. Nelson, R. M. Corn. **2000**, Surface Plasmon Resonance Imaging Measurements of Ultrathin Organic Films. *Annu. Rev. Phys. Chem.* 51, 41–63.

73 Love, J. C., L. A. Estroff, J. K. Kriebel, R. G. Nuzzo, G. M. Whitesides. **2005**, Self-Assembled Monolayers of Thiolates on Metals as a Form of Nanotechnology. *Chem. Rev.* 105, 1103–1169.

74 Fivash, M., E. M. Towler, R. J. Fisher. **1998**, Biacore for Macromolecular Interaction. *Curr. Opin. Biotechnol.* 9, 97–101.

75 Green, R. J., R. A. Frazier, K. M. Shakesheff, M. C. Davies, C. J. Roberts, S. J. B. Tendler. **2000**, Surface Plasmon Resonance Analysis of Dynamic Biological Interactions with Biomaterials. *Biomaterials* 21, 1823–1835.

76 Rich, R. L., D. G. Myszka. **2005**, Survey of the Year 2003 Commercial Optical Biosensor Literature. *J. Mol. Recognit.* 18, 1–39.

77 Barry, R., M. Soloviev. **2004**, Quantitative Protein Profiling Using Antibody Arrays. *Proteomics* 4, 3717–3726.

78 Minke, W. E., C. Roach, W. G. J. Hol, C. Verlinde. **1999**, Structure-Based Exploration of the Ganglioside GM1 Binding Sites of *Escherichia Coli* Heat-Labile Enterotoxin and Cholera Toxin for the Discovery of Receptor Antagonists. *Biochemistry* 38, 5684–5692.

79 Landsteiner, K. **1962**, *The Specificity of Serological Reactions*, Dover, New York.

80 Liang, M. N., S. P. Smith, S. J. Metallo, I. S. Choi, M. Prentiss, G. M. Whitesides. **2000**, Measuring the Forces Involved in Polyvalent Adhesion of Uropathogenic *Escherichia Coli* to Mannose-Presenting Surfaces. *Proc. Natl Acad. Sci. USA* 97, 13092–13096.

81 Qian, X. P., S. J. Metallo, I. S. Choi, H. K. Wu, M. N. Liang, G. M. Whitesides. **2002**, Arrays of Self-Assembled Monolayers for Studying Inhibition of Bacterial Adhesion. *Anal. Chem.* 74, 1805–1810.

82 Williams, D. H., et al. **1991**, Toward the Semiquantitative Estimation of Binding Constants – Guides for Peptide Peptide Binding in Aqueous-Solution. *J. Am. Chem. Soc.* 113, 7020–7030.

83 Gao, J. M., M. Mammen, G. M. Whitesides. **1996**, Evaluating Electrostatic Contributions to Binding with the Use of Protein Charge Ladders. *Science* 272, 535–537.

84 Carbeck, J. D., I. J. Colton, J. Gao, G. M. Whitesides. **1998**, Protein Charge Ladders, Capillary Electrophoresis, and the Role of Electrostatics in Biomolecular Recognition. *Acc. Chem. Res.* 31, 343–350.

85 Carbeck, J. D., I. Gitlin, G. M. Whitesides. **2005**. Why Are Proteins Charged? Networks of Charge-Charge Interactions in Proteins Measured by Charge Ladders and Capillary Electrophoresis. *Angew. Chem. Int. Ed. Eng.* (in press).

86 Liu, J. Y., Z. S. Zhang, X. J. Tan, W. G. J. Hol, C. Verlinde, E. K. Fan. **2005**, Protein Heterodimerization through Ligand-Bridged Multivalent Pre-Organization: Enhancing Ligand Binding toward Both Protein Targets. *J. Am. Chem. Soc.* 127, 2044–2045.

87 Bruinink, C. M., et al. **2005**, Supramolecular Microcontact Printing and Dip-Pen Nanolithography on Molecular Printboards. *Chem. Eur. J.* 11, 3988–3996.

88 Nijhuis, C. A., J. Huskens, D. N. Reinhoudt. **2004**, Binding Control and Stoichiometry of Ferrocenyl Dendrimers at a Molecular Printboard. *J. Am. Chem. Soc.* 126, 12266–12267.

89 Auletta, T., et al. **2004**, Writing Patterns of Molecules on Molecular Printboards. *Angew. Chem. Int. Ed. Eng.* 43, 369–373.

90 Schmid, W., L. Z. Avila, K. W. Williams, G. M. Whitesides. **1993**, Synthesis of Methyl α-Sialosides N-Substituted with Large Alkanoyl Groups, and Investigation of Their Inhibition of Agglutination of Erythrocytes by Influenza-A Virus. *Bioorg. Med. Chem. Lett.* 3, 747–752.

91 Fleming, C., et al. **2005**, A Carbohydrate-Antioxidant Hybrid Polymer Reduces Oxidative Damage in Spermatozoa and Enhances Fertility. *Nat. Chem. Biol.* 1, 270–274.

92 Schneider, G., U. Fechner. **2005**, Computer-Based De Novo Design of Drug-Like Molecules. *Nat. Rev. Drug Discov.* 4, 649–663.

93 Rao, J. H., G. M. Whitesides. **1997**, Tight Binding of a Dimeric Derivative of Vancomycin with Dimeric L-Lys-D-Ala-D-Ala. *J. Am. Chem. Soc.* 119, 10286–10290.

94 Hanabusa, K., A. Kawakami, M. Kimura, H. Shirai. **1997**, Small Molecular Gelling Agents to Harden Organic Liquids: Trialkyl Cis-1,3,5-Cyclohexanetricarboxamides. *Chem. Lett.* 1997, 191–192.

95 Jain, R. K., A. D. Hamilton. **2000**, Protein Surface Recognition by Synthetic Receptors Based on a Tetraphenylporphyrin Scaffold. *Org. Lett.* 2, 1721–1723.

96 Jain, A., G. M. Whitesides, R. S. Alexander, D. W. Christianson. **1994**, Identification of Two Hydrophobic Patches in the Active Site Cavity of Human Carbonic Anhydrase II by Solution-Phase and Solid-State Studies and Their Use in the Development of Tight-Binding Inhibitors. *J. Med. Chem.* 37, 2100–2105.

97 Krishnamurthy, V. M., B. R. Bohall, V. Semetey, G. M. Whitesides. **2005**. The Paradoxical Thermodynamic Basis for the Interaction of Ethylene Glycol, Glycine and Sarcosine Chains with Bovine Carbonic Anhydrase II: An Unexpected Manifestation of Enthalpy/Entropy Compensation. *J. Am. Chem. Soc.* (in press).

98 Wang, G. J., A. D. Hamilton. **2002**, Synthesis and Self-Assembling Properties of Polymerizable Organogelators. *Chem. Eur. J.* 8, 1954–1961.

99 Chapman, W. H., R. Breslow. **1995**, Selective Hydrolysis of Phosphate-Esters, Nitrophenyl Phosphates and UpU, by Dimeric Zinc-Complexes Depends on the Spacer Length. *J. Am. Chem. Soc.* 117, 5462–5469.

100 Schweitzerstenner, R., A. Licht, I. Luscher, I. Pecht. **1987**, Oligomerization and Ring-Closure of Immunoglobulin-E Class Antibodies by Divalent Haptens. *Biochemistry* 26, 3602–3612.

101 Schuler, B., E. A. Lipman, P. J. Steinbach, M. Kumke, W. A. Eaton. **2005**, Polyproline and the "Spectroscopic Ruler" Revisited with Single-Molecule Fluorescence. *Proc. Natl Acad. Sci. USA* 102, 2754–2759.

102 Paar, J. M., N. T. Harris, D. Holowka, B. Baird. **2002**, Bivalent Ligands with Rigid Double-Stranded DNA Spacers Reveal Structural Constraints on Signaling by Fcε RI. *J. Immunol.* 169, 856–864.

103 Semetey, V., D. T. Moustakas, G. M. Whitesides. **2006**, Synthesis and Conformational Study of Water-Soluble, Rigid, Rod-Like Oligopiperidines. *Angew. Chem. Int. Ed. Engl.* 45, 588–591.

104 Wilder, R. L., G. Green, V. N. Schumaker. **1975**, Bivalent Hapten-Antibody Interactions 2. Bivalent Haptens as Probes of Combining Site Depth. *Immunochemistry* 12, 49–54.

105 Kitov, P. I., H. Shimizu, S. W. Homans, D. R. Bundle. **2003**, Optimization of Tether Length in Nonglycosidically Linked Bivalent Ligands That Target Sites 2 and 1 of a *Shiga*-Like Toxin. *J. Am. Chem. Soc.* 125, 3284–3294.

106 Strong, L. E., L. L. Kiessling. **1999**, A General Synthetic Route to Defined, Biologically Active Multivalent Arrays. *J. Am. Chem. Soc.* 121, 6193–6196.

107 Arranz-Plaza, E., A. S. Tracy, A. Siriwardena, J. M. Pierce, G.-J. Boons. **2002**, High-Avidity, Low-Affinity Multivalent Interactions and the Block to Polyspermy in *Xenopus Laevis*. *J. Am. Chem. Soc.* 124, 13035–13046.

108 Disney, M. D., J. Zheng, T. M. Swager, P. H. Seeberger. **2004**, Detection of Bacteria with Carbohydrate-Functionalized Fluorescent Polymers. *J. Am. Chem. Soc.* 126, 13343–13346.

109 Yang, J., I. Gitlin, V. M. Krishnamurthy, J. A. Vazquez, C. E. Costello, G. M. Whitesides. **2003**, Synthesis of Monodisperse Polymers from Proteins. *J. Am. Chem. Soc.* 125, 12392–12393.

110 Spaltenstein, A., G. M. Whitesides. **1991**, Polyacrylamides Bearing Pendant α-Sialoside Groups Strongly Inhibit Agglutination of Erythrocytes by Influenza Virus. *J. Am. Chem. Soc.* 113, 686–687.

111 Woller, E. K., E. D. Walter, J. R. Morgan, D. J. Singel, M. J. Cloninger. **2003**, Altering the Strength of Lectin Binding Interactions and Controlling the Amount of Lectin Clustering Using Mannose/Hydroxyl-Functionalized Dendrimers. *J. Am. Chem. Soc.* 125, 8820–8826.

112 Gestwicki, J. E., C. W. Cairo, L. E. Strong, K. A. Oetjen, L. L. Kiessling. **2002**, Influencing Receptor-Ligand Binding Mechanisms with Multivalent Ligand Architecture. *J. Am. Chem. Soc.* 124, 14922–14933.

113 Saito, R., T. Sato, A. Ikai, N. Tanaka. **2004**, Structure of Bovine Carbonic Anhydrase II at 1.95 Angstrom Resolution. *Acta Crystallogr. Biol. Crystallogr.* 60, 792–795.

114 Newkome, G. R., C. N. Moorefield, F. Vögtle **2001**, *Dendrimers and Dendrons: Concepts, Syntheses, Applications*, Wiley-VCH, Weinheim.

115 Boas, U., P. M. H. Heegaard. **2004**, Dendrimers in Drug Research. *Chem. Soc. Rev.* 33, 43–63.

116 Hecht, S. **2003**, Functionalizing the Interior of Dendrimers: Synthetic Challenges and Applications. *J. Polymer Sci. Polymer Chem.* 41, 1047–1058.

117 Trnka, T. M., R. H. Grubbs. **2001**, The Development of $L_2X_2RU=CHR$ Olefin Metathesis Catalysts: An Organometallic Success Story. *Acc. Chem. Res.* 34, 18–29.

118 Novak, B. M., W. Risse, R. H. Grubbs. **1992**, The Development of Well-Defined Catalysts for Ring-Opening Olefin Metathesis Polymerizations (ROMP). *Adv. Polymer Sci.* 102, 47–72.

119 Kanai, M., K. H. Mortell, L. L. Kiessling. **1997**, Varying the Size of Multivalent Ligands: The Dependence of Concanavalin A Binding on Neoglycopolymer Length. *J. Am. Chem. Soc.* 119, 9931–9932.

120 Mortell, K. H., R. V. Weatherman, L. L. Kiessling. **1996**, Recognition Specificity of Neoglycopolymers Prepared by Ring-Opening Metathesis Polymerization. *J. Am. Chem. Soc.* 118, 2297–2298.

121 Mortell, K. H., M. Gingras, L. L. Kiessling. **1994**, Synthesis of Cell Agglutination Inhibitors by Aqueous Ring-Opening Metathesis Polymerization. *J. Am. Chem. Soc.* 116, 12053–12054.

122 Odian, G. **1991**, *Principles of Polymerization*, John Wiley & Sons, New York.

123 Yang, Z. Q., E. B. Puffer, J. K. Pontrello, L. L. Kiessling. **2002**, Synthesis of a Multivalent Display of a CD22-Binding Trisaccharide. *Carbohydr. Res.* 337, 1605–1613.

124 Broom, A. D. **1989**, Rational Design of Enzyme-Inhibitors – Multisubstrate Analog Inhibitors. *J. Med. Chem.* 32, 2–7.

125 Boriack, P. A., D. W. Christianson, J. Kingery-Wood, G. M. Whitesides. **1995**, Secondary Interactions Significantly Removed from the Sulfonamide Binding Pocket of Carbonic Anhydrase II Influence Binding Constants. *J. Med. Chem.* 38, 2286–2291.

126 Sigal, G. B., G. M. Whitesides. **1996**, Benzenesulfonamide-Peptide Conjugates as Probes for Secondary Binding Sites near the Active Site of Carbonic Anhydrase. *Bioorg. Med. Chem. Lett.* 6, 559–564.

127 Lewis, W. G., L. G. Green, F. Grynszpan, Z. Radic, P. R. Carlier, P. Taylor, M. G. Finn, K. B. Sharpless. **2002**, Click Chemistry in Situ: Acetylcholinesterase as a Reaction Vessel for the Selective Assembly of a Femtomolar Inhibitor from an Array of Building Blocks. *Angew. Chem. Int. Ed. Eng.* 41, 1053–1063.

128 Oltersdorf, T., et al. **2005**, An Inhibitor of Bcl-2 Family Proteins Induces Regression of Solid Tumours. *Nature* 435, 677–681.

129 Parang, K., P. A. Cole. **2002**, Designing Bisubstrate Analog Inhibitors for Protein Kinases. *Pharmacol. Ther.* 93, 145–157.

130 http://www.theravance.com

131 Sorbera, L. A., J. Castaner. **2004**, Televancin Hydrochloride – Glycopeptide Antibiotic. *Drugs Future* 29, 1211–1219.

132 Leadbetter, M. R., et al. **2004**, Hydrophobic Vancomycin Derivatives with Improved ADME Properties: Discovery of

Telavancin (TD-6424). *J. Antibiot.* 57, 326–336.

133 King, A., I. Phillips, K. Kaniga. **2004**, Comparative in Vitro Activity of Telavancin (TD-6424), a Rapidly Bactericidal, Concentration-Dependent Anti-Infective with Multiple Mechanisms of Action against Gram-Positive Bacteria. *J. Antimicrob. Chemother.* 53, 797–803.

134 Goldstein, E. J. C., D. M. Citron, C. V. Merriam, Y. A. Warren, K. L. Tyrrell, H. T. Fernandez. **2004**, In Vitro Activities of the New Semisynthetic Glycopeptide Telavancin (TD-6424), Vancomycin, Daptomycin, Linezolid, and Four Comparator Agents against Anaerobic Gram-Positive Species and *Corynebacterium* spp. *Antimicrob. Agents Chemother.* 48, 2149–2152.

135 Higgins, D. L., et al. **2005**, Telavancin, a Multifunctional Lipoglycopeptide, Disrupts Both Cell Wall Synthesis and Cell Membrane Integrity in Methicillin-Resistant *Staphylococcus aureus*. *Antimicrob. Agents Chemother.* 49, 1127–1134.

136 Stryjewski, M. E., et al. **2005**, Telavancin Versus Standard Therapy for Treatment of Complicated Skin and Soft-Tissue Infections Due to Gram-Positive Bacteria. *Clin. Infect. Dis.* 40, 1601–1607.

137 Woller, E. K., M. J. Cloninger. **2002**, The Lectin-Binding Properties of Six Generations of Mannose-Functionalized Dendrimers. *Org. Lett.* 4, 7–10.

138 Wolfenden, M. L., M. J. Cloninger. **2005**, Mannose/ Glucose-Functionalized Dendrimers to Investigate the Predictable Tunability of Multivalent Interactions. *J. Am. Chem. Soc.* 127, 12168–12169.

139 Cloninger, M. J. **2002**, Biological Applications of Dendrimers. *Curr. Opin. Chem. Biol.* 6, 742–748.

140 http://www.starpharma.com

141 Jiang, Y. H., P. Emau, J. S. Cairns, L. Flanary, W. R. Morton, T. D. McCarthy, C. C. Tsai. **2005**, Spl7013 Gel as a Topical Microbicide for Prevention of Vaginal Transmission of SHIV89.6P in Macaques. *AIDS Res. Hum. Retroviruses* 21, 207–213.

142 Witvrouw, M., et al. **2000**, Polyanionic (i.e., Polysulfonate) Dendrimers Can Inhibit the Replication of Human Immunodeficiency Virus by Interfering with Both Virus Adsorption and Later Steps (Reverse Transcriptase/Integrase) in the Virus Replicative Cycle. *Mol. Pharmacol.* 58, 1100–1108.

143 Karulin, A. Y., B. B. Dzantiev. **1990**, Polyvalent Interaction of Antibodies with Bactericidal-Cells. *Mol. Immunol.* 27, 965–971.

144 Yang, T. L., O. K. Baryshnikova, H. B. Mao, M. A. Holden, P. S. Cremer. **2003**, Investigations of Bivalent Antibody Binding on Fluid-Supported Phospholipid Membranes: The Effect of Hapten Density. *J. Am. Chem. Soc.* 125, 4779–4784.

145 Dembo, M., B. Goldstein. **1978**, Thermodynamic Model of Binding of Flexible Bivalent Haptens to Antibody. *Immunochemistry* 15, 307–313.

146 Baird, E. J., D. Holowka, G. W. Coates, B. Baird. **2003**, Highly Effective Poly(Ethylene Glycol) Architectures for Specific Inhibition of Immune Receptor Activation. *Biochemistry* 42, 12739–12748.

147 Woof, J. M., D. R. Burton. **2004**, Human Antibody – Fc Receptor Interactions Illuminated by Crystal Structures. *Nat. Rev. Immunol.* 4, 89–99.

148 Harris, L. J., S. B. Larson, K. W. Hasel, A. McPherson. **1997**, Refined Structure of an Intact IGG2a Monoclonal Antibody. *Biochemistry* 36, 1581–1597.

149 Bongini, L., D. Fanelli, F. Piazza, P. De los Rios, S. Sandin, U. Skoglund. **2004**, Freezing Immunoglobulins to See Them Move. *Proc. Natl. Acad. Sci. USA* 101, 6466–6471.

150 Green, G., R. L. Wilder, V. N. Schumaker. **1972**, Detection of Antibody Monomers, Dimers and Polymers Upon Interactions of a Homologous Series of Divalent Haptens with Its Specific Antibody. *Biochem. Biophys. Res. Commun.* 46, 738–748.

151 Phillips, M. L., V. T. Oi, V. N. Schumaker. **1990**, Electron-Microscopic Study of Ring-Shaped, Bivalent Hapten, Bivalent Antidansyl Monoclonal-Antibody Complexes with Identical Variable Domains but IGG1, IGG2a and IGG2b Constant Domains. *Mol. Immunol.* 27, 181–190.

152 Posner, R., B. Goldstein, D. Holowka, B. Baird. **1990**, Dissociation Kinetics of Bivalent Haptens Bound to Immunoglobulin-E Antibodies. *Biophys. J.* 57, A295–A295.

153 Posner, R. G., J. W. Erickson, D. Holowka, B. Baird, B. Goldstein. **1991**, Dissociation Kinetics of Bivalent Ligand Immunoglobulin-E Aggregates in Solution. *Biochemistry* 30, 2348–2356.
154 Schweitzerstenner, R., A. Licht, I. Pecht. **1992**, Dimerization Kinetics of the IgE-Class Antibodies by Divalent Haptens 2. The Interactions between Intact IgE and Haptens. *Biophys. J.* 63, 563–568.
155 Subramanian, K., D. Holowka, B. Baird, B. Goldstein. **1996**, The Fc Segment of IgE Influences the Kinetics of Dissociation of a Symmetrical Bivalent Ligand from Cyclic Dimeric Complexes. *Biochemistry* 35, 5518–5527.
156 Wilder, R. L., G. Green, V. N. Schumaker. **1975**, Bivalent Hapten-Antibody Interactions 1. Comparison of Water-Soluble and Water Insoluble Bivalent Haptens. *Immunochem.* 12, 39–47.
157 Carson, D., H. Metzger. **1974**, Separation of Rabbit Anti-Dinitrophenyl IgG Antibodies on Basis of Combining Site Depth. *Immunochem.* 11, 355–359.
158 Holowka, D., B. Baird. **1996**, Antigen-Mediated IgE Receptor Aggregation and Signaling: A Window on Cell Surface Structure and Dynamics. *Annu. Rev. Biophys. Biomol. Struct.* 25, 79–112.
159 Young, S. S., N. X. Ge. **2004**, Design of Diversity and Focused Combinatorial Libraries in Drug Discovery. *Curr. Opin. Drug Discovery Dev.* 7, 318–324.
160 Webb, T. R. **2005**, Current Directions in the Evolution of Compound Libraries. *Curr. Opin. Drug Discovery Dev.* 8, 303–308.
161 Ladner, R. C., A. K. Sato, J. Gorzelany, M. de Souza. **2004**, Phage Display-Derived Peptides as a Therapeutic Alternatives to Antibodies. *Drug Discovery Today* 9, 525–529.
162 Lamanna, A. C., J. E. Gestwicki, L. E. Strong, S. L. Borchardt, R. M. Owen, L. L. Kiessling. **2002**, Conserved Amplification of Chemotactic Responses through Chemoreceptor Interactions. *J. Bacteriol.* 184, 4981–4987.
163 http://www.tevapharm.com.
164 Teitelbaum, D., R. Arnon, M. Sela. **1997**, Copolymer 1: From Basic Research to Clinical Application. *Cell. Mol. Life Sci.* 53, 24–28.
165 Johnson, K. P., et al. **1995**, Copolymer-1 Reduces Relapse Rate and Improves Disability in Relapsing-Remitting Multiple-Sclerosis – Results of a Phase-III Multicenter, Double-Blind, Placebo-Controlled Trial. *Neurology* 45, 1268–1276.
166 Teitelbaum, D., R. Arnon, M. Sela. **1999**, Immunomodulation of Experimental Autoimmune Encephalomyelitis by Oral Administration of Copolymer 1. *Proc. Natl. Acad. Sci. USA* 96, 3842–3847.
167 Teitelbaum, D., et al. **2004**, Oral Glatiramer Acetate in Experimental Autoimmune Encephalomyelitis – Clinical and Immunological Studies. In *Oral Tolerance: New Insights and Prospects for Clinical Application*, ed. Teitelbaum, D., et al.
168 de Seze, J., G. Edan, M. Labalette, J. P. Dessaint, P. Vermersch. **2000**, Effect of Glatiramer Acetate (Copaxone) Given Orally in Human Patients: Interleukin-10 Production During a Phase 1 Trial. *Ann. Neurol.* 47, 686–686.
169 Duncan, R. **2003**, The Dawning Era of Polymer Therapeutics. *Nat. Rev. Drug Discov.* 2, 347–360.
170 Krishnamurthy, V. M., et al. **2006**, Promotion of Opsonization by Antibodies and Phagocytosis of Gram-Positive Bacteria by a Bifunctional Polyacrylamide. *Biomaterials* 27, 3663–3674.

3
Entropic Consequences of Linking Ligands

Christopher W. Murray and Marcel L. Verdonk

3.1
Introduction

A small molecule in solution possesses a considerable amount of rigid body entropy associated with free translation and tumbling motions. On binding to a protein, much of this entropy is lost and this constitutes a rigid body barrier to binding that must be overcome by favorable binding interactions between the ligand and the protein. Estimates of the free energy barrier are important in the interpretation of results from fragment-based drug discovery techniques. In this chapter we describe an analysis of experimental data where two ligands of known binding affinity and binding mode have been linked to yield a molecule for which the binding affinity and binding mode have also been determined. In these circumstances it is possible to obtain a crude approximation of the entropic barrier to binding which we estimate to be 15–20 kJ mol^{-1} at 298 K, that is, around three orders of magnitude in affinity. The consequences of such a large entropic barrier to binding are discussed.

3.2
Rigid Body Barrier to Binding

3.2.1
Decomposition of Free Energy of Binding

Figure 3.1(a) shows a schematic in which a fragment-sized molecule A is bound to a protein. Following Page and Jencks [1–3], the binding affinity for fragment A can be written as:

$$\Delta G^A_{total} = \Delta G^A_{int} + \Delta G_{rigid} \qquad (1)$$

where ΔG_{rigid} is the free energy associated with the loss of rigid body entropy on binding to the enzyme; and ΔG^A_{int} contains other free energy terms that contribute to binding and includes favorable enthalpic and entropic interactions (such as hy-

Fragment-based Approaches in Drug Discovery. Edited by W. Jahnke and D. A. Erlanson
Copyright © 2006 WILEY-VCH Verlag GmbH & Co. KGaA, Weinheim
ISBN: 3-527-31291-9

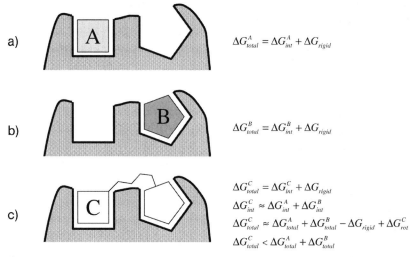

Fig. 3.1
Schematic illustrating the energetics of joining together fragments that exhibit multiple site binding. (a) Fragment A binds to the enzyme and the free energy of binding can be decomposed into an intrinsic affinity for the enzyme, ΔG_{int}^A, and an unfavorable term, ΔG_{rigid}^A, representing the rigid body entropy loss. (b) Fragment B binds to a different site on the enzyme; and (c) molecule C binds with affinity greater than the sum of affinities for fragments A and B when the fragments are joined together ideally.

drogen bonds, lipophilic interactions and the entropy gained by the expulsion of water molecules from the binding site) and unfavorable terms (such as entropy losses associated with freezing rotatable bonds or enthalpy costs associated with fragment A not being bound in its lowest energy state). ΔG_{int}^A is referred to as the *intrinsic* binding affinity associated with fragment A. Note that the decomposition in Eq. (1) is an approximation because it assumes that the two free energy terms are independent [4].

3.2.2
Theoretical Treatment of the Rigid Body Barrier to Binding

Consider the binding of a small rigid fragment to an enzyme pocket. In solution, the fragment has a considerable amount of translational and rigid body rotational entropy associated with its free movement through the solution and with its tumbling motion, and much of this rigid body entropy is lost on specific binding to the larger protein.

Finkelstein and Janin [5] describe a model that accounts for the rigid body movements of a ligand when bound to a protein binding site. Their model assumes that the translational entropy of a ligand, bound to a protein binding site is given by a similar expression to its translational entropy in the gas phase, but that the volume accessible to the ligand is vastly reduced by the interactions it forms with the protein. The translational entropy loss of the ligand upon binding to the protein can then be estimated as:

$$\Delta S_{trans} \approx 3\,R\ln\left(\delta x/v^{1/3}\right) \qquad (2)$$

Where δx represents the average r.m.s. amplitude (Å) in the principal directions of the ligand, when bound to the protein, and v is the volume open to the ligand in solution. Finkelstein and Janin also derive a similar expression for the loss of rotational entropy starting from the gas phase rotational entropy:

$$\Delta S_{rot} \approx 3\,R\ln\left(\delta\alpha/2\pi^{2/3}\right) \qquad (3)$$

where $\delta\alpha$ represents the average r.m.s. amplitude (radians) for rotations about the principal axes of the ligand, when bound to the protein. The model suggests that the loss of translational and rotational entropy is independent of molecular weight; and in another, more detailed, theoretical analysis, Gilson and colleagues [6] come to the same conclusion.

In the analysis below, we likewise assume that the loss of rigid body entropy is independent of molecular weight and additionally that the loss in entropy is approximately constant. This allows us to fit the rigid body barrier to binding, using experimental data associated with fragment linking. It should be noted however that the assumption of a constant barrier is an approximation because, following Searle and Williams [7], one expects the barrier to be larger for ligands forming strong polar interactions [small δx and $\delta\alpha$ values in Eqs. (2) and (3)] compared with ligands forming mainly lipophilic interactions (larger δx and $\delta\alpha$ values). Greater discussion of the approximations we have adopted and comparisons with other approaches can be found in our previous work [8].

3.3
Theoretical Treatment of Fragment Linking

Figure 3.1 (a) and (b) schematically represent two fragments binding to adjacent pockets on a protein. The right-hand side of the figure shows how the total free energy of binding can be decomposed into the intrinsic binding affinities of each fragment (ΔG_{int}^A and ΔG_{int}^B) and the rigid body barrier to binding associated with the fragment's partial loss of translational and rigid body rotational entropy (ΔG_{rigid}).

Imagine joining these two molecules together to form a molecule C [shown schematically in Fig. 3.1(c)] and writing the expression for binding affinity of molecule C as:

$$\Delta G_{total}^C = \Delta G_{int}^A + \Delta G_{int}^B + \Delta G_{rigid} + \Delta G_{rot}^C + \Delta G_{strain}^C + \Delta G_{binding}^{A-B} \qquad (4)$$

The decomposition in Eq. (4) begins with the intrinsic binding affinities of fragments A and B, and the loss of rigid body entropy for molecule C. The next term, ΔG^C_{rot}, takes account of unfavorable entropic terms as a result of introducing new rotatable bonds which become frozen when molecule C binds to the protein. ΔG^C_{strain} takes account of any free energy loss that might have occurred as a result of either not presenting fragments A and B optimally to their respective pockets or as a result of strain introduced through the linker region not being in its lowest energy conformation. The final term, $\Delta G^{A-B}_{binding}$, accounts for any new direct favorable or unfavorable interactions with the enzyme that the linker might form. It also includes indirect interactions facilitated by the linker. For example, the act of joining the two parts together may perturb the solvation energies of A and B, or the presence of fragment A in molecule C can increase the intrinsic binding affinity of fragment B by forming part of the pocket where B binds.

It is useful to replace the first two terms of Eq. (4) using the expressions for the total free energies of binding of A and B shown on the right hand side of Fig. 3.1(a) and (b):

$$\Delta G^C_{total} = \Delta G^A_{total} + \Delta G^B_{total} - \Delta G_{rigid} + \Delta G^C_{rot} + \Delta G^C_{strain} + \Delta G^{A-B}_{binding} \quad (5)$$

We are interested in identifying a number of experimental examples in which the affinity of two fragments and the subsequent joined molecule are known; and this information can be used with Eq. (5) to estimate ΔG_{rigid}, providing that approximations for the final three terms can be derived.

In the subsequent analysis of experimental data, we use the approach of Mammen et al. [9] for the estimation of the entropic penalty associated with the introduction of additional rotatable bonds (that is, ΔG^C_{rot}). The entropic model is based on probabilities derived from torsional energy maps of force fields and yields entropies that vary according to the rotatable bond that is fixed – for example, freezing an sp^3–sp^3 rotatable bond between two carbons gives a free energy loss of 2.2 kJ mol^{-1} at 298 K. More details on our treatment of rotatable bonds and a discussion of alternative approaches are given elsewhere [8].

One of the main difficulties in applying Eq. (5) comes from estimating the strain energy, ΔG^C_{strain}. Most successful examples of joining fragments use flexible linkers and this maximizes the chance that: (a) the fragments A and B will be properly presented when they are joined together in molecule C and (b) that the linker region can adopt a low energy conformation. The only practical way of using Eq. (5) is to restrict the analysis to situations where it is reasonable to assume that the strain energy is approximately zero whilst realising that the calculated values for ΔG_{rigid} could be significant underestimates where these assumptions are erroneous [1].

The final term of Eq. (5) represents additional interactions between the joined molecule and the enzyme (or between the joined molecule and the solvent). These interactions are additional in the sense that they are not represented accurately by the sum of the intrinsic affinities of fragments A and B, and they may arise from a number of sources, as discussed above. In the analysis that follows, it is assumed that $\Delta G^{A-B}_{binding}$ can be neglected. This assumption is more exact for some examples than for others and a few different cases are considered elsewhere [8].

Under these assumptions, Eq. (5) can therefore be simplified to:

$$\Delta G^C_{total} \approx \Delta G^A_{total} + \Delta G^B_{total} - \Delta G_{rigid} + \Delta G^C_{rot} \qquad (6)$$

and should be applicable to situations where fragments, which do not directly interact with each other, are joined together in an ideal way so that they are still correctly presented to the protein using unstrained linkers that form no interactions with the protein. In our subsequent analysis, we restrict ourselves to experimental examples where there is structural information on the binding modes of molecules C and/or molecules A and B, so that the conditions of applicability of Eq. (6) are more likely to be met.

Equation (6) can be used to indicate the entropic consequence of linking ligands in an ideal or approximately ideal manner. Because the rigid body barrier to binding is quite large and because in some circumstances the decrease in entropy associated with freezing rotatable bonds is quite small, the expected free energy of binding of the optimally joined molecule is greater than the sum of the free energies of binding of the two fragments. This was first pointed out by Page and Jencks [1–3] and is illustrated schematically in Fig. 3.1(c). Another way of thinking about this is that the sum of the measured affinities for fragments A and B includes two unfavorable rigid body entropy barriers, whereas the measured affinity for molecule C includes only one such unfavorable term.

3.4
Experimental Examples of Fragment Linking Suitable for Analysis

We wish to estimate the rigid body barrier to binding, using Eq. (6) based on experimental examples in which two fragments, A and B, have been linked to yield a third molecule, C. Clearly we must restrict ourselves to situations where the affinities of all three molecules have been measured accurately using a similar technique and where the actual fragments A and B (rather than related or modified analogues) have been joined together. We also restrict ourselves to situations in which there is experimental information on the binding mode of the molecules because it is important to be able to assess whether: (a) the linker is in a strained conformation, (b) the linker forms strong interactions with the protein and (c) the fragments A and B form the same interactions when they are embedded in molecule C. In a previous study [8], we reviewed several examples of fragment linking and identified four examples that are suitable for evaluation with Eq. (6); and these examples are shown in the first four rows of Table 3.1.

Since that time there have been more examples of fragment linking in the literature, so it is appropriate to assess if they are suitable for inclusion in the analysis. We have gone through the recent reviews by Erlanson et al. [10] and by Rees et al. [11], looking for additional examples where there is structural information

Table 3.1 Examples of fragment linking for stromelysin (1) [19], avidin (2) [20], tryptase (3) [21], vancomycin (4) [22, 23] and glycogen phosphorylase (5) [16].

Example	Fragment A	Fragment B	Molecule C
(1)	K_i = 17 000 µM	K_i = 20 µM	K_i = 0.025 µM
(2)	K_i = 34 µM	K_i = 260 µM	K_i = 0.00000041 µM
(3)	K_i = 17 µM	K_i = 17 µM	K_i = 0.0006 µM (7-mer cyclodextrin scaffold)
(4)	K_i = 4.8 µM	K_i = 4.8 µM	K_i = 0.0011 µM
(5)	IC_{50} = 12.5 µM	IC_{50} = 12.5 µM	IC_{50} = 0.006 µM

and where the affinities of A, B and C appear to be precisely determined. For the agrochemical target, adenylosuccinate synthetase (AdSS), Hanessian et al. [12] linked together hydantocidin 5′-phosphate (HMP, 0.165 µM) and hadacidin (3.5 µM) to yield a 43 nM inhibitor. The tertiary complex of the two fragments in AdSS is available [13] and the fragments have very similar orientations to that exhibited in a complex of the joined molecule. However, in the complex of HMP without hadacidin [14], Arg303 forms good hydrogen bonds to the ligand and occupies a position that would clash with the ligand position of hadacidin observed in the tertiary structure. We therefore reject this example because the intrinsic affinities of HMP and hadacidin separately are unlikely to be a good approximation to the intrinsic affinities of these fragments in the joined molecule.

Another fragment linking case is presented on MMP3 in which 1-naphthyl hydroxamate (50 µM) is joined to various biaryl moieties and there are NMR derived structures for the complexes [15]. This example is rejected, because Hajduk et al. state that the relatively weak inhibitory potency of the linked compounds is due "to suboptimal positioning of the naphthyl and/or biaryl fragments for interaction with the enzyme".

The final example considered in detail here is shown as example 5 in Table 3.1, glycogen phosphorylase. Two individual copies of an indole-based inhibitor, CP-91149, were found bound simultaneously at the dimer interface between two monomers of glycogen phosphorylase [16]. Rath et al. simplified the indole to the fragment shown in Table 3.1 and then linked these fragments together to obtain a potent inhibitor for which the crystal structure has been determined. The difficulty in using this example directly in our analysis is that only the crystal structure of the joined molecule has been determined and, whilst the optimal positioning of the indole carboxamide portion of the fragment is not in doubt because it can be deduced from the crystal structure of the original molecule, CP-91149, the positioning of the flexible ethanolamine tail of the fragment is not known. If anything, the crystal structure of CP-91149 is suggestive of a different placement for the flexible side-chain than that observed in the crystal structure of the linked compound shown in Table 3.1. We decided to include this example but assume that the flexible ethanolamine chain of the fragment is not fixed in the enzyme and becomes fixed only in the joined molecule. (This amounts to assuming that nine rotatable bonds are fixed, rather than three, when these two fragments are joined together.)

3.5
Estimate of Rigid Body Barrier to Binding

Table 3.2 gives the values of ΔG_{rigid} that can be estimated from Eq. (6) using the entries in Table 3.1. The values obtained for ΔG_{rigid} range from 7 kJ mol^{-1} to 29 kJ mol^{-1}, correspondingly from 1.5 orders to 5.0 orders of magnitude in binding affinity. There are two reasons why we obtain this range of ΔG_{rigid} values. First, the assumption that ΔG_{rigid} is a constant is not ideal because the residual rigid body entropy retained by the ligand in the complex depends on the types of interactions formed between protein and ligand. In other words, there genuinely is a range of ΔG_{rigid} values, depending on the system. Second, the approximations in Eq. (6) may lead to under or overestimates of ΔG_{rigid}. For example, in the avidin case, ΔG_{rigid} is probably over-estimated because the intrinsic binding affinity of hexanoic acid (2b) may be much smaller in the absence of molecule 2a than when it is contained in the molecule 2c (this effect is discussed in detail elsewhere

[17]). In the tryptase example, ΔG_{rigid} is probably over-estimated because we decided that only four rotational bonds are introduced upon linking fragments A and B. However, the cyclodextrin linker is very large and not completely rigid and may also be forming favorable interactions with the enzyme.

Table 3.2 Estimates of from Eq. (6) for the examples shown in Table 3.1. Δn_{rot}^C is the number of extra rotatable bonds in molecule C that are frozen upon binding to the protein, compared to fragments A and B. ΔG_{rot}^C was calculated as outlined in the text using Mammen et al. [9].

Example	ΔG_{total}^A (kJ mol^{-1})	ΔG_{total}^B (kJ mol^{-1})	ΔG_{total}^C (kJ mol^{-1})	ΔG_{rot}^C (kJ mol^{-1})	Δn_{rot}^C	ΔG_{rigid} (kJ mol^{-1})
Stromelysin (1)	−10.1	−26.8	−43.4	+7.6	3.0	+14.1
Avidin (2)	−25.5	−20.5	−70.7	+4.2	2.0	+28.9
Tryptase (3)	−27.2	−27.2	−52.6	+8.5	4.0	+6.7
Vancomycin (4)	−30.3	−30.3	−51.1	+30.9	13.0	+21.4
Glycogen phosphorylase (5)	−28.0	−28.0	−46.9	+18.5	9.0	+9.5

In conclusion, ΔG_{rigid} varies with the system, but the actual range may well be smaller than that seen in Table 3.2, because the spread in Table 3.2 is also affected by the approximations of Eq. (6). On the whole, we think it is best to adopt 15–20 kJ mol^{-1}, i.e. about 3.0 orders of magnitude in the affinity, as a crude estimate of the rigid body rotational and translational barrier to binding, but to realize that it may vary, depending on the types of interactions formed in the complex.

3.6
Discussion

Unfortunately, only a handful of examples that are suitable for our analysis are available from the literature. However, in spite of the limited number of examples and the rather crude approximations we need to make in our analysis, we believe that the estimate we derive for ΔG_{rigid}, 15–20 kJ mol^{-1}, can be used to rationalize experience with fragment based drug discovery approaches.

In the context of the relatively large barrier to binding of approximately 3.0 orders of magnitude in affinity, it is useful to readdress the concept of 'ligand efficiency', LE, which was introduced recently by Hopkins et al [18], and who defined it simply as:

$$LE = -\Delta G_{total}/HAC \approx -RT \ln(IC_{50})/HAC \tag{7}$$

where HAC is the number of heavy atoms in the ligand and IC_{50} represents the measured potency of the ligand for the protein. Ligand efficiency provides a useful way to assess the quality of the interactions formed between a compound and a target, and it can be used to prioritize compounds to be taken forward for optimization. Taking into account the barrier to binding associated with the loss of translational and rotational entropy [Eq. (1)] allows us to define the *intrinsic* ligand efficiency as follows:

$$LE_{int} = -\Delta G_{int}/HAC \approx [\Delta G_{rigid} - RT\ln(IC_{50})]/HAC \tag{8}$$

As ΔG_{int} only reflects the interactions between protein and ligand, the intrinsic ligand efficiency may be a more appropriate estimate of the per-atom contribution to the affinity.

Compared to larger compounds, fragments need to be highly efficient binders, that is, form high-quality interactions with the target. This is a direct consequence of a high barrier to binding that is independent of molecular size. In order for a compound to bind to a target, it needs to compensate for the loss of rigid body entropy by forming favorable interactions with the target. Smaller compounds obviously have fewer atoms to form these favorable interactions than larger compounds and hence the interactions formed by fragments need to be more efficient.

The high entropic barrier to binding is also consistent with our observation that there are very few suitable examples of the binding of multiple fragments to adjacent pockets on a target. Only very efficiently binding fragments can be expected to overcome the entropic barrier. This means that sufficient affinity to overcome this substantial barrier needs to be available in both the adjacent pockets.

Even though fragment-sized hits are often weak binders (typically 30 µM to 3 mM), in our experience, optimizing them into low-nanomolar leads is generally tractable. This too can be explained in terms of a constant, and relatively high, barrier to binding. Consider a fragment A that binds to an enzyme with an activity of 100 µM (–23 kJ mol^{-1}). According to Eq. (1), fragment A is actually contributing approximately –40 kJ mol^{-1} of favorable interactions (using 17 kJ mol^{-1} for ΔG_{rigid}). Now assume that fragment A is developed into a much larger, potent drug molecule, C, that still contains fragment A and has a potency of 3 nM (–49 kJ mol^{-1}). Molecule C must overcome a similar rigid body entropic barrier to binding and forms roughly –66 kJ mol^{-1} of favorable interactions. Note that the majority of favorable interactions are provided by fragment A, despite the fact that molecule C is 33 000 times more potent than fragment A. This indicates that optimizing a weak-binding fragment hit into a nanomolar lead may not be as daunting a task as it at first appears.

It is interesting to analyse what happens if, in the above example, fragment A is removed altogether from molecule C. The new molecule will form –26 kJ mol^{-1} of favorable interactions to offset against the barrier to binding, leading to an affinity of approximately 29 mM. This would be considered inactive in any drug program that already had molecules in the nanomolar range. It follows that molecule C would exhibit hypersensitive SAR when changes are made to fragment A. We believe that this kind of analysis is consistent with much of medicinal chemistry experience on drug design projects. It suggests that activity in drugs is not evenly distributed through the molecule and this has important implications in the best strategies for drug discovery.

This uneven distribution of the activity throughout the molecule is related to the fact that the majority of the affinity available in a binding site is often concentrated in a key recognition pocket. If, in the above example, fragment A binds to such a key recognition pocket, then it is critical that fragment A is a perfect match for that pocket and that the rest of molecule C allows the optimal presentation of fragment A to its pocket. Historically, high-throughput screening (HTS) has focused on screening "drug-like" libraries and a typical HTS hit might be a drug-sized molecule with low-micromolar affinity. It is our belief that such a hit is most likely to derive its potency from low-quality interactions spread evenly through the molecule, rather than contain the perfect key fragment, optimally presented to its pocket. For this reason, HTS hits often prove to be difficult to optimize whereas the optimization of fragment hits is more tractable because they form the desired interactions in the key recognition pocket. In our experience, structure-based optimization of such fragment hits generally leads to potent, but smaller and simpler lead compounds (typically with MW 300–400). Hence, if additional molecular weight needs to be added to the compound in order to address issues that may arise at a later stage, then this is possible without generating compounds that are outside the "drug-like" space.

3.7
Conclusions

This chapter outlines an analysis to allow the calculation of the rigid body entropic barrier to binding, using examples from the literature where two fragments that bind to adjacent sites on the protein are linked together. Following Finkelstein and Janin [5], our analysis assumes that this barrier is independent of molecular size. From the analysis, the barrier to binding is estimated to be 15–20 kJ mol^{-1} (about 3.0 orders of magnitude in affinity). This estimate is crude, because of the limited number of suitable examples from the literature and because of the many approximations we have to make in our analysis. Nevertheless, we believe the estimate is useful in rationalizing experience with fragment based drug discovery approaches. We use the relatively high and constant barrier to explain that even low-affinity fragments are often highly efficient binders and that the optimization of a weak-binding fragment hit into a nanomolar lead compound is tractable.

References

1 Jencks, W. P. **1981**, On the attribution and additivity of binding energies, *Proc.Natl. Acad.Sci.USA* 78, 4046–4050.
2 Page, M. I., Jencks, W. P. **1971**, Entropic contributions to rate accelerations in enzymic and intramolecular reactions and the chelate effect, *Proc. Natl. Acad. Sci.USA* 68, 1678–1683.
3 Page, M. I. **1973**, The energetics of neighbouring group participation, *Chem. Soc. Rev.* 2, 295–323.
4 Dill, K. A. **1997**, Additivity principles in biochemistry, *J.Biol.Chem.* 272, 701–704.
5 Finkelstein, A. V., Janin, J. **1989**, The price of lost freedom: entropy of bimolecular complex formation, *Protein Eng.* 3, 1–3.
6 Gilson, M. K., Given, J. A., Bush, B. L., Mccammon, J. A. **1997**, The statistical-thermodynamic basis for computation of binding affinities: a critical review, *Biophys. J.* 72, 1047–1069.
7 Searle, M. S., Williams, D. H. **1992**, Partitioning of free energy contributions in the estimation of binding constants: residual motions and consequences for amide-amide hydrogen bond strengths, *J.Am.Chem.Soc.* 114, 10690–10704.
8 Murray, C. W., Verdonk, M. L. **2002**, The consequences of translational and rotational entropy lost by small molecules on binding to proteins, *J.Comput.Aided Mol.Des.* 16, 741–753.
9 Mammen, M., Shakhnovich, E. I., Whitesides, G. M. **1998**, Using a convenient, quantitative model for torsional entropy to establish qualitative trends for molecular processes that restrict conformational freedom, *J.Org.Chem.* 63, 3168–3175.
10 Erlanson, D. A., McDowell, R. S., O'Brien, T. **2004**, Fragment-based drug discovery, *J. Med. Chem.* 47, 3463–3482.
11 Rees, D. C., Congreve, M., Murray, C. W., Carr, R. **2004**, Fragment-based lead discovery, *Nat. Rev. Drug Discov.* 3, 660–672.
12 Hanessian, S., Lu, P. P., Sanceau, J. Y., Chemla, P., Gohda, K., Fonne-Pfister, R., Prade, L., Cowan-Jacob, S. W. **1999**, An enzyme-bound bisubstrate hybrid inhibitor of adenylosuccinate synthetase, *Angew. Chem. Int. Ed. Engl.* 38, 3159–3162.
13 Poland, B. W., Lee, S. F., Subramanian, M. V., Siehl, D. L., Anderson, R. J., Fromm, H. J., Honzatko, R. B. **1996**, Refined crystal structure of adenylosuccinate synthetase from *Escherichia coli* complexed with hydantocidin 5'-phosphate, GDP, HPO4(2-), Mg2+, and hadacidin, *Biochemistry* 35, 15753–15759.
14 Fonne-Pfister, R., Chemla, P., Ward, E., Girardet, M., Kreuz, K. E., Honzatko, R. B., Fromm, H. J., Schar, H. P., Grutter, M. G., Cowan-Jacob, S. W. **1996**, The mode of action and the structure of a herbicide in complex with its target: binding of activated hydantocidin to the feedback regulation site of adenylosuccinate synthetase, *Proc. Natl. Acad. Sci.USA* 93, 9431–9436.
15 Hajduk, P. J., Shuker, S. B., Nettesheim, D. G., Craig, R., Augeri, D. J., Betebenner, D., Albert, D. H., Guo, Y., Meadows, R. P., Xu, L., Michaelides, M., Davidsen, S. K., Fesik, S. W. **2002**, NMR-based modification of matrix metalloproteinase inhibitors with improved bioavailability, *J. Med. Chem.* 45, 5628–5639.
16 Rath, V. L., Ammirati, M., Danley, D. E., Ekstrom, J. L., Gibbs, E. M., Hynes, T. R., Mathiowetz, A. M., McPherson, R. K., Olson, T. V., Treadway, J. L., Hoover, D. J. **2000**, Human liver glycogen phosphorylase inhibitors bind at a new allosteric site, *Chem. Biol.* 7, 677–682.
17 Olejniczak, E. T., Hajduk, P. J., Marcotte, P. A., Nettesheim, D. G., Meadows, R. P., Edalji, R., Holzman, T. F., Fesik, S. W. **1997**, Stromelysin inhibitors designed from weakly bound fragments: effects of linking and cooperativity, *J. Am. Chem. Soc.* 119, 5828–5832.
18 Hopkins, A. L., Groom, C. R., Alex, A. **2004**, Ligand efficiency: a useful metric for lead selection, *Drug Discov. Today* 9, 430–431.
19 Hajduk, P. J., Sheppard, G., Nettesheim, D. G., Olejniczak, E. T., Shuker, S. B., Meadows, R. P., Steinman, D. H., Carrera, G. M., Jr., Marcotte, P. A., Severin, J., Walter, K., Smith, H., Gubbins, E., Simmer, R., Holzman, T. F., Morgan, D. W., Davidsen, S. K., Summers, J. B.,

Fesik, S. W. **1997**, Discovery of potent nonpeptide inhibitors of stromelysin using SAR by NMR, *J. Am. Chem. Soc.* 119, 5818–5827.

20 Green, N. M. **1975**, Avidin, *Adv. Protein Chem.* 29, 85–133.

21 Schaschke, N., Matschiner, G., Zettl, F., Marquardt, U., Bergner, A., Bode, W., Sommerhoff, C. P., Moroder, L. **2001**, Bivalent inhibition of human beta-tryptase, *Chem. Biol.* 8, 313–327.

22 Rao, J., Whitesides, G. M. **1997**, Tight binding of a dimeric derivative of vancomycin with dimeric L-Lys-D-Ala-D-Ala, *J. Am. Chem. Soc.* 119, 10286–10290.

23 Rao, J., Lahiri, J., Isaacs, L., Weis, R. M., Whitesides, G. M. **1998**, A trivalent system from vancomycin.D-ala-D-Ala with higher affinity than avidin.biotin, *Science* 280, 708–711.

4
Location of Binding Sites on Proteins by the Multiple Solvent Crystal Structure Method

Dagmar Ringe and Carla Mattos

4.1
Introduction

One of the greater challenges of structural biology is to characterize proteins in enough detail so that the structures can be used for computational prediction of function, and for the design of compounds that can be used to modulate or to interfere with that function. In general, the function of a protein is associated with its interaction with other molecules, either small molecules or other large biological molecules, such as nucleic acids or proteins. Invariably, the function can be disrupted when that interaction is prevented. A complete industry exists solely for the purpose of finding such disruptors – they are called drugs. The discovery of drugs had been a serendipitous process, in an age when very little was known about the structures of the small molecules or their targets. The process of discovery has become more sophisticated as such structures became available.

The success of this process relies on knowing where an interaction site is located on a protein and characterizing the properties of that site. The first step is therefore the location of such a site. The properties of the site can then be used to design a molecule that fits it perfectly in terms of size, shape and chemical properties. One method to obtain such properties is to map a site with chemical probe molecules. Ideally these probe molecules can be connected to form a larger molecule that takes advantage of as many of the potential interaction partners in the site as possible. Finally, when the process of finding and characterizing becomes robust enough, the process can be accomplished computationally.

At this moment in time, locating a binding site on a protein without any extraneous information, such as a fortuitously bound ligand in the crystal structure, is only marginally successful. The characterization of a binding site is still very laborious when done experimentally and is not robust when done computationally. Thus, mapping such a site with chemical probes could help characterize the site, the map then being used to design a specific ligand unique for the site. In addition, such a map can provide the data from which computational approaches can be calibrated. One way in which mapping can be achieved is to use a variety of small probe molecules and observe where they bind to the surface of the protein.

Fragment-based Approaches in Drug Discovery. Edited by W. Jahnke and D. A. Erlanson
Copyright © 2006 WILEY-VCH Verlag GmbH & Co. KGaA, Weinheim
ISBN: 3-527-31291-9

The probes can represent a diverse set of chemical functionalities, such as those found in ligands, except as individual entities rather than linked into a single molecule. Any one probe could bind to the surface of the protein in a region that is an energy minimum for the affinity of the chemical functionality of the compound.

4.2
Solvent Mapping

The process of solvent mapping involves the use of small organic probe molecules to map protein surfaces. The method, called multiple solvent crystal structures (MSCS), was first suggested by Allen et al. [1] in an application to porcine pancreatic elastase. The method is discussed in detail by Mattos and Ringe [2]. In brief, the method depends on the structure determination of a protein in the presence of an organic solvent and the analysis of the surface binding sites at which the organic molecules appear. When repeated with a different organic solvent, a new pattern of solvent-binding sites emerges. Clustering of solvent molecules at a single site seems to identify ligand-binding sites, both catalytic and recognition. This method essentially provides an experimental approach to determine the binding site and key interactions of the fragments with the target protein. The mapping of binding sites in terms of the interactions that such sites make with other molecules is crucial to an understanding of the function of a protein and to the ability to design specific ligands that can interfere with that function.

The function of a protein depends on the ability of other proteins or small molecules to interact specifically with a portion of the surface. These interactions are localized to distinct sites, very often only one per protein. In the cases of enzymes, the critical sites are active sites, usually designed to recognize a specific substrate and to promote a chemical reaction with a substrate that is unique. Alternatively, the purpose of the site may be for recognition only, in which case the ligand may induce a function that is lacking in the target protein alone. In either case, the interactions are unique, both in terms of the chemical entities recognized and in the orientation of the ligand on the surface of the protein. Since these interactions occur only at specified sites on a protein, the notion exists that these sites have special properties designed to recognize the precise characteristics of the ligands. Therefore, if these properties can be predicted, it should be possible to identify the sites themselves. The characterization of binding sites can be used to develop specific ligands, particularly inhibitors that can interfere with the natural action of a protein by blocking access to important binding sites. The MSCS method described here enables one to design such inhibitors.

4.3
Characterization of Protein–Ligand Binding Sites

The MSCS method was not possible before the age of structural biology. Rather, the characterization of protein–ligand binding sites depended on complex biochemical investigations. Primarily through the use of kinetic experiments, these investigations sought to discover several types of information: the identities of the residues that constitute the binding site, their arrangement on the protein surface and the affinity of a ligand to that binding site. The identity of residues involved in the interaction can be determined using chemical probes that derivatize, and thereby inactivate, a site. The arrangements of residues in a binding site can only be determined indirectly, by using a variety of inhibitors to probe the site and measure their affinities. Mapping a binding site in terms of the ligand can thus be accomplished by using comparisons of similar, but systematically different, compounds that bind to a site. The properties of the ligands help to define the properties of the interaction sites. The mapping of binding sites by kinetic experiments can thus be a very powerful method, but requires the basic assumption that the geometry of binding is the same for substrates and their analogs, and that their binding behaviour is therefore analogous. If that assumption does not hold, the results can be conflicting and misleading. In general, kinetic methods only detect binding sites that influence rates of reactions, but they give no clues as to the existence and properties of other binding sites not located in close proximity to the active site.

Other approaches are also available to determine the arrangements of residues in a binding site. One involves a laborious process of systematic derivatization and mutation of residues of a peptide or protein ligand, in order to determine the effects of these changes on a biological response, which then provide spatial information about the binding surface and the characteristics of that surface. One version of this process that is widely used for detecting and analyzing binding interfaces is that of alanine scanning mutagenesis [40, 41, 42, 43]. This technique, which was first applied to the human growth hormone and its receptor, consists of the systematic mutation of surface residues to Ala, with concomitant measurement of the effect on the binding constant for the protein–ligand interaction [3]. The seminal work on the human growth hormone/receptor complex showed that a few hydrophobic interactions contribute a significant portion of the binding affinity in this largely hydrophilic interface. These areas were termed "affinity hot spots" or simply hot spots.

Once structures were available for proteins the task became easier, especially if the structure of a ligand–protein complex could be determined. Given a known inhibitor of the reaction of an enzyme, especially if the compound in question bears a resemblance to the substrate, the structure of the complex identifies the binding site, the residues that make important interactions, and potentially, the position and orientation of the substrate. Such information provides a template from which the chemist can model viable compounds. If the inhibitor binds in a position or orientation that does not mimic the binding of the substrate, the information can be used to design ligands that extend beyond the active site and reach into other grooves on the surface of the protein. This information could be very important in the development of inhibitors that serve as drugs, since such compounds could potentially have fewer undesirable side-effects. The MSCS method makes it possible to find such binding sites that are further removed from the active site and therefore are likely to be unique to a protein in a structural family of homologous members. This provides specificity of the ligand for the target protein.

The process whereby a ligand binds to a protein is thought to involve recognition by the ligand of a site on the protein, followed by conformational rearrangements that optimize packing at the interface [4, 5, 6]. Data base analyses show that ligand-binding sites do not follow any general patterns of hydrophobicity, shape or charge [7]. Although a few guiding principles have been uncovered, we have not yet discovered a single parameter that can be used to distinguish a binding site from the rest of the protein surface [8], nor have we observed a general pattern for all interfaces [6]. Therefore, in order to effectively design specific ligands for a specified site, the structural biologist must fully understand the characteristics of a binding site that distinguish a ligand-binding site from any other site [9]. The two most important goals to be considered when designing specific ligands are tight binding and specificity. These properties can be optimized by using the geometry of the interactions between ligand and protein as a variable. Finally, it should be added that it is becoming clear that plasticity may play a major role in protein–ligand interactions [10].

The past two decades have witnessed considerable progress in the study of protein interactions, and on ligand design in particular. Experimental approaches involving enzyme kinetics, structure determination [11] and combinatorial chemistry [12], combined with computational strategies used to analyze charge distribution, surface shape and energetics of ligand binding have contributed greatly to elucidating the important factors involved in ligand recognition, binding and specificity. For example, kinetic data on the inhibitors of a target protein become extremely useful when accompanied by a study of the crystal structures of the inhibitor/protein complexes. A prime example of this sort of study involves the extensive analysis on the complexes of elastase with trifluoracetyl-dipeptide-analide inhibitors [11]. This has resulted in the design of a new ligand that simultaneously combined functional groups of at least two of the original inhibitors [13]. Inhibitors can also be optimized to make new drugs by using the trial and error method to modify previously known small molecule ligands. This has recently been coupled with combinatorial chemistry methods in the exploration of ligand–protein interactions [12].

A number of ligand design strategies have been developed that take advantage of the latest advances in the basic understanding of protein-binding surfaces. Of particular interest is the idea that functional groups can be optimized independently for different regions of a protein-binding site [14, 15]. These functional groups can then be linked to form a ligand with high affinity and specificity to the target protein [16, 17]. The protein-binding affinity of the resulting molecule will be, in principle, the product of the binding constants for the individual fragments plus a term that accounts for changes in binding affinity due to the linker portion of the larger ligand [18].

4.4
Functional Characterization of Proteins

The function of a protein is always based on the ability to interact with a ligand; and the site at which the interaction occurs is usually unique. If we understand the characteristics of a binding site, we can predict its location and subsequently the precise ligand that is recognized. The recent explosion of genomic information has made numerous new targets available for therapeutic intervention. Most of this information comes in the form of sequences of gene products. Unfortunately, the functions of many of these gene products are unknown and the sequences often do not contain clues to help us understand them.

In order to help define the functions of these proteins, three-dimensional structures are being determined at a rapid rate. This has caused structural biologists to focus on three major areas: the determination of protein function, the locations of the sites of protein–protein and/or protein–ligand interactions and the problem of ligand design aimed at the control of disease. The factors that determine these areas are ultimately controlled at the physico-chemical level by the surface features of the protein. These features include surface shape, electrostatics, hydrophobicity, dynamics and solvation. In particular, the crucial role that solvent molecules play in the dynamics and function of proteins has become increasingly apparent [19, 20]. Thus, the current challenge in structural biology is to understand the general features that underlie protein–protein and protein–ligand interactions.

Under ideal conditions, we would be able to determine function computationally. However, a primary challenge in the computational approach is the identification and characterization of binding sites and the difficulty in predicting optimal binding modes for probe fragments. Until recently, there had been little experimental data for a comparison with computational results. The MSCS method provides these experimental data, as well as primary data for molecule-building from fragments. In fact, experimental and computational methods can, in principle, be complementary to each other, as can be seen in the following discussion.

4.5
Experimental Methods for Locating the Binding Sites of Organic Probe Molecules

Examples of organic solvent molecules bound to the surface of a protein have been observed before, sometimes by accident. For instance, the use of an organic solvent as precipitant in crystallization of a protein leads to such solvent molecules appearing in the electron density map. In a study that was not directed at the effect of organic solvents on protein structure, ribonuclease A was crystallized from

50% (v/v) iso-propanol, with the result that eight molecules of the solvent were detected [21]. An alternative study aimed at determining the effect of acetonitrile on the structure of subtilisin showed discrete organic solvent molecules bound to the surface of the protein along with a shell of water molecules, even though the crystal of the enzyme had been transferred from an aqueous solution to one of neat acetonitrile [22, 23]. A similar experiment involved crystals of gamma-chymotrypsin grown in aqueous solution soaked in n-hexane. Seven hexane molecules along with a shell of water molecules were found in the electron density maps. Two of the seven hexane molecules are found near the active site; and the rest are close to hydrophobic regions on or near the surface of the enzyme [24, 25].

4.6
Structures of Elastase in Nonaqueous Solvents

In addition to these fortuitous observations, the first and most comprehensive intentional study of solvent binding to the surface of a protein was conducted with elastase, with surprising results [57]. The crystal structures of elastase were solved in the presence of neat acetonitrile (ACN) [1], 95% acetone (ACE; red code 2FO9), 55% dimethylformamide (DMF; yellow, 2FOC), 80% 5-hexene1,2-diol (HEX; purple, 2FOE), 80% isopropanol (IPR; blue, 2FOF), 80% ethanol (ETH; green, 2FO9) and two structures solved in 40% trifluoroethanol (TFE1 and TFE2; magenta/pink, 2FOG, 2FOH). In addition, crystal structures were solved in a mixture of 40% benzene, 50% isopropanol and 10% water (IBZ; blue, 2FOA) and one in 40% cyclohexane, 50% isopropanol and 10% water (ICY; cyan, 2FOB). Isopropanol was bound to elastase in these last two structures, but no bound benzene or cyclohexane molecules were observed in the electron density maps. By overlaying these ten structures and the structure of cross-linked elastase solved in aqueous solution [1], patterns of the bound organic solvent molecules on the surface of the protein could be discerned. In addition, patterns of hydration and plasticity of the protein structure in response to the changing solvent environments became apparent.

Overall, the structures of the protein have not undergone major global changes. However, the structure of a protein can respond to changes in solvent environment locally. When the C-alpha traces of the 11 elastase models are superimposed, main chain rmsd values range from 0.11 Å^2 (between IBZ and ICY models) to 0.33 Å^2 (between ACE and the aqueous model). These overall average deviations are misleading, because the variations in backbone structure are greater than elsewhere on the protein in four areas. For instance, the deviations in C-alpha positions around Asn 25, Ser 39, Ser 122 and Ser 177 are greater then 1 Å. Of these, Ser 39 and Ser 177 have residue B-factors greater then 40 Å^2 in both the aqueous structure and in the organic solvent structures. Furthermore, their positions in the superimposed models are continuous within the 1 Å range, which together with the high B-factors, is consistent with disorder in the structure. In contrast, residues 24–27 (RNSW) and 122–123 (SY) have B-factors around 20 Å^2 and adopt one of two very distinct backbone and side-chain conformations. Such changes in secondary structure, although subtle, were also observed in the structure of subtilisin in neat acetonitrile. The peptide from residues 156–166 occurs as a loop of indeterminate structure in aqueous solution, but forms a tight turn in the organic solvent.

When the comparison of the elastase structures includes all atoms (rmsd values range from 0.39 Å^2 for IBZ and ICY models to 0.87 Å^2 for ACN and IBZ models), most of the differences between them occur in the conformations of the side-chains. A few notable exceptions occur where the changes in conformations of the side-chains are correlated with changes in backbone configuration. An example of the correlation between secondary structural change and side chain occurs in the region around Asn 25. The peptide bond between Arg 24 and Asn 25 adopts two conformations, depending on whether the Asn is oriented towards the higher dielectric constant solvents or away from the solvent, when the Asn is tucked into the protein (as in the case of solvents with lower dielectric constants). This difference in peptide bond conformation leads the backbone through two different paths that meet again at the C-alpha atom of Trp 27. In order for Asn 25 to pack closer to the protein surface in the low dielectric constant solvents, Tyr 123 must rotate nearly 40 degrees about the chi1 dihedral angle so that the C-gamma atom moves about 1 Å and the side-chain oxygen atom moves about 3.5 Å from its original position. In this new conformation, the H-bonding interactions of the Asn side-chain are with protein atoms rather than with solvent. Ser 26 is also turned toward the solvents in the group of models containing the aqueous structure and buried in the protein solved in the solvents with low dielectric constants. In the model determined from crystals in aqueous solution and others in the higher dielectric constant group, the O-gamma of Ser 122 makes an OH aromatic interaction with the aromatic ring of Tyr 123. With Ser 122 in the orientation observed in the structures solved in low dielectric constant solvents, this polar–aromatic interaction is no longer possible and Ser 122 is oriented toward the solvent.

Thus the extent to which parts of the models deviate from each other can be used to place the models into two distinct groups. One contains the aqueous model, ACN, DMF, HEX and TFE2 models, obtained from elastase in a mother liquor containing a solvent of dielectric constant ≥ 27. The other includes TFE1, ETH, ACE, IPR, IBZ and ICY models, the elastase structures solved from crystals in a mother liquor containing a solvent of dielectric constant < 27. Changes in secondary structure observed for elastase can therefore be correlated with the type of organic solvent used in the mother liquor.

4.7
Organic Solvent Binding Sites

If the locations of solvent molecules on the surface of the protein are to be used for mapping surface properties that can be associated with functional sites, the locations of these solvent-binding sites should be discrete and unique. The MSCS experiments confirm this expectation. In general, only a few crystallographically visible organic solvent molecules were found in each of the models. In each of the ten experiments mentioned above, no great difference in the number of organic solvent molecules was found, despite the fact that the fraction of organic solvent varies from >99% for ACN to 40% for TFE1 and TFE2. The important point is that, even at very high organic solvent concentrations, only a few molecules of the solvent are observed binding to elastase at positions that are not randomly distributed on the protein surface. Nevertheless the number of sites occupied by a particular solvent has been shown to be concentration-dependent [26], consistent with the MSCS method for studying the relative affinities of binding sites for a particular solvent or functional group. This elastase study is not concerned with relative binding affinities of particular solvent molecules, but rather focuses on the structures of elastase at the highest achievable concentrations of each solvent in order to locate the maximum number of binding sites for the solvents. This strategy illustrates that the MSCS method is a way of locating and mapping binding surfaces.

Altogether, a total of 60 organic molecules observed in 31 unique sites can be identified bound to the surface of the protein. The majority (35) of the organic molecules cluster in one region of the protein surface and specifically the active site. The remaining molecules are found distributed over the protein surface without any obvious clustering. In general, the positions of the probe molecules are stabilized through hydrogen bonds to polar groups, if possible, and van der Waal's interactions with the hydrophobic parts of protein side-chains. The implication is that these probe molecules bind to regions of the protein surface that are more hydrophobic than hydrophilic. When the models of elastase in different organic solvents are superimposed, a pattern becomes apparent in which a few binding sites on the protein surface bind a variety of different types of organic solvent molecules, while most other sites are individual niches for a particular type of molecule. Clustering of different types of organic solvent molecules on the surface of elastase occurs prominently in the active site, where five of the known subsites on the protease are clearly delineated by the organic solvents (Fig. 4.1).

The active site of elastase was extensively characterized by kinetic experiments in the mid-1970s with substrates of different length and different amino acid composition [27]. This study identified five subsites on the acyl enzyme intermediate side of the active site (S_1–S_5) and three on the leaving group side (S'_1–S'_3) that play an important role in the catalytic efficiency of peptide hydrolysis. Small aliphatic amino acid residues were shown to be favored in the S_1, and S_4 subsites, with lysine preferred at S_2. The S_3 subsite is highly exposed to solvent and has a less distinct preference for any given amino acid residue.

The active site binding pockets in which organic solvents are found in the elastase structures are the S_1, S_4, the oxyanion hole and two sites on the leaving group side of the catalytic triad, likely to be the S'_1 and S'_3 binding pockets. Since in most of the existing crystal structures of elastase/inhibitor complexes, the inhibitor binds in the S subsites, little is known about the exact location of the S' sites. These five pockets are each observed to bind at least three different types of organic solvent molecules.

In addition, there is a single acetonitrile molecule at the entrance to the S_2 pocket [1], making a total of six binding sites for organic solvents in the active site.

The clustering and distributions of the probe molecules within these sites overlays precisely the possible binding modes of inhibitors of elastase (Fig. 4.2). For instance, trifluoroacetyl-Lys-Pro-*p*-isopropylanilide (gray; 1ELA) binds with the trifluoroacetyl group in the S1 subsite, the lysine side chain at S2, the proline side chain at S3 and the anilide group at S4, while trifluoroacetyl-Lys-Phe-*p*-isopropylanilide (black; 1ELC) binds with the trifluoroacetyl group in the oxyanion hole, the phenyl side chain at S1, the lysine side chain at S2, and the anilide group at a subsite that had previously been observed in only one inhibited structure [2]. That subsite was subsequently utilized in an inhibitor (Fig. 4.3; black; 1BMA) designed specifically to place a group into it [13].

In addition to locating binding sites, solvent mapping also gives information about the plasticity of the protein. As already mentioned, the protein may respond locally to the presence of ligands by adjusting conformationally to their presence. For instance, the S1 subsite of elastase is a site at which the residues Gln200 and Ser203 show such adjustments. In the superposition of all solvent exposed structures, the positions of the backbone atoms of the strand containing these residues are indistinguishable from one structure to the other. The side-chains, however, are in slightly different positions in each model and provide key areas of plasticity on either side of the probe ligand within the binding pocket. Thus the probes provide not only an identification of the active site, but also the potential plasticity of this region of the protein.

The remaining binding sites can be characterized by the type of location on or near the surface of the protein where they are found: at crystal contacts, in a channel or randomly distributed on the surface away from the active site or from a crystal contact. Most of these sites do not exhibit the type of clustering that is observed in the active site. For instance, organic solvent molecules binding in crystal contacts are found near each other, defining four areas where different protein molecules interact in the crystal. In fact, this type of clustering can be used to identify crystal contacts and to potentially distinguish them from active sites.

4.8
Other Solvent Mapping Experiments

A more comprehensive study of crystals exposed to organic solvents, similar to that of elastase, was aimed at mapping the binding surface of thermolysin. The crystal structures of thermolysin in different concentrations of isopropanol [26] and in three other organic solvents [28] were determined, the authors observing that the organic solvents cluster in the known substrate-binding site of the enzyme and appear in some areas of crystal contacts. In the first set of experiments, crystals soaked in 2–100% isopropanol caused only minor changes to the conformation of the protein. An increasing number of isopropanol interaction sites could be identified as the solvent concentration was increased. Isopropanol occupied all four of the main subsites in the active site, although this was only observed at very high concentrations of isopropanol for three of the four subsites.

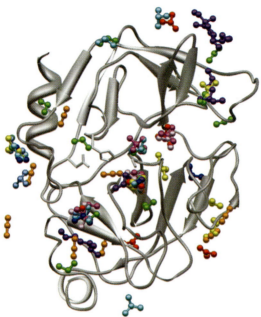

Fig. 4.1
Solvent molecule distribution on the surface of elastase.

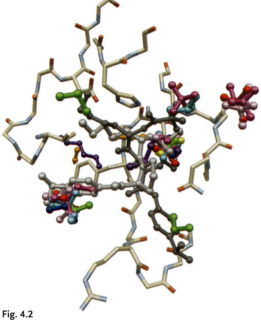

Fig. 4.2
Clustering of organic solvent probe molecules in the active site region of elastase overlaid with the models of two inhibitors.

4.8 *Other Solvent Mapping Experiments* | 77

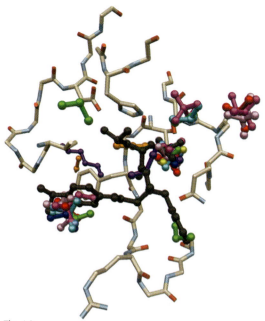

Fig. 4.3
Overlay of a designed inhibitor with solvent molecule clusters.

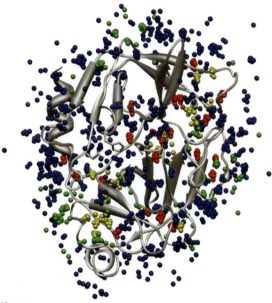

Fig. 4.4
Water molecule distribution on elastase. A water is identified-
according to the site on the structure in which it was observed.

Such experimentally determined positions of isopropanol are consistent with the structures of known protein–ligand complexes of thermolysin [26].

In the second set of experiments, high-resolution crystal structures of thermolysin were determined with crystals soaked in 50–70% acetone, 50–80% acetonitrile and 50 mM phenol. The structures of the protein in the aqueous–organic mixtures are essentially the same as the native enzyme; and a number of solvent interaction sites were identified for each probe molecule. After superposition of the individual models, clusters in the main specificity pocket of the active site and a buried subsite were observed for the distribution of probe molecules. The experimentally determined solvent positions were compared with predictions from two computational functional group mapping techniques, GRID and multiple copy simultaneous search (MCSS). Analysis of the probe positions, computed for isopropanol, shows little correlation with interaction energy computed using a molecular mechanics force field. Again, the experimentally determined clusters of small molecule positions are consistent with the structures of known protein–ligand complexes of thermolysin [28].

In addition to crystallography, NMR can also be used to map protein-binding sites using organic probe molecules. The original demonstration of this technique, called SAR by NMR [18], is a linked-fragment approach and was used to discover ligands that bind tightly to the protein that forms a complex with the potent immunosuppressant FK506, the FK506-binding protein (FKBP). Small organic molecules that bind to proximal subsites of the protein were identified, optimized and linked together to produce high-affinity ligands. NMR spectroscopy has also been used to map protein-binding sites using organic solvent probes in aqueous solution [29]. The specificity-determining substrate-binding sites of hen egg-white lysozyme, for example, could be readily identified in aqueous solution using small organic solvent molecules as probes [29]. Recently, the surface of the *Escherichia coli* peptide deformylase was also mapped using organic solvents by NMR [30]. One of the advantages of NMR over crystallography is the fact that surface regions blocked by crystal contacts are free. However, the NMR method cannot be used with higher solvent concentrations due to the instability of the protein in such solutions.

4.9
Binding of Water Molecules to the Surface of a Protein

The binding of organic probe molecules to a protein does not preclude the binding of water molecules to the surface [31]. In any one crystallographic experiment, somewhere between 100 and 200 electron density peaks are interpreted as water molecules in the electron density maps of the structures of elastase, either in aqueous solution or in the presence of an organic solvent. The locations of these water molecules are not constant from one structure determination to another. Some of the identified waters are found in the same location in every experiment and some are in different locations. Some regions of the surface do not have any electron density identifiable as water molecules at all. However, the complete surface of the protein is covered with water molecules when all of the structures being compared in this study are superimposed [32] (Fig. 4.4).

The selectivity for water molecules on the surface is independent of the fraction of water present in each of the 11 experimental conditions used to obtain the models discussed in the elastase study. For instance, approximately the same number of water molecules are found in the ACN and ACE models (<1% water and 5% water, respectively) as in the alcohol structures (20% water) and in the DMF model (45% water). Many of these waters are observed at the same or at a very close site on the protein, leading to a clustering when all of the waters are superimposed on one structural background. When counting all of these waters, no matter how often they are observed, there are 247 water molecule clusters (within 1 Å of each other) and 178 waters that are observed in only one structure. Similar to the classification of solvent molecule-binding sites, these waters can be classified on the basis of their locations relative to the protein and each other. This classification is based on location within the protein, on the surface of the protein and in the active site, although these last water molecules are technically on the surface of the protein.

4.10
Internal Waters

Two types of internal waters can be characterized: individual molecules in the interior (Fig. 4.4; red) and those found in channels (yellow). An internal water is defined as a water molecule in the interior of the protein and not in direct contact with the surface of the protein, or as a water molecule that has at least three strong hydrogen bonds (~2.8 Å) to the protein and, typically, a fourth hydrogen bond to either the protein or to another crystallographic water. The characterization of water molecules in the interior of the protein is based on the following crystallographic criteria: these waters show little displacement within a cluster (generally less than 0.5 Å between the farthest lying members) and those with low average B-factors close to that for main-chain atoms. In the reference structure (aqueous model), the internal waters are characterized as having at least three strong hydrogen bonds to the protein and an additional interaction to either the protein or a neighboring water molecule (Fig. 4.5). In the solvent exchanged structures, these interactions are generally conserved. However, an individual structure containing an interaction that is not conserved typically replaces the lost interaction with another one, thereby satisfying the requirement for the number of interactions of an internal site, but lowering the average number of conserved interactions within this classification.

The second type, channel waters, are those in which the water molecule is within hydrogen-bonding distance to a buried water or another channel water. The water molecules that occupy these sites typically have fewer hydrogen bonds than buried waters and are in hydrogen-bonding distance to at least two other water molecules. These waters typically extend inward from the surface. Such waters are often conserved in comparing multiple structures of the same protein.

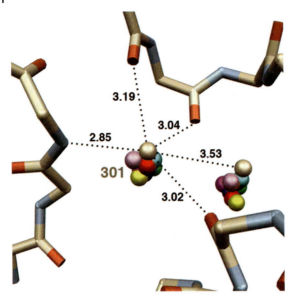

Fig. 4.5
Example of an internal water molecule, water 301. Note the potential hydrogen bonding interactions available to the molecule.

Thus, eight of the ten channel waters that are observed in all of the structures determined from crystals in aqueous medium are observed in all of the structures determined from solvent-exchanged crystals. Four waters are observed to exist only in the structures determined from crystals exchanged into organic solvents. All others are sometimes observed in any one structure but not in others.

In addition to the observation of water molecules within these channel sites, there are also solvent molecules clearly defined, including a hexenediol molecule in one of the channels. For example, one hexenediol molecule points its polar end towards the center of elastase inside a deep channel and thereby replaces a number of water molecules. This channel is also occupied by acetonitrile, which replaces two water molecules. Within another channel, a dimethylformamide molecule displaces one water molecule. Thus, although the binding sites at which these waters occur are well conserved, the water molecules that occupy these sites are exchangeable with organic solvents in at least two different channels.

4.11
Surface Waters

Several types of surface waters can be characterized. One consists of crystal contact waters (green) in which there is a hydrogen bond (< 3.5 Å) to the reference protein molecule and a polar interaction or a Van der Waals contact (< 4.0 Å) to a polar protein atom in a symmetry-related protein in the asymmetric unit of the crystal.

Another consists of active site waters (blue), in the case of elastase, defined as lying within 3.5 Å of any residue known to be in the S_1–S_4 subsites of elastase. Interestingly, most of these water molecules are not well conserved from one structure to another and have slightly higher than average B-factors. This is a general observation for water molecules in the active site regions of most enzymes. A total of 21 unique sites at which a water molecule is observed are found in the active site of elastase; and most are occupied in only two to four of the solvent structures. In fact, either a water molecule or a solvent molecule occupies only two protein sites in all of the solvent structures. One is in the S_1 sub-site, and the other is located at the position of a new sub-site that had not been identified in any of the inhibitor structures. These two sites are actually very highly conserved. A water molecule is always observed in these sites except when displaced by an inhibitor.

Finally, many waters are on the surface (blue) with only one hydrogen bond to a polar protein group. These sites are not well conserved and any one may be observed in only one of the structures. A comparison of the solvent exchanged structures with the two structures of elastase with inhibitors bound shows that most of the active site water molecules lie in regions occupied by one of the two inhibitors (Fig. 4.2; trifluoroacetyl-Lys-Pro-p-isopropylanilide (1ELA) or trifluoroacetyl-Lys-Phe-p-isopropylanilide (1ELC), respectively) [11]. In other words, most of the waters in the active site area are displaced by the incoming inhibitors. In fact, it is a general observation that these waters are displaced by incoming ligands, be they inhibitors or solvent molecules. Thus, these waters seem to be easily displaced, a requirement for facile binding to and recognition of a ligand by the protein.

4.12
Conservation of Water Binding Sites

Several general features of water binding sites can be used to characterize them semiquantitatively. One is the frequency with which a water molecule is observed in a particular water binding site. In other words, in how many of the structures we are analyzing is a water molecule observed at a specific site. Second, does frequency of observation at a site correlate with the type of water molecules as characterized previously (i.e. in terms of location in or on the protein). In other words, are internal waters found more often in an internal water binding site than are surface waters at a surface water binding site. Finally, what type of interactions, and how many, can a water molecule make with a water binding site? A site will have a number of nonconserved hydrogen-bonding partners that are available to interact with a water molecule (or an organic solvent molecule). Since water can only make a maximum of four such interactions, two hydrogen bond-donating and two hydrogen bond-accepting, these interaction partners must be complementary. However, any one site may have more than this number of potential partners, allowing any one water molecule to have a set of interactions that may not be identical every time the site is observed. A subset of these interactions may indeed be made every time a water molecule interacts with a particular site and can be considered conserved (Fig. 4.5).

In general, the sites at which water molecules are observed most frequently (in 9–11 of the structures) correlate with internal and channel sites. Waters at these sites tend to have B-factors similar to those of the surrounding protein atoms, indicating that either the water site is fully occupied in a crystallographic sense, or the water position is well defined crystallographically, or both. These sites tend to have higher numbers of conserved interactions between protein and water and tend to cluster tightly, in terms of the distances between member water molecules. Sites at which a water molecule is found least frequently (in 1–4 of the structures) correlate with surface sites. These sites are often associated with side-chains that are found in multiple positions. Surprisingly, these sites often correlate with active site positions.

4.13
General Properties of Solvent and Water Molecules on the Protein

Solvent mapping of elastase was originally used to map potential binding sites on the protein with a view toward aiding rational drug design. This information not only proves useful to aid drug design, but it also proves valuable in developing a greater understanding of the solvent in terms of the protein structure. The importance of waters for the stability of a protein is well known, due in part to the ability of a water molecule to act as a hydrogen bond donor or acceptor. However, it has become increasingly apparent that other solvent molecules can substitute for this purpose. A crystal of a functional protein mildly cross-linked with glutaraldehyde and transferred into an organic solvent retains both the crystalline lattice and protein structure with very little change in either. There are a significant number of water molecules that remain associated with the protein, even when the crystal is transferred into high concentrations of water-miscible organic solvent. In addition, there are a significant number of solvent molecules that appear and even replace water molecules from their binding sites. Most of the solvents that have been used in this elastase MSCS study cannot substitute equally for the hydrogen-bonding capacity of water, despite their polar functional groups. In fact, most of the solvents used here have some hydrophobic portion as part of the molecular structure. All of the solvents are much larger than water and have different steric properties than water.

The common binding sites shared between water and other organic solvent molecules appear in almost all parts of the protein, although solvent molecules are excluded from most buried sites. It is proposed that the waters in these very tightly bound buried sites cannot be replaced while still retaining an active protein conformation. The MSCS analysis of a protein, however, successfully maps the surface of a protein, identifying the locations of specific binding sites, the distribution of water-binding sites and regions of plasticity in the protein. In the case of elastase, the organic solvents cluster in the binding site pockets that have hydrophobic exposed areas. These hydrophobic anchor regions are surrounded by more polar residues that exhibit plasticity, allowing a diversity of local interactions within each subsite. The water molecules in the active site of elastase collectively trace the shape of known ligands and outline the places in the active site where polar interactions may occur. Together, the solvent-binding sites and water-binding sites provide a detailed map that is consistent with the binding of known inhibitors. Because the properties being probed have been shown to be component features of binding sites, MSCS is likely to be a powerful method for locating and analyzing the active sites of enzymes in terms of their essential properties. In addition, binding sites that are not part of an active site, or are part of a protein that does not have an enzymatic function can be identified and analyzed in this way. Ultimately, if all probes in all binding sites, both solvent and water, are modeled as functional groups, their stereoelectronic features can be used to build larger molecules by connecting the individual groups. This process could be useful in rational drug design.

4.14
Computational Methods

Because experimental methods such as MSCS are time-consuming, mapping of the protein surface computationally in terms of functional groups would simplify the process considerably. In fact, computational methods play a key role in deriving structural information for designing compounds that fit a particular site on a protein. These methods fall into a number of categories, each distinguished by the level of detail and the complexity of the problem [33, 34]. The process of designing ligands for a protein site involves two major factors: the physical compatibility of the ligand for the receptor and the interaction energy between the ligand and the residues of the interaction site. Depending on the approach being taken, the balance between these factors may be different. For instance, docking programs tend to emphasize complimentary shapes [34], whereas *de novo* design programs tend to focus on interaction energy. Ideally, both programs need to be considered simultaneously. One of the difficulties with these approaches is the role of water in mediating protein–ligand or protein–protein interactions, since waters are, in general, not included in such calculations owing to the complexity of the resulting computation.

Nevertheless, the process of computational mapping of the surface of a protein in terms of probe molecules or groups has become more reliable. Computational mapping methods place molecular probes (small molecules or functional groups) on a protein surface to identify the most favorable binding positions by calculating an interaction potential. Such calculations actually predate the first solvent mapping experiments. For instance, Goodford mapped a receptor active site in the drug design program GRID, followed by a fragment assembly process in which some of the favorable positions found for individual molecular fragments were connected into a single viable molecule [14]. This active site mapping and fragment assembly strategy has been implemented in a number of drug design programs. [14–16, 35–39]. An interesting approach to mapping is the MCSS method, which optimizes the free energy of numerous ligand copies simultaneously, each transparent to the others but subject to the full force of the receptor [15, 39] (chapter 7).

The most extensive use of computational probe mapping has been developed in the laboratory of Sandor Vajda. Algorithms developed to perform solvent mapping computationally find the consensus site that binds the highest number of different probes. Invariably, the consensus site found by the mapping is always a major subsite of the substrate-binding site. This result can be compared with enzyme structures determined experimentally in organic solvents that show that most organic molecules cluster in the active site, delineating the binding pocket. Calculations on ligand-bound and apo-structures of enzymes show that the mapping results are not very sensitive to moderate variations in the protein coordinates. The computational probe–protein interactions at this site can then be compared to the intermolecular interactions seen in the known complexes of the enzyme with various ligands (substrate analogs, products and inhibitors). For example, experimental mapping results are available for thermolysin [26, 28], which has also been mapped computationally with solvent molecules as probes by Vajda's group. The consensus site is located at a major subsite of the substrate-binding site. In addition, the probes at this location make hydrogen bonds and nonbonded interactions with the same residues that interact with the specific ligands of the enzyme. Thus, computational solvent mapping can provide detailed and reliable information on substrate-binding sites. [44, 45, 46, 47].

Vajda's method relies on the computation of free energies of ligand–protein binding through thermodynamic integration and free energy perturbation via molecular dynamics and Monte Carlo simulations. Such free energy calculations typically focus on a single molecule or a series of molecules in a pre-designated binding site. These methods have generated high-quality results, but they are time-consuming, especially when attempting to locate all possible binding sites. A second basic approach predicts free energies using ensemble average data from Monte Carlo simulations in parameterized equations. The "linear interaction energy" (LIE) method of Aqvist [48] has been used by several groups to produce very good results for large numbers of HIV inhibitors [49] and for thrombin [50]. However these methods require calibration using a set of known ligand–protein binding energies, as well as prior knowledge of the binding site.

An alternative approach in computational mapping is a derivation of a method used by Guarnieri and Mezei to explore water binding to DNA [51, 52]. It uses Monte Carlo sampling to construct grand canonical ensembles based on ligand–protein free energies [53]. The method was used to illustrate the identification of hydrophobic ligands and the rank ordering of ligand affinities to an artificial cavity of T4 lysozyme [54, 55]. The results are in general agreement with other free energy computation methods. In addition, the method was used in a computational version of MSCS, to map the binding of hydrophilic ligands to the surface of thermolysin. Comparison with the experimental results of English et al. [26] shows good agreement [56].

4.15
Conclusion

In the pharmaceutical industry, these types of experimental and computational methods have been successfully applied. They have been focused on active sites of enzymes, protein–protein interaction sites and allosteric sites that were previously unknown in some cases. These latter cases, exploiting binding sites that are not used by a substrate or known inhibitor, pose a more formidable challenge, while the process of locating and characterizing such interaction sites is, in principle, the same as for the others. First, a site needs to be identified and then characterized so that chemically complementary compounds can be designed, synthesized and optimized for specificity and affinity. Since there are no known ligands that bind to such sites, there are no templates from which to start. The combination of solvent mapping and computational design provides an avenue to identify these sites and to identify compounds that could interact with these sites. Standard methods can then be used to optimize a compound to provide affinity and specificity.

Acknowledgments

Chery Kreinbring is acknowledge for assistance with the figures.

References

1 K.N. Allen, C.R. Bellamacina, X. Ding, C.J. Jeffrey, C. Mattos, G.A. Petsko, D. Ringe **1996**, "An Experimental Approach to Mapping the Binding Surfaces of Crystalline Proteins", *J. Phys. Chem.* 100, 2605–2611.

2 C. Mattos, D. Ringe **1996**, "Locating and Characterizing Binding Sites on Proteins," *Nat. Biotechnol.* 14, 595–599.

3 T. Clackson, J.A. Wells **1995**, "A hot spot of binding energy in a hormone–receptor interface", *Science* 267, 383–386.

4 T. Selzer, S. Albeck, G. Schreiber **2000**, "Rational design of faster associating and tighter binding protein complexes", *Nat. Struct. Biol.* 7, 537–541.

5 D. Rajamani, S. Thiel, S. Vajda, C. J. Camacho **2004**, "Anchor residues in protein-protein interactions", *Proc. Natl. Acad. Sci. USA* 101, 11287–11292

6 O. Keskin, B. Ma, R. Nussinov **2005**, "Hot regions in protein–protein interactions: the organization and contribution of structurally conserved hot spot residues", *J. Mol. Biol.* 345, 1281–1294.

7 S. Jones, J.M. Thornton **1996**, "Principles of protein–protein interactions", *Proc. Natl. Acad. Sci. USA* 93, 13–20.

8 W. L. DeLano **2002**, "Unraveling hot spots in binding interfaces: progress and challenges", *Curr. Opin. Struct. Biol.* 12, 14–20.

9 D. Ringe **1995**, "What makes a binding site a binding site?" *Curr. Opin. Struct. Biol.* 5, 825–829.

10 R. Najmanovich, J. Kuttner, J., Sobolev,V., M. Edelman **2000**, "Side chain flexibility in proteins upon ligand binding", *Proteins* 39, 261–268.

11 C. Mattos, D.A. Giammona, G.A. Petsko, D. Ringe **1995**, "Structural analysis of the active site of porcine pancreatic elastase based on the x-ray crystal structures of complexes with trifluoroacetyl-dipeptide-anilide inhibitors". *Biochemistry* 34, 3193–3203.

12 I. Huc, J.M. Lehn **1997**, "Virtual combinatorial libraries: dynamic generation of molecular and supramolecular diversity by self-assembly" [published erratum appears in *Proc. Natl. Acad. Sci. USA* 94, 8272]. *Proc. Natl. Acad. Sci. USA* 94, 2106–2110.

13 E. Peisach, D. Casebier, S.L. Gallion, P. Furth, G.A. Petsko, J.C. Hogan, Jr., D. Ringe **1995**, "Interaction of a Peptidomimetic Aminimide Inhibitor with Elastase", *Science* 269, 66–69.

14 P.J. Goodford **1985**, "A computational procedure for determining energetically favorable binding sites on biologically important macromolecules". *J. Med. Chem.* 28, 849–857.

15 A. Miranker, M. Karplus **1991**, "Functionality maps of binding sites: a multiple copy simultaneous search method (MCSS)". *Proteins Struct. Funct. Genet.* 11, 2934.

16 C.L. Verlinde, G. Rudenko, W. G. Hol **1992**, "In search of new lead compounds for trypanosomiasis drug design: a protein structure-based linked-fragment approach". *J. Comput. Aided Mol. Des.* 6, 131–147.

17 M.B. Eisen, D.C. Wiley, M. Karplus, R.E. Hubbard **1994**, "HOOK: a program for finding novel molecular architectures that satisfy the chemical and steric requirements of a macromolecule binding site". *Proteins* 19, 199–221.

18 S.B. Shuker, P.J. Hajduk, R.P. Meadows, S.W. Fesik **1996**, "Discovering high-affinity ligands for Proteins: SAR by NMR", *Science* 274, 1531–1534.

19 D. Vitkup, D. Ringe, G.A. Petsko, M. Karplus **2000**, "Solvent Mobility and the Protein 'Glass' Transition", *Nat. Struct. Biol.* 7, 1–5.

20 C. Mattos, D. Ringe **2001**, "Proteins in organic solvents". *Curr. Opin. Struct. Biol.* 11, 761–764.

21 L. Esposito, L. Vitagliano, F. Sica, G. Sorrentino, A. Zagari, L. Mazzarella **2000**, "The ultrahigh resolution crystal structure of ribonuclease A containing an isoaspartyl residue: hydration and sterochemical analysis". *J. Mol. Biol.* 297, 713–732.

22 P.A. Fitzpatrick, D. Ringe, A.M. Klibanov **1994**, "X-ray crystal structure of crosslinked subtilisin carlsberg in water vs. acetonitrile", *Biochem. Biophys. Res. Comm.* 198, 675–681.

23 J.L. Schmitke, L.J. Stern, A.M. Klibanov **1997**, "The crystal structure of subtilisin Carlsberg in anhydrous dioxane and its comparison with those in water and acetonitrile", *Proc. Natl. Acad. Sci. USA* 94, 4250–4255.

24 N.H. Yennawar, H.P. Yennawar, G.K. Farber **1994**, "X-ray crystal structure of gamma-chymotrypsin in hexane, *Biochemistry* 33, 7326–7336.

25 G.K. Farber **1999**, "Crystallographic analysis of solvent-trapped intermediates of chymotrypsin", *Methods Enzymol.* 308, 201–215.

26 A.C. English, S. H. Done, L.S. Caves, C.R. Groom, R.E. Hubbard **1999**, "Locating interaction sites on proteins: the crystal structure of thermolysin soaked in 2% to 100% isopropanol", *Proteins* 37, 628–640.

27 D. Atlas **1975**, "The active site of porcine elastase". *J. Mol. Biol.* 93, 39–53.

28 A.C. English, C.R. Groom, R.E. Hubbard **2001**, "Experimental and computational mapping of the binding surface of a crystalline protein", *Protein Eng.* 14, 47–59.

29 E. Liepinsh, G. Otting **1997**, "Organic solvents identify specific ligand binding sites on protein surfaces". *Nat. Biotechnol.* 15, 264–268.

30 D.W. Byerly, C.A. McElroy, M.P. Foster **2002**, "Mapping the surface of *Escherichia coli* peptide deformylase by NMR with organic solvents". *Protein Sci.* 11, 1850–1853.

31 C. Mattos **2002**, "Protein–water interactions in a dynamic world". *Trends Biochem. Sci.* 27, 203–208.

32 C. Mattos, D. Ringe **2001**, "Solvent Structure in International Tables for Crystallography", Vol. F, ed. M.G. Rossman, E. Arnold, Kluwer Academic, Dordrecht, p. 623–640.

33 T. Kortvelyesi, S. Dennis, M. Silberstein, L. Brown, 3rd, S. Vajda **2003**, "Algorithms for computational solvent mapping of proteins", *Proteins* 51, 340–351.

34 M.D. Cummings, R.L. DesJarlais, A.C. Gibbs, V. Mohan, E. P. Jaeger **2005**, "Comparison of automated docking programs as virtual screening tools". *J. Med. Chem.* 48, 962–976.

35 H.J. Bohm **1992**, "The computer program LUDI: a new method for the de novo design of enzyme inhibitors", *J. Comput. Aided Mol. Des.* 1, 61–78.

36 M.C. Lawrence, P.C. Davis **1992**, "PC. CLIX: a search algorithm for finding novel ligands capable of binding proteins of known three-dimensional structure". *Proteins* 12, 31–41.

37 S.H. Rothstein, M.A. Murcko **1993**, "GroupBuild: a fragment-based method for de novo drug design". *J. Med. Chem.* 36, 1700–1710.

38 B.L. King, S. Vajda, C. DeLisi **1996**, "Empirical free energy as a target function in docking and design: application to HIV-1 protease inhibitors". *FEBS Lett.* 384, 87–91.

39 A. Caflisch, A. Miranker, M. Karplus **1993**, "Multiple copy simultaneous search and construction of ligands in binding sites: application to inhibitors of HIV-1 aspartic proteinase". *J. Med. Chem.* 36, 2142–2167.

40 K.S. Thorn, A. A. Bogan **2001**, "ASEdb: a database of alanine mutations and their effects on the free energy of binding in protein interactions". *Bioinformatics* 17, 284–285.

41 A.A. Bogan, K.S. Thorn **1998**, "Anatomy of hot spots in protein interfaces". *J. Mol. Biol.* 280, 1–9.

42 B. Ma, T. Elkayam, H. Wolfson, R. Nussinov **2003**, "Protein–protein interactions: structurally conserved residues distinguish between binding sites and exposed protein surfaces". *Proc. Natl. Acad. Sci. USA* 100, 5772–5777.

43 L. Lo Conte, C. Chothia, J. Janin **1998**, "The atomic structure of protein–protein recognition sites". *J. Mol. Biol.* 285, 2177–2198.

44 M. Silberstein, S. Dennis, L. Brown T. Kortvelyesi, K. Clodfelter, S. Vajda **2003**, "Identification of substrate binding sites in enzymes by computational solvent mapping". *J. Mol. Biol.* 332:1095–113.

45 S.H. Sheu, T. Kaya, D.J. Waxman, S. Vajda **2005**, "Exploring the binding site structure of the PPAR gamma ligand-binding domain by computational solvent mapping". *Biochemistry* 44, 1193–1209.

46 T. Kortvelyesi, M. Silberstein, S. Dennis, S. Vajda **2003**, "Improved mapping of protein binding sites. *J. Comput. Aided Mol. Des.* 17, 173–186.

47 S. Dennis, T. Kortvelyesi, S. Vajda **2002**, "Computational mapping identifies the binding sites of organic solvents on proteins". *Proc. Natl. Acad. Sci. USA* 99, 4290–4295.

48 J. Aqvist, J. Marelius **2001**, "The Linear Interaction Energy Method for Predicting Ligand Binding Free Energies". *Comb. Chem. High Throughput Screening* 4, 613–626.

49 R. C. Rizzo, M. Udier-Blagovic, D. P. Wang, E. K. Watkins, M.B. Kroeger, M. B. Smith, R.H. Smith Jr., J. Tirado-Rives, W. L. Jorgensen **2002**, "Estimation of Binding Affinities for HEPT and Nevirapine Analogues with HIV-1 Reverse Transcriptase via Monte Carlo Simulations". *J. Med. Chem.* 45, 2970–2987.

50 A. C. Pierce, W. L. Jorgensen **2001**, "Estimation of Binding Affinities for Selective Thrombin Inhibitors via Monte Carlo Simulations". *J. Med. Chem.* 44, 1043–1050.

51 F. Guarnieri, M. Mezei **1996**, "Simulated Annealing of Chemical Potential: A General Procedure for Locating Bound Waters. Application to the Study of the Differential Hydration Propensities of the Major and Minor Grooves of DNA". *J. Am. Chem. Soc.* 118, 8493–8494.

52 H. Resat, M. Mezei **1994**, "Grand Canonical Monte Carlo Simulation of Water Positions in Crystal Hydrates". *J. Am. Chem. Soc.* 116, 7451–7452.

53 F. Guarnieri **2004**, "Computational Protein Probing to Identify Binding Sites". US patent 6,735,530.

54 A. Morton, W. A. Baase, B.W. Matthews **1995**, "Energetic Origins of Specificity of Ligand Binding in an Interior Nonpolar Cavity of T4 Lysozyme". *Biochemistry* 34, 8564–8575.

55 A. Morton, B.A. Matthews **1995**, "Specificity of Ligand Binding in a Buried Nonpolar Cavity of T4 Lysozyme: Linkage of

Dynamics and Structural Plasticity". *Biochemistry* 34, 8576–9888.

56 M. Clark, F. Guarnieri, I. Shkurko, J. Wiseman **2006**, "Grand canonical Monte Carlo simulation of ligand–protein binding". *J. Chem. Inf. Model* 46, 231–242.

57 C. Mattos, C. R. Bellamacina, E. Peisach, D. Vitkup, G. A. Petsko, D. Ringe **2006**, "Multiple solvent crystal structures: probing binding sites, plasticity and hydration". *J. Mol. Biol.* 357, 1471–1482.

Part 2: Fragment Library Design and Computional Approaches

5
Cheminformatics Approaches to Fragment-based Lead Discovery

Tudor I. Oprea and Jeffrey M. Blaney

5.1
Introduction

For most of its century-old history, drug discovery has been a science-driven process that benefited from serendipity [1]. Notable serendipitous discoveries include Penicillin V (the first antibiotic), discovered by Sir Alexander Fleming in 1929, and Sildenafil (Viagra), discovered by Andrew Bell, Nicholas Terrett and David Brown in 1992. Originally developed as an antianginal agent [2], Sildenafil was later [3] found to be more effective in the treatment of male erectile disfunction [4]. Among the science-driven discoveries, we note Captopril (the first angiotensin-converting enzyme inhibitor) by Miguel Ondetti and David Cushman in 1976 [5] and Esomeprazole (the first enantiopure proton pump inhibitor) by Sverker von Unge in 1993 [6]. While every effort is spent to replace serendipity with a rational workflow in drug discovery and development, the top 50 pharmaceutical companies have spent, on average, U.S.$ 750×10^6 for each of the 21 truly novel drugs launched during the 1991–2000 decade [7]. Since pharmaceutical companies need to be run as businesses, i.e., they require positive levels of profitability, major pharmaceutical houses no longer consider a new drug for market launch unless its sales are expected to top U.S.$ 800×10^6, and that number is soon expected to pass U.S.$ 1×10^9. To complicate the economic aspects of this science-driven workflow, most major pharmaceutical companies focus on "first world" diseases. Overall, these companies attempt to address the same (first world) clinical needs and are, in most cases, competing on the same targets, as marketing research is done on (similar) population samples from the industrialized world [8]. Their aim is to develop single-dose, orally available compounds with no side-effects (if possible) and low dosage (e.g., not exceeding 500 mg day^{-1}).

Thus, there is a strong impetus to minimize drug discovery costs and to reduce the time-lapse between idea (e.g., target and lead identification) and drug, in order to maximize the rate of success of preclinical R&D projects. The preclinical research workflow is constantly reevaluated, and stronger emphasis is placed on rational efforts, rather than serendipity. However, as major pharmaceutical houses

Fragment-based Approaches in Drug Discovery. Edited by W. Jahnke and D. A. Erlanson
Copyright © 2006 WILEY-VCH Verlag GmbH & Co. KGaA, Weinheim
ISBN: 3-527-31291-9

compete on the same therapy area, this leads to steep competition in the intellectual property (IP) arena, increased costs in clinical research (due to, e.g., limited clinical research capacity and increased regulatory demands). The unreliability of marketing analyses forecasts has also been a contributing factor to the increased pressure being placed on preclinical research; and all these factors – external to preclinical R&D – play an important role in the decision making process at the earliest level. In particular, the IP portfolio, a commodity by which most companies (continue to) attract investors, forces the entire industry to make decisions when important issues are, too often, unquantifiable variables because of the secretive nature of preclinical research. Hence, there is a justified effort to explore novel chemistries and novel chemical spaces, as this type of research is likely to ensure freedom of operation as long as that region of chemical space remains IP-protected.

This chapter begins with a theoretical exploration of chemical space, as we discuss an effort to sytematically enumerate chemical graphs. To a chemist, these graphs are single-bond carbon–hydrogen molecules (alkanes). To a computer, they are mathematical constructs based on edges ("bonds") and vertices ("atoms"). The reason we address this issue relates to the efforts of seeking novel IP: How vast is chemical space? How can we, at least from a theoretical (cheminformatics) perspective, begin to explore yet-unmapped regions of the chemical space of potential drugs? We then focus on the chemical space of leads, i.e., structures that are part of a series with known activity that are amenable for further chemistry optimization and exhibit favorable IP situation [9, 10]. By understanding the limitations of the chemical space for leads [11], we can further relate to the concept of fragment-based lead discovery [12] and show how cheminformatics can provide practical suggestions for the practicing medicinal chemist.

5.2
The Chemical Space of Small Molecules (Under 300 a.m.u.)

The vastness of chemical space cannot be better illustrated than by discussing the total number of chemical graphs (Csp^3 and H only) under 300 a.m.u. [13] (see Table 5.1). This effort came to address a challenge posed by W.G. Richards, namely to provide a complete list of C, N and O containing molecules with molecular weight (MW) between 150 a.m.u. and 250 a.m.u. This list is required for the screensaver-based drug discovery effort [14] that Richards initiated – where volunteers can download software that computationally evaluates the binding of small molecules to certain drug targets of known structure, using software that runs in the "screensaver" mode (i.e., when the volunteer's desktop computer is idle). To generate such complete lists, one needs to systematically enumerate these molecules; we used GENSMI [13] to exhaustively enumerate chemical graphs with less than 21 atoms. GENSMI [13] performs constructive enumeration of canonical non-isomeric SMILES [15, 16]: Each time a new atom is added, connectivity is explored (e.g., branching, rings) in all possible ways until constraints are violated.

Table 5.1 Complete scaffold enumeration by GENSMI. #C is the number of carbons; only rows with 10–20 atoms and 0–5 rings are shown. Highlighted cells indicate molecules of possible interest for medicinal chemistry (complete enumeration only). Cells that end in thousands ('000) represent estimates, not actual enumeration.

#C	0 Rings	1 Ring	2 Rings	3 Rings	4 Rings	5 Rings
10	75	475	1,792	4,875	10,162	16,461
11	159	1,231	5,533	17,978	45,282	90,111
12	355	3,232	16,977	64,720	192,945	460,699
13	802	8,506	51,652	227,842	790,849	2,222,549
14	1,858	22,565	156,291	787,546	3,138,808	10,216,607
15	4,347	60,077	470,069	2,678,207	12,116,550	45,076,266
16	10,359	160,629	1,407,264	8,982,754	45,675,153	191,989,014
17	24,894	430,724	4,193,977	29,761,361	168,615,086	800,000,000
18	60,523	1,158,502	12,451,760	97,557,854	593,000,000	3,000,000,000
19	148,284	3,122,949	36,838,994	317,000,000	2,000,000,000	10,000,000,000
20	366,319	8,437,289	102,733,261	1,000,000,000	7,000,000,000	35,000,000,000

GENSMI relies on SMILES, a linear notation for chemical structures, as this is a very convenient and systematic way to store chemical structures; SMILES canonization is needed to ensure uniqueness, as there are multiple ways to write valid SMILES (i.e., chemical structures), even for molecules as simple as isopentane. Canonical non-isomeric SMILES refer to the fact that we store unique chemical structures without taking into account chiral information.

As of August 2005, 1 503 201 444 unique non-isomeric SMILES have been stored. We evaluated these alkanes for "drug-likeness" from a topological standpoint only, since enumeration was unrestricted. We monitored the following topological properties (besides number of "nodes" – or atoms, and RNG, the number of rings): *branching*, or the number of bonds per non-terminal atom; *cyclization*, or the number of cycles per ring atom, and *ring* size. Of the six different databases evaluated when restricted to 20 atoms or less and eight rings or less, the World Drug Index (WDI) [17] was by far the most topologically diverse (covering a wider distribution, while having comparable median values). We found 48 771 867 graphs that match the median ±2 SD (95% of the distribution for each property) from WDI (see also Table 5.2).

While seemingly intractable from a synthetic chemical perspective, these SMILES are stored in a database that effectively becomes a resource for mining the entire virtual space of small molecule scaffolds. From a computational and logistic viewpoint, over 10^{16} molecules are anticipated. This effort is supported by our observation that available chemicals significantly under-sample chemical space at MW >300 [11]; in other words there are (potentially) IP-free regions that need to be explored. Therefore we are interested in systematically mapping all

Table 5.2 Topologically „drug-like" scaffold enumeration by GENSMI; see also Table 5.1.

#C	0 Rings	1 Ring	2 Rings	3 Rings	4 Rings	5 Rings
10	73	124	115	56	0	0
11	146	351	449	337	50	0
12	346	970	1,650	1,785	691	0
13	798	2,711	5,802	8,544	6,083	208
14	1,821	7,433	19,593	37,306	41,017	3,463
15	4,326	20,615	64,475	152,299	232,931	42,982
16	10,184	56,130	205,565	586,821	1,168,284	240,358
17	24,790	154,750	647,676	2,158,602	3,856,934	
18	59,680	419,620	2,000,657	7,662,010		
19	147,722	1,152,845	6,077,679			
20	362,008	3,147,135	17,930,155			

possible chemical scaffolds for such fragment-like molecules. This area is of interest because of the recent trends to explore fragment-based drug discovery [12], itself rooted in the concept of lead-likeness [11] discussed below.

5.3
The Concept of Lead-likeness

The concept of lead-likeness is central to drug discovery [11, 18]. The pharmaceutical industry has done a poor job in documenting the decision process, e. g., why certain chemical steps were followed to reach a particular compound, despite its century-old history [19]. This turns out to be a particularly important question today, since the industry is under pressure to reduce costs, increase productivity and provide high quality leads in early preclinical discovery. Chemical aspects of the history of drug discovery history are discussed in the three-volume series *Chronicles in Drug Discovery* [20–22], in Walter Sneader's *Drug Prototypes and their Exploitation* [23] and in *Integration of Pharmaceutical Discovery and Development: Case Histories* [24]. These accounts, while rich in historical chemical information, do not focus explicitly on the choice of lead structures in drug discovery.

The importance of restricting small molecule synthesis in the pharmacokinetics-friendly property space was first emphasized by the "rule-of-five" (RO5), which significantly changed our perception regarding leads, lead properties and lead discovery. RO5 is a stepwise descriptor proposed by Lipinski et al. [25], based on ClogP [26] (the calculated logarithm of the octanol–water partition coefficient), MW, the number of hydrogen-bond donors (HDO) and acceptors (HAC). The 90th percentile [25] for the distribution of MW (≤ 500), ClogP (≤ 5), HDO (≤ 5)

and the sum of nitrogen and oxygen (accounting for HAC ≤10) defines "RO5 compliance", based on the property distribution of these parameters for 2245 compounds from WDI [17] that had reached phase II clinical trials or higher. If any one of the four properties is higher than the above limits, RO5 = 1; and if any two properties are higher, RO5 = 2, etc. Any RO5 >1 is likely to yield molecules that are not orally available [25].

Following the RO5 criteria introduction in 1997, many library design programs based on combinatorial chemistry or compound acquisition enforced RO5 compliance. The existence of a drug-like space was established [27, 28] recently thereafter (1998). This provided the ability to discriminate between "drugs" and "non-drugs" based on chemical fingerprints, offering a computer-based discrimination between "drugs", as represented by WDI [17] or MDDR [29], and "non-drugs", represented by ACD [29]. Although this result was reproduced by other groups [30–32], it has yet to become accepted by the chemistry community as a decision-enabling scheme. If it was truly effective, it could assist chemists to quickly evaluate, for example, what other chemists before them have considered worthy of synthesis (and patenting).

A good drug-like score does not make a molecule a drug; rather, it indicates that more of its features are encountered in other molecules from MDDR and WDI, and less of its features in ACD. It is often assumed that Lipinski's RO5 criteria define drug-likeness, but the distinction between the "drug-like property space", to which the "rule-of-five" applies [25], and the "drug-like chemical space", defined by fingerprint discrimination models [27, 28], is often overlooked. More ACD compounds, i.e., "non-drugs", are RO5-compliant, compared to MDDR compounds, or "drugs" [33]; thus, most ACD compounds are drug-like in property space, but clearly not in chemical space. Vieth et al. [34] looked at the differences in the properties of drugs having a variety of routes of administration and confirmed that oral drugs have properties associated with lower MW, fewer HAC and HDO and fewer rotatable bonds (RTB), compared to drugs that have other routes of administration (chapter 6). Despite this extension to RO5 criteria, there remains a gulf between these rules of thumb and true discriminating power for specific design purposes. It is therefore more appropriate to think of the RO5 type criteria as necessary, but not sufficient to create an oral drug-like molecule, and of the drug-like scores as guidelines for filtering chemical libraries, not as a tool to evaluate the quality of individual compounds. Since they are complementary, the two filters should be combined during chemical library analysis.

Unlike drug-like scores, based on statistical analyses for large numbers ($\sim 2 \times 10^5$) of chemical structures, the lead-like concept [35] relies on significantly smaller datasets [36–38] that when combined amounts only to 894 structures (*vide infra*). Despite this, the concept of lead-likeness has a significant impact in the design of chemical libraries [39–41]. This is, in part, because the concepts and methods related to lead-likeness are very intuitive and fit with the current experience of what typically happens [42] in lead optimization. Based on current data, it appears that, on average, effective leads have lower molecular complexity [36], when compared to drugs, have fewer rings (RNG) and rotatable bonds [37], have a lower MW and are more polar [35].

The following computational criteria should be applied [11] for lead-like libraries: MW ≤460, −4.0 ≤ ClogP ≤ +4.2 (−4.0 ≤ LogD$_{7.4}$ ≤ +4.0), LogS$_w$ −5, HDO ≤5 and HAC ≤9 (where LogD$_{7.4}$ is the ClogP value corrected for pH 7.4; and LogS$_w$ is the predicted aqueous solubility [43]), as well as RTB ≤ 9, RNG ≤ 4. These computational criteria, derived from the RO5 cut-off values, are also rooted into a distribution analysis of drug-like databases [33], and in the property distribution of leads versus drugs [19]. Additional experimental criteria [11], related to *in vivo* properties (e.g., in rat), should be applied individually to lead-like compounds: Oral bioavailability above 30%, low clearance (e.g., below 30 ml min^{-1} kg^{-1}), measured LogD$_{7.4}$ between 0 and 3, poor (or no) inhibition of drug-metabolizing cytochrome P450 isozymes, plasma protein binding below 99.5%, lack of acute toxicity at the expected therapeutic window (e.g., assuming 500 mg day^{-1} P.O. regimen for 7 days), as well as poor binding to hERG. hERG is a K$^+$ channel implicated in sudden cardiac death and possibly responsible for the cardiac toxicity of a wide range of compounds [44], now withdrawn from the market.

5.4
The Fragment-based Approach in Lead Discovery

The lead-like concept served as the basis for developing screening strategies that are complementary to more traditional screening methods. Some companies, e.g., Astex [45, 46], Plexxikon [47], Vertex [48] and SGX Pharmaceuticals [49] have implemented the concept of screening fragments or very small lead like entities (in connection with X-ray crystallography or NMR) as their main lead generation paradigm. The general approach is to find start points for the LI phase which are "lead-like" and less complex than those derived solely on "drug-like" criteria.

The "reduced complexity" criteria [11] for a fragment-based screening set are mostly computational, e.g., average values for MW <350, RTB ≤6, heavy atoms ≤22, HDO ≤3, HAC ≤8 and ClogP ≤2.2. These criteria are consistent with the "rule of three" (RO3) criteria [50], MW ≤300, ClogP ≤3, RTB ≤3, HDO ≤3, HAC ≤3 and PSA ≤60 Å^2, where PSA is the polar surface area. Examples of known actives (potency above 100 nM) that are RO3-compliant on a variety of targets are given in Fig. 5.1. In this and the following figures and tables, all biological potency values are converted to −log$_{10}$(activity). The examples in Fig. 5.1 are extracted from WOMBAT 2005.1.

WOMBAT (World of Molecular BioAcTivity) [51] is a bioactivity database [52] that contains 117 007 entries (104 230 unique SMILES [15, 16]), with over 230 000 biological activities on 1021 unique targets in the 2005.1 release [53]. This release covers 4786 series from 4773 papers published in medicinal chemistry journals between 1975 and 2004. There are 5321 RO3-compliant molecules (807 "generics") in WOMBAT 2005.1; and, of these, 941 entries (242 "generics") have potency ≤10 nM. The "generics" terminology covers mostly compounds that have a specific (typically unique) designation, which means they are either launched drugs, or natural products, or otherwise in an advanced stage of development. The exam-

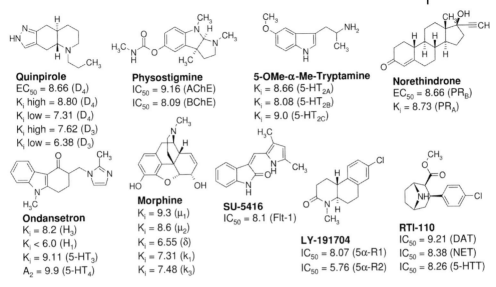

Fig. 5.1
Examples of RO3 compliant molecules. Target names: D_3, D_4 – dopaminergic receptor types 2 and 3; AChE, BChE – acetyl- and butyryl-choline esterases; PR_A, PR_B – progesterone receptor types A and B; H_1, H_3 – histamine receptor types 1 and 3; $5\text{-}HT_{2A}$, $5\text{-}HT_{2B}$, $5\text{-}HT_{2C}$, $5\text{-}HT_3$, $5\text{-}HT_4$ – serotonin receptor subtypes 2A, 2B, 2C and types 3 and 4; DAT, NET, 5-HTT – dopamine, norepinephrine and serotonin transporter proteins; μ_1, μ_2, δ, k_1, k_3 – opioid receptor types mu-1, mu-2, delta, kappa-1 and kappa-3; 5α-R1, 5α-R2 – 5-alpha-reductase isozymes 1 and 2; Flt-1 – fms-like tyrosine kinase receptor.

ples given in Fig. 5.1 illustrate the chemotype, target and activity diversity that can be found in RO3-compliant molecules: Neurotransmitter and nuclear hormone receptor agonists (EC_{50}) and antagonists (K_i, IC_{50} and A_2), neurotransmitter transporters as well as enzyme inhibitors are present, most of them with multiple activities. Based on the WOMBAT 2005.1 entries, it appears that there is a number of interesting RO3-compliant chemotypes.

Of equal interest is the question whether lead-like (i.e., ignoring potent ligands such as NO and CO) molecules with MW ≤ 200 (but not necessarily RO3-compliant) can have potency above 10 nM. Such examples, also extracted from WOMBAT 2005.1, are given in Fig. 5.2. We found 1460 activities better than 10 nM, from 342 unique structures, acting on 98 targets: 41 enzymes, 42 receptors, six ion channels, and nine proteins (where "proteins" are targets that cannot be classified as enzymes, receptors or ion channels). What is equally interesting is that, at pH 7.4, 80% of these small molecules are likely to be charged: There are 269 cations, 22 anions and 15 zwitterions in this list. This indicates that potent low-MW compounds are likely to require salt bridge interactions with the receptor in order to achieve high activity.

Andrews et al. were influenced by this observation, based on a limited set of 200 structures, in their efforts to define the "average intrinsic binding energy" of drugs.

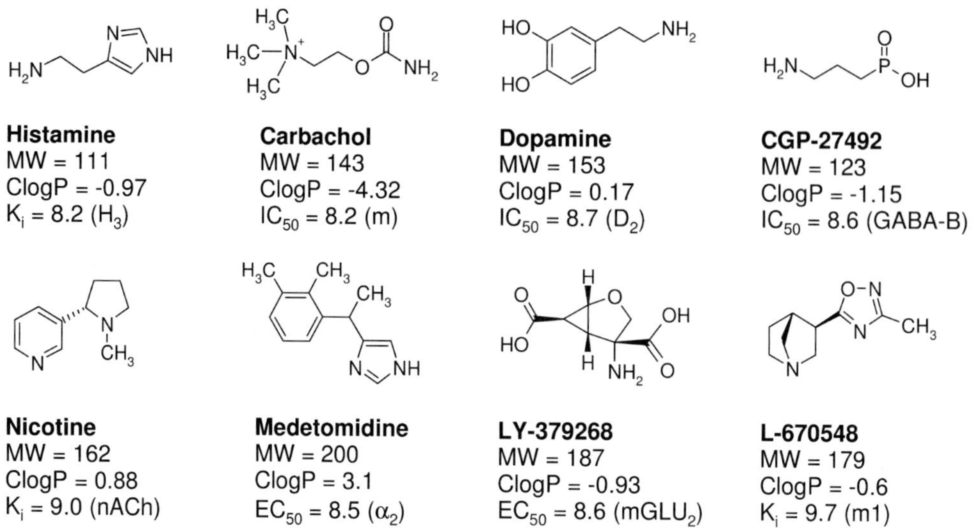

Histamine
MW = 111
ClogP = -0.97
K_i = 8.2 (H_3)

Carbachol
MW = 143
ClogP = -4.32
IC_{50} = 8.2 (m)

Dopamine
MW = 153
ClogP = 0.17
IC_{50} = 8.7 (D_2)

CGP-27492
MW = 123
ClogP = -1.15
IC_{50} = 8.6 (GABA-B)

Nicotine
MW = 162
ClogP = 0.88
K_i = 9.0 (nACh)

Medetomidine
MW = 200
ClogP = 3.1
EC_{50} = 8.5 (α_2)

LY-379268
MW = 187
ClogP = -0.93
EC_{50} = 8.6 ($mGLU_2$)

L-670548
MW = 179
ClogP = -0.6
K_i = 9.7 (m1)

Fig. 5.2
Examples of small molecules (MW ≤200) that have biological activity better than 10 nM. Under each molecule, the following information is included: molecule name, MW, ClogP, biological activity type, value and target. Target names are as follows: H_3 – histamine receptor subtype 3; m – muscarine receptor; D_2 – dopaminergic receptor type 2; nACh – nicotine receptor; α_2 – alpha adrenergic receptor subtype 2; $mGLU_2$ – metabotropic glutamate receptor subtype 2; GABA-B – gamma-amino butyric acid receptor subtype B; m1 – muscarinic receptor subtype 1.

They found that coefficients for charged atom types are approximately ten times higher compared to the coefficients for neutral atom types based on 200 molecules [54]. However, the median MW of their dataset is below 300; and hence the "average intrinsic binding energy" [54], an empirical scheme better known as the Andrews binding energy (ABE), is likely to be biased towards such interactions. Used to estimate the bioactivity potential of chemical libraries [35, 40], the ABE scoring scheme parallels the increase in MW (R^2 = 0.72), but not the increase in activity (R^2 = 0.01) when tested on N = 55 658 activities better than 0.1 µM from WOMBAT [19]. However, as MW increases from 150 a.m.u. to, e.g., 1200 a.m.u., the affinity of ligands might increase from, e.g., 1 µM to 1 pM, or between 8.5 kcal mol^{-1} and 17 kcal - mol^{-1}. Thus, while MW may increase nine-fold, the free energy of binding might only double its value. Kuntz et al. surveyed the strongest-binding ligands for a large number of targets and concluded that binding affinity improves very little once the number of heavy atoms increases above 15 [55]. Since the ABE scoring scheme makes no correction for higher MW, its use to evaluate chemical libraries should be restricted to small molecules (MW <300), where the presence of charged moieties can indeed improve activity, as illustrated in Fig. 5.2.

Unlike the situations described in Figs. 5.1 and 5.2, the effort to identify small MW molecules for screening is likely to result in fragment hits that have significantly lower potency (low mM to low μM range). Such hits would not always be identifiable using more traditional screening methods, where the final compound concentration is in the order of 10 μM or less. The obvious solution is to screen compounds at higher concentrations, e.g., 50 μM. This introduces problems related to solubility, purity and interference with readout, e.g. by fluorescence quenching. As the expected affinity of fragments is significantly lower and the physico-chemical properties need to be significantly different compared to drugs, the selection criteria that focus on yet smaller molecules need to be redefined and perhaps maintained more flexible, as illustrated by the RO3 or "reduced complexity" criteria above. With careful selection of compounds and robust screens, one can design screening approaches for biomolecular targets (mainly enzymes) using up to 1 mM concentrations and still extract useful information.

5.5
Literature-based Identification of Fragments: A Practical Example

Mining literature-derived data from, e.g., WOMBAT, can provide some interesting insights into how to assemble chemical libraries of fragments. The following examples (summarized in Tables 5.3–5.7) provide suggestions for the design of fragment-based libraries related to several biochemical enzyme classes. Table 5.3 provides background information on the type of unique WOMBAT entries related to four classes of enzymes, those having Enzyme Commission (EC) designations [56] starting with 1, 2, 3 and 4, respectively. There was very little data related to EC 5 and EC 6, therefore these were not included in the analysis. All entries in Table 5.3 were filtered to remove any activities on receptors, ion channels and proteins, and were further filtered to remove entries tested on another EC category. Thus, all entries and scaffolds reported below were filtered to show biological activity for that respective EC category only.

Table 5.3 Distribution of active WOMBAT entries and related scaffolds for the four EC categories.

Category	Entries with activity better than 100 nM	Unique scaffolds	Unique scaffolds with occurrence	Unique scaffolds[a] compared to all other EC classes
EC1	3501	236	64	32
EC2	4550	265	93	54
EC3	10209	460	142	95
EC4	1254	53	15	8
Total	19514	1014	238	189

a) Only scaffolds with occurrence > 10 are reported.

Table 5.4 Fragments, occurrence, enzyme and parent compound information for EC 1-specific WOMBAT entries.

Structure	Occurrence	MW Fragment	Reference	Enzyme E.C. Number	Enzyme Family	Enzyme Name	Generic Name	Activity	Activity Type	Parent Structure	MW Parent
	113	245.4	J. Med. Chem. 40(9)-1997 1293-1315	EC 1.3.99.5	steroid 5-alpha-reductase	3-oxo-5-alpha-steroid 4-dehydrogenase 2		8.495	IC50		331.5
	65	175.3	J. Med. Chem. 37(8)-1994 1153-1164	EC 1.13.11.34	lipoxygenase; dioxygenase	arachidonate 5-lipoxygenase	L-699333	8.155	IC50		627.2
	61	245.4	J. Med. Chem. 45(16)-2002 3406-3417	EC 1.3.99.5	steroid 5-alpha-reductase	3-oxo-5-alpha-steroid 4-dehydrogenase 1	finasteride	8.301	IC50		372.6
	56	100.1	J. Med. Chem. 37(20)-1994 3240-3246	EC 1.1.1.34	HMG-CoA reductase; oxidoreductase	3-hydroxy-3-methylglutaryl-coenzyme A reductase	lovastatin	8.155	IC50		404.6
	51	132.1	J. Med. Chem. 37(14)-1994 2167-2174	EC 1.5.1.3	oxidoreductase	dihydrofolate reductase	methotrexate	7.951	IC50		454.4
	49	145.2	Bioorg. Med. Chem. Lett. 13(3)-2003 543-546	EC 1.1.1.205	IMPDH/GMPR; oxidoreductase	inosine-5'-monophosphate dehydrogenase 2		8.301	IC50		332.4
	34	258.4	J. Med. Chem. 37(14)-1994 2198-2205	EC 1.14.14.1	cytochrome P450; oxidoreductase	estrogen synthetase; cytochrome P450 19A1; aromatase	androstenedione	7.699	Ki		286.4

5.5 Literature-based Identification of Fragments: A Practical Example | 101

29	244.2	J. Med. Chem. 43(6)- 2000 1062-1070	EC 1.1.1.21	aldo/keto reductase; oxidoreductase	aldose reductase	WAY-121365	7.136	IC50	276.2
26	184.2	J. Med. Chem. 39(20)- 1996 3951-3970	EC 1.13.11.34	lipoxygenase; dioxygenase	arachidonate 5-lipoxygenase		8.301	IC50	456.5
21	244.3	J. Med. Chem. 41(12)- 1998 2118-2125	EC 1.4.3.4	flavin monoamine oxidase	amine oxidase [flavin-containing] A		7.222	IC50	258.3
20	232.3	J. Med. Chem. 41(12)- 1998 2118-2125	EC 1.4.3.4	flavin monoamine oxidase	amine oxidase [flavin-containing] A		7.523	IC50	246.3
10	133.2	J. Med. Chem. 39(9)- 1996 1924-1927	EC 1.1.1.21	aldo/keto reductase; oxidoreductase	aldose reductase	epalrestat	7.678	IC50	319.4

Table 5.5 Fragments, occurrence, enzyme and parent compound information for EC 2-specific WOMBAT entries.

Structure	Occurrence	MW Fragment	Reference	Enzyme E.C. Number	Enzyme Family	Enzyme Name	Generic Name	Activity	Activity Type	Parent Structure	MW Parent
	159	195.3	J. Med. Chem. 40(26)- 1997 4290-4301	EC 2.5.1.58	protein prenyl-transferase	protein farnesyl-transferase type I alpha subunit	SCH-56580	7.398	IC50		444.0
	139	146.1	J. Med. Chem. 43(21)- 2000 3837-3851	EC 2.1.1.45	thymidylate synthase; transferase; methyl-transferase	thymidylate synthase	ZD-1694; tomudex	6.000	IC50		458.5
	98	212.2	J. Med. Chem. 41(16)- 1998 2960-2971	EC 2.7.7.49	reverse transcriptase	RNA-directed DNA polymerase; DNA nucleotidyl-transferase; revertase	nevirapine	7.097	IC50		266.3
	97	114.1	J. Med. Chem. 37(20)- 1994 3274-3281	EC 2.5.1.21	phytoene/squalene synthetase; transferase; oxidoreductase	squalene synthetase; farnesyl-diphosphate farnesyl-transferase	squalestatin H1	7.585	IC50		538.5
	90	112.1	J. Med. Chem. 46(2)- 2003 207-209	EC 2.4.2.4	thymidine/pyrimidine-nucleoside phosphorylase; transferase	thimidine phosphorylase; platelet-derived endothelial cell growth factor; PD-ECGF; gliostatin	TPI	7.699	IC50		242.7
	76	133.2	J. Med. Chem. 45(26)- 2002 5687-5693	EC 2.7.1.112	tyrosine-protein kinase; transferase	vascular endothelial growth factor receptor 1; fms-like tyrosine kinase 1	SU-5416; semaxanib	7.367	IC50		238.3
	73	147.1	J. Med. Chem. 43(24)- 2000 4606-4616	EC 2.7.1.37	serine/threonine-protein kinase; transferase	cyclin-dependent kinase 4		7.495	IC50		378.5

71	131.1	J. Med. Chem. 39(9)-1996 1823-1835	EC 2.7.1.112	tyrosine-protein kinase; transferase	epidermal growth factor receptor	PD-158780	11.097	IC50	330.2
67	148.2	J. Med. Chem. 43(10)-2000 2019-2030	EC 2.7.7.49	reverse transcriptase	RNA-directed DNA polymerase; DNA nucleotidyl-transferase; revertase	DPC-963	7.745	IC50	316.2
63	119.1	J. Med. Chem. 36(22)-1993 3424-3430	EC 2.7.1.20	carbohydrate kinase pKB; transferase	adenosine kinase; adenosine 5'-phospho-transferase	iodotubercidin	7.523	IC50	392.2
62	169.2	Bioorg. Med. Chem. Lett. 12(10)-2002 1361-1364	EC 2.7.1.112	tyrosine-protein kinase; transferase	proto-oncogene tyrosine-protein kinase Lck		8.046	IC50	308.8

Table 5.6 Fragments, occurrence, enzyme and parent compound information for EC 3-specific WOMBAT entries.

Structure	Occurrence	MW Fragment	Reference	Enzyme E.C. Number	Enzyme Family	Enzyme Name	Generic Name	Activity	Activity Type	Parent Structure	MW Parent
	206	98.1	J. Med. Chem. 44(14)-2001 2319-2332	EC 3.4.23.16	peptidase A2; hydrolase; aspartyl protease	HIV-1 retropepsin; HIV-1 protease	CI-1029	9.959	IC50		483.7
	190	114.1	J. Med. Chem. 40(25)-1997 4079-4088	EC 3.4.23.16	peptidase A2; hydrolase; aspartyl protease	HIV-1 retropepsin; HIV-1 protease	DMP-450	9.509	IC50		536.7
	167	185.2	J. Med. Chem. 35(14)-1992 2672-2687	EC 3.1.4.17	cyclic nucleotide phosphodiesterase; hydrolase	cGMP-inhibited PDE	BMY-20844	8.000	IC50		213.2
	106	96.1	Bioorg. Med. Chem. Lett. 12(20)-2002 2925-2930	EC 3.4.21.5	peptidase S1; serine protease	prothrombin; coagulation factor II	L-375378	9.000	Ki		406.5
	102	134.2	J. Med. Chem. 46(5)-2003 685-690	EC 3.4.21.6	peptidase S1; serine protease	coagulation factor Xa	RPR-200095	8.886	Ki		463.0
	76	183.3	J. Med. Chem. 46(6)-2003 954-966	EC 3.1.1.7	carboxylesterase / lipase B-type; serine esterase; hydrolase; peptidase M10A	acetylcholinesterase	tacrine	6.821	Ki		198.3
	67	150.2	J. Med. Chem. 46(10)-2003 2008-2016	EC 3.1.4.17	cyclic nucleotide phosphodiesterase; hydrolase; peptidase M10A	cAMP-specific phosphodiesterase (PDE)		7.900	IC50		362.4

5.5 Literature-based Identification of Fragments: A Practical Example

65	147.1	J. Med. Chem. 46(18)-2003 3840-3852	EC 3.4.24.35	gelatinase B; peptidase M10A; matrix metallo-peptidase M10A; protease-9 hydrolase		7.921	IC50	485.6
59	160.2	J. Med. Chem. 46(6)-2003 954-966	EC 3.1.1.8	carboxyl-esterase/ lipase B-type; serine esterase; hydrolase; peptidase M10A	physostigmine	7.640	IC50	Chiral 275.4
49	152.1	Bioorg. Med. Chem. Lett. 12(21)-2002 3149-3152	EC 3.1.4.17	cyclic nucleotide phosphodiestera se; hydrolase	cGMP-specific PDE; PDE5	9.086	IC50	415.5
49	118.1	J. Med. Chem. 46(26)-2003 5663-5673	EC 3.2.2.16/EC 3.2.2.9	PNP/UDP phosphorylase	5'-methylthio-adenosine nucleosidase/ S-adenosyl-homocysteine nucleosidase	7.284	IC50	456.0

Table 5.7 Fragments, occurrence, enzyme and parent compound information for EC 4-specific WOMBAT entries.

Structure	Occurrence	MW Fragment	Reference	Enzyme E.C. Number	Enzyme Family	Enzyme Name	Generic Name	Activity	Activity Type	Parent Structure	MW Parent
	66	101.1	J. Med. Chem. 46(11)- 2003 2187-2196	EC 4.2.1.1	eukaryotic-type carbonic anhydrase; lyase	carbonic anhydrase II	methazolamide	7.854	Ki		236.3
	16	208.3	J. Med. Chem. 39(14)- 1996 2745-2752	EC 4.6.1.1	lyase	adenylyl cyclase type I	forskolin	7.387	IC50		410.5

The unique scaffolds analysis was performed using the ChemoSoft program [57]. Care was given to extract only those scaffolds that did not overlap between the various EC categories; and, in the extraction procedure, all single-bond substitutions and all carbon-only cycles were ignored. Only non-overlapping scaffolds are illustrated in Tables 5.4–5.7. The degree of overlap between various chemotypes can be inferred from comparing columns 4 and 5 in Table 5.3. These scaffolds are likely to reflect a certain bias, not only in terms of the current targets of interest in medicinal chemistry literature (as expected), but also in terms of the amount of effort in conducting such research and reporting such results. For example, unique EC 1 scaffolds are dominated by those active on dihydrofolate reductase (EC 1.5.1.3) and 5-alpha reductase (EC 1.3.99.5), whereas unique EC 4 scaffolds are dominated by those active on carbonic anhydrase (EC 4.2.1.1). Whenever possible, parent structures having generic names (e.g., methotrexate and methazolamide) are reported, together with the corresponding literature reference and the biological activity, preferably on an enzyme of human origin. This explains why not all examples given have activity better than 100 nM – however, such activities have been reported for the same enzymes, in different species.

The types of scaffolds given here are not necessarily linked to specificity for a certain enzyme. However, they illustrate the type of chemistry that has been, at least in the past decade, *reported* specifically for, e.g., class 1 compared to class 2 etc., in medicinal chemistry literature. While not exhaustive (since we did not cover patents), this type of search is likely to reflect certain biases derived either from experimental reasons, e.g., the azasteroid scaffold for 5α-reductase or the cyclic urea for HIV-1 protease. Therefore, in the context of chemical genomics, one could envision such a library of fragments, or lead-like libraries containing these fragments, to be used in phenotypic screening. Given their (apparent?) specificity, their observed binding could offer mechanistic insights in various biochemical pathways and perhaps assist in target identification.

5.6
Conclusions

The initial publications on lead-like properties suggested that small, simple molecules would have higher hit rates and lower affinities than the typical larger, more complex molecules used in classic high-throughput screening (HTS) discovery programs. This concept has now been demonstrated by SGX Pharmaceuticals [49], who reported typical fragment screening hit rates of 15% for several different target classes. Such high hit rates suggest that it is not necessary to rely on previously known chemotypes, such as those described in the earlier section, to discover hits. Indeed, one of the potential advantages of fragment-based screening is to discover new chemotypes which might be missed by screening larger, more complex molecules. The cheminformatic analyses presented above provide a useful approach for suggesting molecules that are compatible with lead-like properties (e.g., RO3-compliant); and they further demonstrate a surprising diversity of known biologically active compounds that are compatible with the rather restrictive rules proposed for fragments. These results (specifically Fig. 5.2) also suggest that fragment-screening libraries which are enriched in charged compounds may increase the probability of finding higher affinity hits – since salt-bridges are likely to contribute to binding. The target-class breakdown for the 342 unique molecules, discussed earlier, indicates an equal amount of molecules binding to enzymes and G protein-coupled receptors, with surprisingly fewer numbers of ion channels and proteins. Since multiple aminergic receptor-targeted drugs are protonated amines and entire classes of drugs (e.g., non-steroidal anti-inflammatory) are acids, it is likely that oral bioavailability will not be significantly influenced by the presence of one such center.

The experimentally observed high fragment screening hit rates also suggest that fragment-screening libraries can be very carefully pruned to include only the most desirable compounds. Further work defining the properties of such "most desirable" compounds beyond lead-like, reduced screening set, or RO3 properties will be needed and will develop as we gain more experience with fragment-based drug discovery. It is quite likely that, as new companies enter this arena, different flavors of "fragment-based libraries design" are likely to exist – not only in terms of particular chemistry make-up (for intellectual property reasons), but also in terms of screening methodology, overall molecular complexity, physico-chemical profile, etc.

Cheminformatics and computational chemistry approaches need further development to address the challenging problem of predicting which fragments or HTS hits are most likely to be optimized into high activity (e.g., low-digit nM IC_{50}) leads. For example, can a classic HTS hit with IC_{50} = 10 µM which is not compliant with lead-like criteria (e.g., MW = 550) be a better starting point for optimization than a 200-MW, RO3-compliant fragment with IC_{50} = 500 µM? This is a key question posed by fragment-based methods. A systematic analysis of the calculable physicochemical properties and diversity of virtual libraries which could rapidly be derived in two or three synthetic steps from such fragment-based hits could assist with the decision-making process. When using structure-based virtual screening [58], the number of compounds in such virtual libraries might be prohibitive, even for high-throughput docking, so sampling methods (e.g., pooling representatives with various chemotypes that match certain physico-chemical, or perhaps pharmacophoric profiles) will be required to dock and score subsets of the virtual libraries. The issue of correctly evaluating, *in silico*, the binding affinity for "weak binders" remains open – as scoring functions have been known to evaluate better nanomolar, as opposed to millimolar, binding affinities. Overall, the challenges raised by developing fragment-based approaches have yet to be solved. Given the results so far, their integration into the evolutionary process of lead development is likely to direct us towards interesting and stimulating new discoveries.

Acknowledgments

This work was supported by New Mexico Tobacco Settlement funds for Biocomputing and by the New Mexico Molecular Library Screening Center, MH074425–01 (T.I.O.).

References

1 Kubinyi, H. **1999**, Chance favors the prepared mind – from serendipity to rational drug design. *J. Rec. Signal Transduction Res.* 19, 15–39.
2 Bell, A. S., Terrett, N. K., Brown, D. **1992**, Pyrazolopyrimidinone antianginal agents. EU Patent EP0463756, available at: http://l2.espacenet.com/espacenet/viewer? PN=EP0463756.
3 Ellis, P., Terrett, N. K. **1994**, Pyrazolopyrimidininones for the treatment of impotence. EU Patent EP0702555, available at: http://l2.espacenet.com/espacenet/viewer? PN=WO9428902.
4 Terrett, N. K., Bell, A. S., Brown, D., Ellis, P. **1996**, Sildenafil (VIAGRA™), a potent and selective inhibitor of type 5 cGMP phosphodiesterase with utility for the treatment of male erectile dysfunction, *Bioorg. Med. Chem. Lett.* 6, 1819–1824.
5 Ondetti, M. A., Cushman, D. W. **1976**, Compounds and method for alleviating angiotensin related hypertension. US Patent 4,053,651.
6 Von Unge, S. **1995**, Novel ethoxycarbonyloxymethyl derivatives of substituted benzimidazoles. EU Patent WO9532957, available at: http://l2.espacenet.com/espacenet/viewer?PN=WO9532957.
7 Drews, J. **1998**, Innovation deficit revisited: reflections on the productivity of pharmaceutical R&D. *Drug Discov. Today*, 3, 491–494.
8 Horrobin, D. F. *J. Royal Soc. Med.* **2000**, Innovation in the pharmaceutical industry. 93, 341–345.
9 DeStevens, G. **1986**, Serendipity and structured research in drug discovery. *Prog. Drug. Res.* 30, 189–203.
10 Sneader, W. **1996**, *Drug Prototypes and their Exploitation*, Wiley, Chichester.
11 Hann, M. M., Oprea, T. I. **2004**, Pursuing the leadlikeness concept in pharmaceutical research. *Curr. Opin. Chem. Biol.* 8, 255–263.
12 Fattori, D. **2004**, Molecular recognition: the fragment approach in lead generation. *Drug Discov. Today* 9, 229–238.
13 Kappler, M. A., Allu, T. K., Bologa, C., Oprea, T. I. **2005**, Exhaustive Enumeration of Small Molecules. 1. Simple Graphs. *J. Chem. Inf. Model.* (in preparation).
14 Screensaver Lifesaver **2005**, available at: http://www.chem.ox.ac.uk/curecancer.html.
15 Weininger, D. **1988**, SMILES 1. Introduction and encoding rules. *J. Chem. Inf. Comput. Sci.* 28, 31–36.
16 Weininger, D., Weininger, A., Weininger, J. L. **1989**, SMILES 2. Algorithm for generation of unique SMILES notation. *J. Chem. Inf. Comput. Sci.* 29, 97–101.
17 World Drug Index database **2005**, Derwent Publications, available at: http://www.derwent.com/products/lr/wdi/; and Daylight Chemical Information Systems, available at: http://www.daylight.com/products/databases/WDI.html.
18 Hann, M. M., Leach, A., Green, D. V. S. **2005**, Computational chemistry, molecular complexity and screening set design. In: *Cheminformatics in Drug Discovery.* ed. Oprea, T. I., Wiley–VCH, New York, p. 43–57.
19 Oprea, T. I. **2004**, Cheminformatics in Lead Discovery. In: *Cheminformatics in Drug Discovery.* ed. Oprea, T. I., Wiley–VCH, New York, p. 23–41.
20 Bindra, J. S., Lednicer D. (eds.) **1982**, *Chronicles of Drug Discovery, Vol. 1*, Wiley–Interscience, New York.
21 Bindra, J. S., Lednicer D. (eds.) **1983**, *Chronicles of Drug Discovery, Vol. 2*, Wiley–Interscience, New York.

22 Lednicer, D. (ed.) **1993**, *Chronicles of Drug Discovery*, Vol. 3, ACS Publishers, Washington D.C.

23 Sneader, W. **1996**, *Drug Prototypes and Their Exploitation*, Wiley, Chichester.

24 Borchardt, R. T., Freidinger, R. M., Sawyer, T. K., Smith P. L. (eds.) **1998**, *Integration of Pharmaceutical Discovery and Development: Case Histories*, Plenum Press, New York.

25 Lipinski, C. A., Lombardo, F., Dominy, B. W., Feeney, P. J. **1997**, Experimental and computational approaches to estimate solubility and permeability in drug discovery and development settings. *Adv. Drug Deliv. Rev.* 23, 3–25.

26 Leo, A., Weininger, D. **2005**, *ClogP ver. 4.0*, Daylight Chemical Information Systems, Santa Fe, available at: http://www.daylight.com/.

27 Ajay, Walters, W. P., Murcko, M. A. **1998**, Can we learn to distinguish between "drug-like" and "nondrug-like" molecules? *J. Med. Chem.* 41, 3314–3324.

28 Sadowski, J., Kubinyi, H. **1998**, A scoring scheme for discriminating between drugs and nondrugs. *J. Med. Chem.* 41, 3325–3329.

29 MDDR and ACD **2005**, MDL Information Systems, available at: http://www.mdli.com/products/finders/database_finder/; MDDR is developed in cooperation with Prous Science Publishers, available at: http://www.prous.com/index.html.

30 Wagener, M., Van Geerenstein, V. J. **2000**, Potential drugs and nondrugs: prediction and identification of important structural features. *J. Chem. Inf. Comput. Sci.* 40, 280–292.

31 Frimurer, T. M., Bywater, R., Naerum, L., Lauritsen, L. N., Brunak, S. **2000**, Improving the odds in discriminating "druglike" from "nondruglike" compounds. *J. Chem. Inf. Comput. Sci.* 40, 1315–1324.

32 Brüstle, M., Beck, B., Schindler, T., King, W., Mitchell, T., Clark, T. **2002**, Descriptors, physical properties and drug-likeness. *J. Med. Chem.* 45, 3345–3355.

33 Oprea, T. I. **2000**, Property distribution of drug-related chemical databases. *J. Comput.-Aided Mol. Design* 14, 251–264.

34 Vieth, M., Siegel, M. G., Higgs, R. E., Watson, I. A., Robertson, D. H., Savin, K. A., Durst, G. L., Hipskind, P. A. **2004**, Characteristic physical properties and structural fragments of marketed oral drugs. *J. Med. Chem.* 47, 224–232.

35 Teague, S. J., Davis, A. M., Leeson, P. D., Oprea, T. I. **1999**, The design of leadlike combinatorial libraries. *Angew. Chem. Int. Ed.* 38, 3743–3748 (in German: *Angew. Chem.* 1999, 111, 3962–3967).

36 Hann, M. M., Leach, A. R., Harper, G. **2001**, Molecular complexity and its impact on the probability of finding leads for drug discovery. *J. Chem. Inf. Comput. Sci.* 41, 856–864.

37 Oprea, T. I., Davis, A. M., Teague, S. J., Leeson, P. D. **2001**, Is there a difference between leads and drugs? A historical perspective. *J. Chem. Inf. Comput. Sci.* 41, 1308–1315.

38 Proudfoot, J. R. **2002**, Drugs, leads, and drug-likeness: An analysis of some recently launched drugs, *Bioorg. Med. Chem. Lett.* 12, 1647–1650.

39 Davis, A. M., Teague, S. J., Kleywegt, G. J. **2003**, Application and limitations of X-ray crystallographic data in structure-based ligand and drug design. *Angew. Chem. Int. Ed.* 42, 2718–2736 (in German: *Angew. Chem.* 2003, 115, 2822–2841).

40 Goodnow Jr., R. A., Gillespie, P., Bleicher, K. **2005**, Chemoinformatic tools for library design and the hit-to-lead process: A user's perspective. In: *Cheminformatics in Drug Discovery*. Oprea, T. I. (ed.), Wiley–VCH, New York, p. 381–435.

41 Sanders, W. J., Nienaber, V. L, Lerner, C. G., McCall, J. O., Merrick, S. M., Swanson, S. J., Harlan, J. E., Stoll, V. S., Stamper, G. F., Betz, S. F., Condroski, K. R., Meadows, R. P. Severin, J. M., Walter, K. A., Magdalinos, P., Jakob, C. G., Wagner, R., Beutel, B. A. **2004**, Discovery of Potent Inhibitors of Dihydroneopterin Aldolase Using CrystaLEAD High-Throughput X-ray Crystallographic Screening and Structure-Directed Lead Optimization. J. Med. Chem. 47, 1709–1718.

42 Kenakin, T. **2003**, Predicting therapeutic value in the lead optimization phase of drug discovery. *Nat. Rev. Drug Discov.* 2, 429–438.

43 Tetko, I. V., Tanchuk, V. Y. **2002**, Application of associative neural networks for prediction of lipophilicity in AlogPS 2.1 program. *J. Chem. Inf. Comput. Sci.* 42, 1136–1145.

44 Vandenberg, J. I., Walker, B. D., Campbell, T. J. **2001**, HERG K+ channels: friend or foe. *Trends Pharm. Sci.* 22, 240–246.

45 Carr, R., Jhoti, H. **2002**, Structure-based screening of low-affinity compounds. *Drug Discov. Today* 7, 522–527.

46 Hartshorn, M. J., Murray, C. W., Cleasby, A., Frederickson, M., Tickle, I. J., Jhoti, H. **2005**, Fragment-Based Lead Discovery Using X-ray Crystallography. *J. Med. Chem.* 48, 403–413.

47 Milburn, M. V. **2002**, Drug discovery on a proteomic scale. *Abstr. Pap. ACS Nat. Meet.* 224, COMP-042.

48 Moore, J., Abdlu-Manan, N., Fejzo, J., Jacobs, M., Lepre, C., Peng, J., Xie, X. **2004**, Leveraging structural approaches: applications of NMR-based screening and X-ray crystallography for inhibitor design. *J. Synchrotron Radiation* 11, 97–100.

49 Blaney et al. **2005**, Chapter 11 in this book

50 Congreve, M., Carr, R., Murray, C., Jhoti, H. **2003**, A "Rule of Three" for fragment-based lead discovery? *Drug Discov. Today* 8, 876–877.

51 Olah, M., Mracec, M., Ostopovici, L., Rad, R., Bora, A., Hadaruga, N., Olah, I., Banda, M., Simon, Z., Mracec, M., Oprea, T. I. **2005**, WOMBAT: World of Molecular Bioactivity. In: *Chemoinformatics in Drug Discovery*. ed. Oprea, T. I., Wiley–VCH, New York, p. 223–239.

52 Olah, M., Oprea, T. I. **2006**, Informatics and databases: Bioactivity databases. In: *Comprehensive Medicinal Chemistry II, Vol. 3*, ed. Taylor, J., Triggle, D., Elsevier, New York (in press).

53 WOMBAT **2005**, Sunset Molecular Discovery, Santa Fe, available at: http://www.sunsetmolecular.com/.

54 Andrews, P. R., Craik, D. J., Martin, J. L. **1984**, Functional group contributions to drug-receptor interactions. *J. Med. Chem.* 27, 1648–1657.

55 Kuntz, I. D., Chen, K., Sharp, K. A., Kollman, P. A. **1999**, The maximal affinity of ligands. *Proc. Natl. Acad. Sci. USA* 96, 9997–10002.

56 ExPASy **2005**, *Enzyme Nomenclature Database*, available at: http://www.expasy.org/enzyme/.

57 ChemoSoft **2005**, available at: http://www.chemosoft.com/.

58 Oprea, T. I., Matter, H. **2004**, Integrating virtual screening in lead discovery. *Curr. Opin. Chem. Biol.* 8, 349–358.

6
Structural Fragments in Marketed Oral Drugs

Michal Vieth and Miles Siegel

6.1
Introduction

In this chapter we present a view on the chemically and pharmacologically relevant fragments/building blocks of marketed drugs, with particular emphasis on the fragments contained in oral drugs. Several recent papers have emphasized the evolution of the physical properties of oral drugs [1–3]. Some authors have suggested that the evolution of the target and chemistry space has caused small but meaningful changes in the mean properties of newer drugs as compared with older drugs [2, 3]. However, we have found little evidence in the literature to link these apparent differences to specific chemical fragments that are the pharmacophorically important building blocks of drugs. This chapter describes an analysis of the fragments present in drugs and other groups of compounds, to investigate similarities and differences between these groups. In particular, we look into the frequency of occurrence of fragments in groups of compounds, to emphasize and understand physical property similarities and differences between sets. By taking this approach, we have found that the distribution of chemical fragments in injectable drugs differs significantly from that of oral drugs, mostly in the scaffolds connecting these fragments. We also use a similar comparison to look into the fragment-based similarities between oral drugs and compounds in clinical and preclinical development. In addition to speculating on the interesting implications of these findings, we present some practical directions on how the available data can be utilized to guide compound development in the direction of increasing probability of pharmacokinetic success.

6.2
Historical Look at the Analysis of Structural Fragments of Drugs

Retrosynthetic combinatorial analysis procedure, or RECAP [4], was the first tool used to generate, analyze, and suggest the utilization of a set of pharmacologically relevant fragments from the set of molecules in a drug database. The significance

Fragment-based Approaches in Drug Discovery. Edited by W. Jahnke and D. A. Erlanson
Copyright © 2006 WILEY-VCH Verlag GmbH & Co. KGaA, Weinheim
ISBN: 3-527-31291-9

of RECAP in comparison to other fragmentation approaches [5, 6] was the almost immediate and clear-cut utilization of the knowledge present in the database of biologically relevant compounds in the design and potential synthesis of compound libraries. It is worth noting the RECAP approach came before the concept of targeted libraries [7–9] and inspired a number of approaches for *de novo* generation of chemically feasible and pharmacologically relevant structures [10–13]. The visionary character of RECAP can be best described by the number of follow-up approaches and companies funded around the concept, including De Novo Pharmaceuticals (www.denovopharma.com) and Locus Discovery (www.locusdiscovery.com).

Interestingly, the RECAP analysis of the World Drug Index [14] gave significantly different fragments from our recent analysis of marketed drugs using a similar fragmentation approach. For example, of the 15 most common side-chains in oral and injectable drugs found in our work, shown in Fig. 6.1, only six were exemplified in 35 presented using the RECAP method; simple phenyl, the side-chain we found to be most common in both sets of drugs, was not exemplified by RECAP.

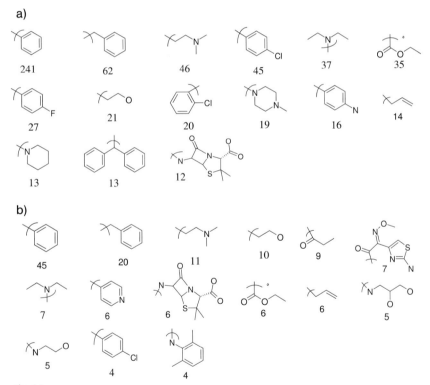

Fig. 6.1
Comparison of the most frequent side-chains in oral (a) and injectable (b) drugs. The numbers indicate the count of the drugs containing that fragment. The means of properties are not significantly different for a and b. Reproduced from Vieth at al. [1].

6.3
Methodology Used in this Analysis

The structural fragments analyzed in this chapter were generated using an internally developed tool, Molecular Slicer (MS), which we routinely use to analyze the common fragments of biologically targeted compounds [1]. MS is similar to other retrosynthetic algorithms previously described in the literature, such as RECAP [4] and REOS [15]. Our tool decomposes compounds into core and side-chain fragments, using a sequential set of 15 pre-assigned rules. Side-chain fragments are characterized by having only one "break-point", while scaffolds have two or more. The breakpoints are not always based on a logical retro-synthetic step, but are defined in a manner that allows us to analyze the composition of large volumes of compounds. The rules encoded and used in this study to generate these fragments are detailed in Table 6.1. After the molecules are decomposed, the resulting fragments (with the addition of explicit hydrogens) are then used to perform substructure searches to determine the frequency of occurrence of these specific fragments in each set of analyzed molecules (Fig. 6.2). We feel that, as long as the same process is used to dissect and assemble compounds, internal consistency insures transferability of statistics.

Five different data sets were used to perform the fragment analysis: marketed oral drugs (1192 compounds), marketed injectable drugs (308 compounds), compounds in clinical trials as listed in MDDR [16] (1943 compounds; marketed drugs were substracted from this set), and compounds from MDDR described as in "biological testing". (137 550). In addition, we subdivided the marketed oral drugs for which the approval date was available into a set of 332 "new oral drugs" approved after 1982, and a set of 859 "old oral drugs" approved before 1982. The final list of 139 126 unique molecules was then subjected to the MS process. As a result of the fragmentation, 52 131 side-chains and 15 652 scaffold fragments were identified. 19 914 fragments with four or more atoms (14 330 side-chains and 5584 scaffolds) were present in two or more structures and were therefore considered meaningful and used in the subsequent analysis. For each group, we computed the mean frequency of all fragments present and used it to assign importance for each fragment. Table 6.2 summarizes the fragment statistics for each group.

In order to compare the fragment distribution between two groups, we used all fragments present with frequencies greater than the standard deviation from the mean for each group. Thus, for comparison of frequencies of fragments in oral and injectable drugs, we took 54 fragments from oral drugs and 91 fragments from injectable drugs, for a total of 122 unique fragments (145 total with 23 in common). For the comparisons of new vs old drugs, oral vs clinical, and oral vs MDDR biological testing, we ended up with 52, 174, and 335 unique and significant fragments, respectively. In order to compare the frequencies between groups, we used the following difference function, shown here for the oral vs injectable comparison:

$$Diff_i = 100 \cdot \frac{abs\left(freq_i^{oral} - freq_i^{injectable}\right)}{max\left(freq_i^{oral}, freq_i^{injectable}\right)} \qquad (1)$$

6 Structural Fragments in Marketed Oral Drugs

Table 6.1 Sequentially applied SMARTS queries used in Molecular Slicer algorithm. The bond breakages occur between the indicated pair of query atoms. The SMARTS contain locally developed extensions to the SMARTS language, most notably the relational operator. Isotopic labels are applied to designate previously perceived bond breakages, so the queries require non-isotopic atoms for new breakages. Adapted from Vieth at al. [1].

Sequence	SMARTS	Query atoms bond break	Isotopic label for break atoms	Query name
1	a-[CH,CH2;R00*]-[N,O;R00*]-[CHR0,CH2R0]-a	1 2	+1	Heteroatom beta to a ring
2	[C0*D>1,c0*D>1]-[NH,ND2,ND3,OD2;R00*]-[CH,CH2;R00*]-a	1 2	+1	Heteroatom beta to a ring
3	[R;0*]-[CH2R0,NHR0,OR0;0*]-[R]	0 1	+2	Separate two rings connected by an atom
4	*-[CD3H,ND2;R00*](-a)-a	0 1	+3	Separate biphenyl
5	[a]-&!@[a]	0 1	+4	Break ring–ring non–ring bonds
6	[NR;0*]-[CD3R0;0*](=O)-[R]	0 1	+5	Separate biphenone from ring N
7	[NR;0*]-[CD2R0;0*]-[R]	0 1	+6	Ring nitrogen beta to a ring
8	[N0*,n0*;!H2]-[SR0;0*]-[CD>1,cD>1,ND>1]	0 1	+18	Break N-S bonds
9	[NR;0*]-[CD2R0;0*]-[CD2,CD3,OD2,ND2,ND3,aD2,aD3]	0 1	+9	Break Ring nitrogen – no ring carbon
10	a-[NHR0]-[CR0;0*](=O)-[OR0,NR0;0*]	0 1	+10	Separate anilines
11	[CRD>1;0*]-[NH,O;0*]-[CR0;0*](=O)-[NH,O;0*]	1 2	+11	Break N-C in ureas
12	[OD1H0]=[CD3R0;0*](-[ND2,ND3;0*]-[CD>1,aD>1;0*])-[CD>1,OD2,aD>1,ND>1;0*]	1 2	+11	Break amides
13	[OD1H0]=[CD3R0;0*](-[CD>1;0*])-[CH2;R00*]-[CH,CH2;R00*]-a	1 3	+12	Gamma carbonyls
14	[a;0*]-[CD3R0;0*](=O)-[D2,D3,D4;0*]-[D<4]-[D<4]	0 1	+6	Break phenone
15	[CR,NR]=[CR]-&!@[a]	1 2	+13	Another ring breakage connected by no-ring bond

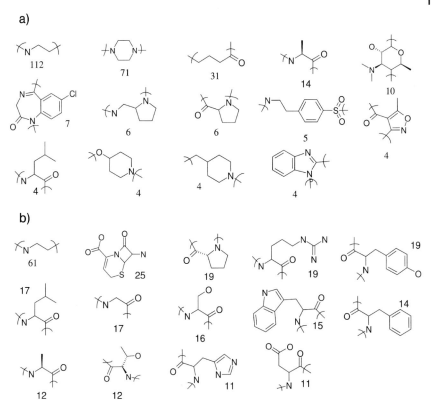

Fig. 6.2
Comparison of most frequent scaffolds in oral (a) and injectable (b) drugs. The numbers indicate the number of drugs containing the fragment. The means of physical properties (CLOGP, ON, rotbond) are significantly different for a and b. Reproduced from Vieth et al. [1].

If a fragment is found only in the oral drugs, the difference is 100%, however if the fragment is found in both groups with the same frequencies, the difference is 0%. For each of the significant fragments we computed the difference function and binned the differences into ten bins. The plots of the difference functions are shown in Fig. 6.3a–d for oral vs injectable, new vs old, oral vs clinical and oral vs MDDR biological testing compounds. The specific comparisons were chosen to understand whether the property differences between these sets would translate into differences in the nature and frequency of the fragments derived from each set.

Table 6.2 General statistics for fragments generated from different drug data sets.

Group	Number of molecules	Number of fragments with match in at least one molecule	Mean fragment frequency	STD	Mean + STD	Number of fragments with frequencies greater than 1 STD away from the mean
Marketed oral drugs	1192	1630	0.27	0.70	0.97	54
Marketed injectable drugs	308	762	1.19	1.67	2.86	91
Clinical compounds	1943	3065	0.28	0.71	0.99	157
MDDR biological compounds	137 550	20 577	0.02	0.25	0.28	330
New oral drugs	332	779	0.59	0.98	1.57	39
Old oral drugs	859	1204	0.31	0.84	1.14	48

6.4
Analysis of Similarities of Different Drug Data Sets Based on the Fragment Frequencies

The distribution of frequency differences gives a breakdown of the number of fragments (both side-chains and scaffolds) between groups (Fig. 6.3). For example, fragments having large differences in frequencies between the oral and injectable drugs dominate the distribution, with median and mean differences of 90% and 74%, respectively. A similar distribution is observed for the differences between fragments in oral and clinical compounds. The median difference is slightly lower at 85%, with the mean identical and indistinguishable from the oral vs injectable group at 75%. In contrast, the distribution of differences for the new vs old marketed drugs is more uniformly distributed, with no particular preference to any difference bin. This is substantiated by the median value of 53% and the mean value of 52%. Interestingly, the distribution of frequency differences of 335 significant fragments compared between marketed oral drugs and biological testing shows a largely uniform character, with a slight bias towards higher difference bins. The median of this distribution 68% and mean of 60% suggests more oral-like than injectable-like fragment character.

Inspection of the side-chain and scaffold structures giving the largest frequency difference between oral drugs and the three groups injectable, clinical, and biologically active molecules reveals that many of these side-chains likely derive from

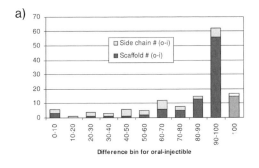
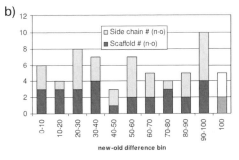
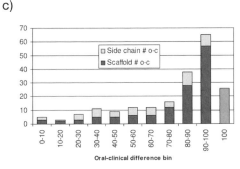
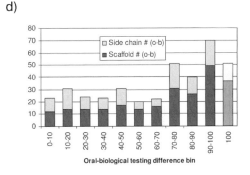

Fig. 6.3
Distribution of the fragment frequency differences as a function of difference bins (Eq. 1). We added 100 difference bin (included in 90–100 bin) to separately view the structures which occur only in one group. (a) The total count of 122 side-chains and scaffold fragments in oral vs injectable comparison. (b) The total count of 52 side-chains and scaffold fragments in new (post-1982) vs old (pre-1982) marketed oral drug comparison. (c) The total count of 174 side-chains and scaffolds for oral vs clinical comparison. (d) The total count of 335 side-chains and scaffolds for oral vs MDDR biological comparison.

peptides, which are present in injectable, clinical, and biologically active molecules but are largely absent in oral drugs. As it is difficult to determine the intended route of administration in either the clinical or the biologically active sets, these sets include compounds not intended for oral delivery and/or for which ADME properties are not optimized.

In order to shed light on the specific differences and similarities between fragments from different sets, we looked into the most and least different fragments for each comparison groups. The oral vs injectable 100% difference bin contains 15 significant scaffolds and two side-chains that occur in one group but not in the other, representatives of which are shown in Fig. 6.4a. As pointed out in our original work, the majority of scaffolds (14) present only in the injectable drugs are amino acids. Six fragments (three side-chains and three scaffolds) are found with almost identical frequencies in oral and injectable drugs. These six are exemplified in Fig. 6.4b. Interestingly, ethanolamine scaffolds and selected tertiary amines are present with very similar frequencies in oral and injectable drugs.

6 Structural Fragments in Marketed Oral Drugs

a)

Structure	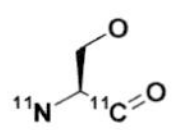 [img]	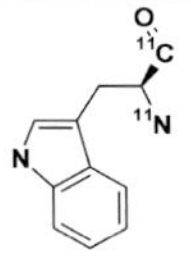 [img]	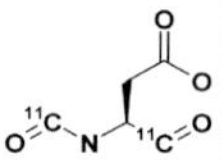 [img]	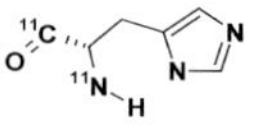 [img]	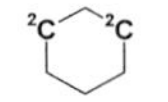 [img]	 [img]	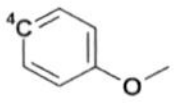 [img]
injectable matches (% of total)	16 (5.2%)	12 (3.9%)	11 (3.6%)	12 (3.9%)	0 (0%)	0 (0%)	0 (0%)
oral matches (% of total)	0 (0%)	0 (0%)	0 (0%)	0 (0%)	22 (1.8%)	20 (1.7%)	12 (1%)

b)

Structures	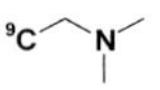 [img]	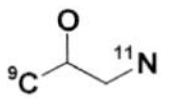 [img]	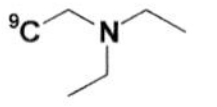 [img]	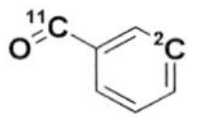 [img]	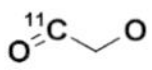 [img]
injectable matches (% of total)	11 (3.6%)	7 (2.3%)	6 (1.9%)	4 (1.4%)	4 (1.4%)
oral matches (% of total)	46 (3.9%)	25 (2.1%)	23 (1.9%)	14 (1.2%)	14 (1.2%)

Fig. 6.4
Exemplification of significant fragments most different (a) and most similar (b) between oral and injectable drugs. The counts and percentages of structures in which each fragment is present are given for both groups.

When comparing the fragments with similar and dissimilar frequencies in new and old oral drugs, one finds some penicillin and sugar structures that are absent in the new drugs, with unnatural amino acid present with much greater frequency in the new drugs (Fig. 6.5). These differences reflect to some degree evolution of targets trends within the pharmaceutical industry. The pencillin-derived beta-lactam framework, common in many older antibiotics, has not been actively pursued for some time due to several factors, including developing drug resistance and the advent of alternative anti-infectives such as the fluoroquinolones. By contrast, three of the five fragments appearing in newer drugs that are not reflected in older drugs are related to homophenylalanine, a fragment common to many ACE inhibitors, such as enalapril, that have been launched in the past 20 years. As we and others have shown [1, 17], however, the physical property differences for new vs old oral drugs are generally very small. These fragment differences, notably smaller than for the other comparisons, likely reflect target-related differences. The fragments common to both new and old drugs, such as 2-substituted pyridine, isopropyl and phosphate side-chains, and propylamine, butyl, and butylamine, are perhaps more generic in helping to confer favorable PK/PD properties and biological activity on oral drugs. These conclusions, although similar in spirit to the ones presented by Leeson [2], come from the analysis of building blocks and thus may have more direct application particularly in the area of drug-like library design and construction.

a)

Structure	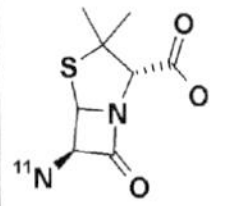	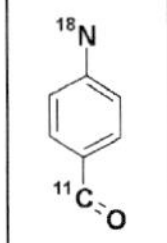	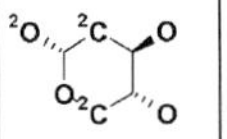	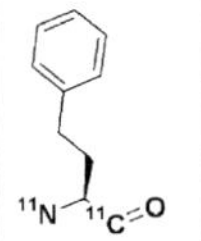	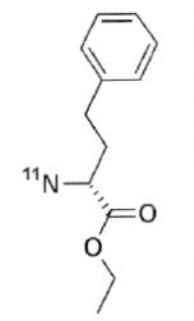	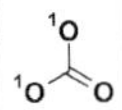	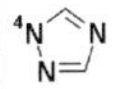
new drug matches (%)	0 (0%)	0 (0%)	0 (0%)	11 (3.3%)	9 (2.7%)	6 (1.8%)	6 (1.8%)
old drug matches (%)	12 (1.4%)	11 (1.3%)	4 (0.4%)	1 (0.1%)	1 (0.1%)	1 (0.1%)	1 (0.1%)

b)

Structure	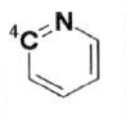	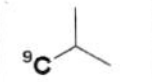	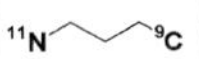	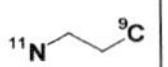	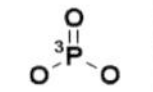	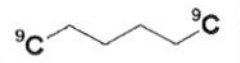
new drug matches (%)	6 (1.8%)	7 (2.1%)	4 (1.2%)	10 (3%)	4 (1.2%)	5 (1.5%)
old drug matches (%)	16 (1.9%)	19 (2.2%)	10 (1.2%)	26 (3%)	11 (1.3%)	13 (1.5%)

Fig. 6.5
Exemplification of significant fragments most different (a) and most similar (b) between new (post-1982) and old (pre-1982) drugs. The counts and percentages of structures in which each fragment is present are given for both groups.

a)

Structure	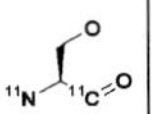	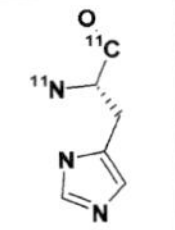	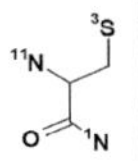	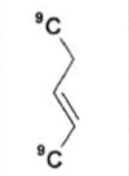	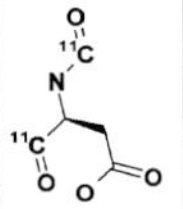	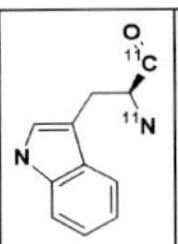	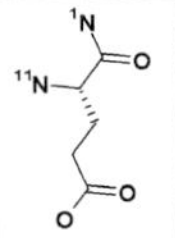	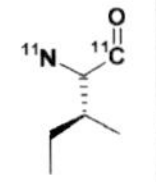	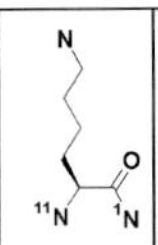	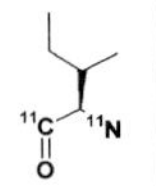
chiral clinical matches	23 (1.2%)	23 (1.2%)	23 (1.2%)	21 (1.1%)	17 (0.9%)	15 (0.7%)	5 (0.3%)	4 (0.2%)	13 (0.7%)	11 (0.6%)
oral matches	0 (0%)	0 (0%)	0 (0%)	0 (0%)	0 (0%)	0 (0%)	0 (0%)	0 (0%)	0 (0%)	0 (0%)

b)

Structures	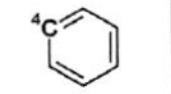	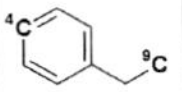	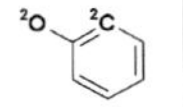	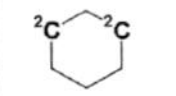	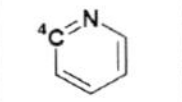
clinical matches (%)	361 (18.6%)	61 (3.1%)	86 (4.4%)	35 (1.8%)	33 (1.7%)
oral matches (%)	240 (20.2%)	36 (3%)	48 (4%)	22 (1.8%)	22 (1.8%)

Fig. 6.6
Exemplification of significant fragments most different (a) and most similar (b) between oral and clinical drugs. The counts and percentages of structures in which each fragment is present are given for both groups.

As noted above, fragments with the greatest oral vs clinical differences are similar to those from oral vs injectable differences (amino acid scaffolds; Fig. 6.6). Fragments with almost identical frequencies in marketed oral and clinical compounds include phenyl and 2-substituted pyridine side-chains, and para-substituted benzyl, 2-alkoxy-substituted phenyl, and 1,3 di-substituted cyclohexyl scaffolds.

The fragments with the largest differences between oral drugs and biologically active compounds also include amino acid side-chains (similar to clinical and injectable compounds), while the notable fragments with similar frequencies in the

a)

Structure

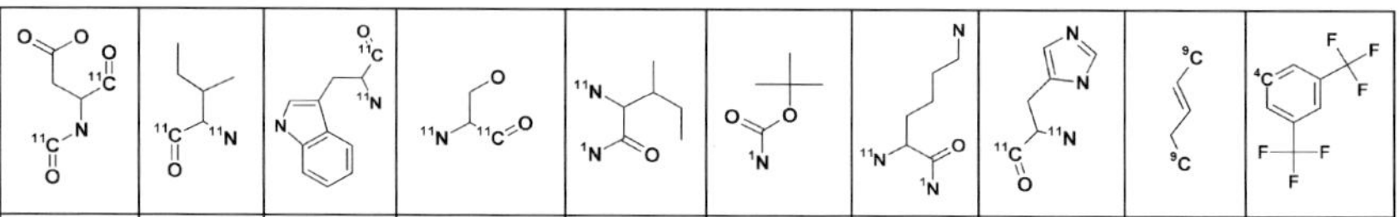

biological matches (%)	1552 (1.1%)	1285 (0.9%)	1271 (0.9%)	1159 (0.8%)	1052 (0.8%)	1001 (0.7%)	977 (0.7%)	952 (0.7%)	906 (0.7%)	840 (0.6%)
oral matches (%)	0 (0%)	0 (0%)	0 (0%)	0 (0%)	0 (0%)	0 (0%)	0 (0%)	0 (0%)	0 (0%)	0 (0%)

Structure

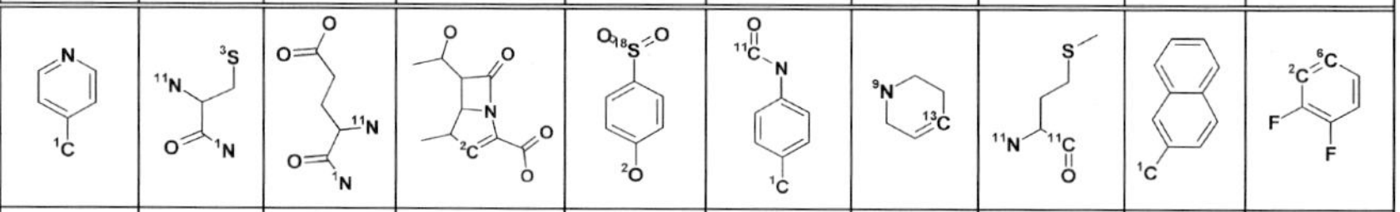

biological matches (%)	826 (0.6%)	807 (0.6%)	771 (0.6%)	705 (0.5%)	699 (0.5%)	635 (0.5%)	601 (0.4%)	589 (0.4%)	567 (0.4%)	557 (0.4%)
oral matches (%)	0 (0%)	0 (0%)	0 (0%)	0 (0%)	0 (0%)	0 (0%)	0 (0%)	0 (0%)	0 (0%)	0 (0%)

b)

Structure

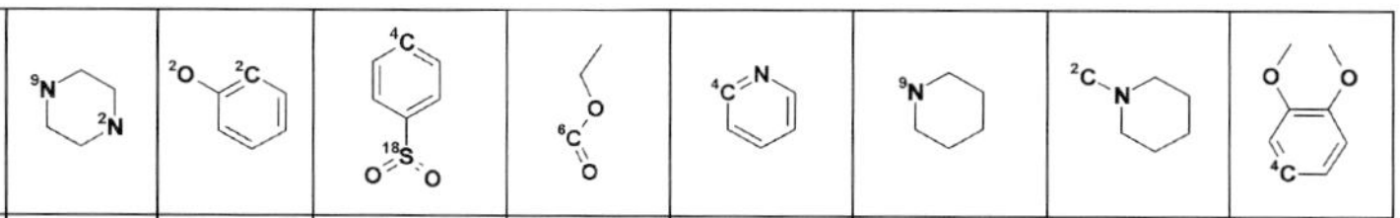

biological matches (%)	7744 (5.6%)	5909 (4.3%)	4610 (3.4%)	3703 (2.7%)	2633 (1.9%)	1578 (1.1%)	1017 (0.7%)	985 (0.7%)
oral matches (%)	71 (6%)	48 (4%)	36 (3%)	35 (2.9%)	22 (1.8%)	13 (1.1%)	8 (0.7%)	8 (0.7%)

Structure

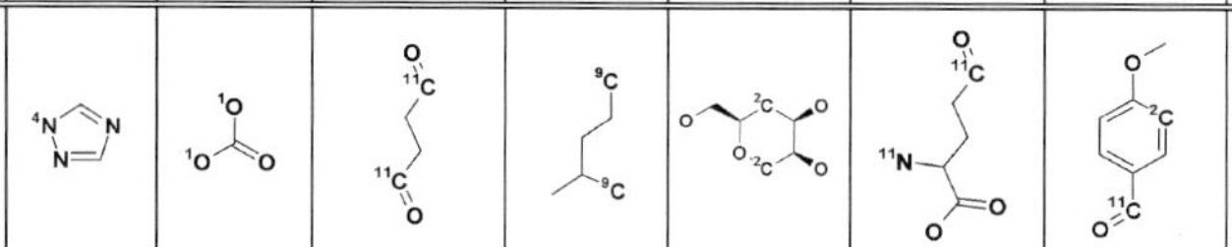

biological matches (%)	775 (0.6%)	632 (0.5%)	641 (0.5%)	531 (0.4%)	545 (0.4%)	545 (0.4%)	509 (0.4%)
oral matches (%)	7 (0.6%)	6 (0.5%)	5 (0.4%)	5 (0.4%)	5 (0.4%)	5 (0.4%)	4 (0.3%)

Fig. 6.7
Exemplification of significant fragments most different (a) and most similar (b) between oral and MDDR biological testing drugs. The counts and percentages of structures in which each fragment is present are given for both groups.

two groups include piperidine and pyridine scaffolds, in addition to several already exemplified in earlier comparisons (Fig. 6.7).

6.5
Conclusions

As we pointed out in our earlier communication, there is some association of properties of drug types and their derived fragments, in particular when comparing oral and injectable drugs. The apparent difference in some properties (increased MW by 10%) in new vs old drugs observed by Lesson does not appear to be related to the properties of the most commonly occurring fragments in these groups. The 39 significant fragments in new drugs have a mean MW of 90 Da, while the 48 significant fragments in old drugs have a slightly higher mean MW of 93 Da. As Vieth and Leeson pointed out, there is no significant difference in year to year mean MW of approved drugs, which seems to be consistent with small changes in frequently used fragments in the whole groups of old and new marketed oral drugs. It may be that the most significant differences in fragments between older and newer drugs reflect a shifting biological target emphasis over the past 25 years [1, 2]. Our analysis suggests that building compounds from marketed oral drug fragments should increase the bioavailability of compounds, compared to random selection. It is intriguing that there are many fragments common to nearly all groups and routes of administration, suggesting they may be generally useful for obtaining biological activity while also proving beneficial for, or at least not detrimental to, reasonable oral exposure. By contrast, many of the fragments that appear frequently in injectable, clinical, and biologically active compounds but not oral drugs are reflective of peptide substructures. These fragments clearly have biological relevance as well. Peptides themselves frequently display poor PK/PD properties; this however may be more a reflection of the properties of the molecules as a whole rather than the individual fragments. At the same time, many of the particular amino acid-related side-chains that appear frequently in injectable but not oral drugs derive from amino acids such as tryptophan and histidine that may suffer from metabolic liabilities on oral dosing.

We believe that, as researchers continue to analyze fragment differences between classes of drugs and between drugs and biologically active molecules or non-drugs, these differences will help to shed light on property differences between the molecules as a whole. In addition, as we demonstrated here, the significant fragments could tell the story about the resulting molecules constructed from them.

Acknowledgments

The authors would like to thank Dan Robertson, Ken Savin, Phil Hipskind, and Ian Watson for their contribution to Molecular Slicer. Dr. Tudor Oprea is acknowledged for helpful comments on the chirality of fragments and molecules in the database.

References

1 Vieth, M., Siegel, M. G., Higgs, R. E., Watson, I. A., Robertson, D. H., Savin, K. A., Durst, G. L., Hipskind, P. A. **2004**, Characteristic physical properties and structural fragments of marketed oral drugs. *J. Med. Chem.* 47, 224–232.

2 Leeson, P. D., Davis, A. M. **2004**, Time-related differences in the physical property profiles of oral drugs. *J. Med. Chem.* 47, 6338–6348.

3 Proudfoot, J. R. **2005**, The evolution of synthetic oral drug properties. *Bioorg. Med. Chem. Lett.* 15, 1087–1090.

4 Lewell, X. Q., Judd, D. B., Watson, S. P., Hann, M. M. **1998**, RECAP–retrosynthetic combinatorial analysis procedure: a powerful new technique for identifying privileged molecular fragments with useful applications in combinatorial chemistry. *J. Chem. Inform. Comput. Sci.* 38, 511–522.

5 Bemis, G. W., Murcko, M. A. **1996**, The properties of known drugs. 1. Molecular frameworks. *J. Med. Chem.* 39, 2887–2893.

6 Bemis, G. W., Murcko, M. A. **1999**, Properties of known drugs. 2. Side chains. *J. Med. Chem.* 42, 5095–5099.

7 Aronov, A. M., Bemis, G. W. **2004**, A minimalist approach to fragment-based ligand design using common rings and linkers: application to kinase inhibitors. *Proteins Str. Funct. Bioinform.* 57, 36–50.

8 Chen, G., Zheng, S., Luo, X., Shen, J., Zhu, W., Liu, H., Gui, C., Zhang, J., Zheng, M., Puah, C. M., Chen, K., Jiang, H. **2005**, Focused combinatorial library design based on structural diversity, druglikeness and binding affinity score. *J. Combin. Chem.* (in press).

9 Todorov, N. P., Buenemann, C. L., Alberts, I. L. **2005**, Combinatorial ligand design targeted at protein families. *J. Chem. Inform. Mod.* 45, 314–320.

10 Stahl, M., Todorov, N. P., James, T., Mauser, H., Boehm, H.-J., Dean, P. M. A. **2002**, Validation study on the practical use of automated de novo design. *J. Comp. Aided Mol. Des.* 16, 459–478.

11 Rees, D. C., Congreve, M., Murray, C. W., Carr, R. **2004**, Fragment-based lead discovery. *Nat. Rev. Drug Disc.* 3, 660–672.

12 Erlanson, D. A., McDowell, R. S., O'Brien, T. **2004**, Fragment-based drug discovery. *J. Med. Chem.* 47, 3463–3482.

13 DeSimone, R. W., Currie, K. S., Mitchell, S. A., Darrow, J. W., Pippin, D. A. **2004**, Privileged structures: applications in drug discovery. *Combin. Chem. HTS* 7, 473–493.

14 Derwent 2001, *Derwent World Drug Index, 2001–2002 edn*, Derwent, London.

15 Walters, W. P., Stahl, M. T., Murcko, M. A. **1998**, Virtual screening – an overview. *Drug Discov. Today* 3, 160–178.

16 MDL Information Systems **2002**, *MDL Drug Data Report 2002.2 edn*, MDL Information Systems, London.

17 Leeson, P. D., Davis, A. M., Steele, J. **2004**, Drug-like properties: guiding principles for design – or chemical prejudice? *Drug Discov. Today: Tech.* 1, 189–195.

7
Fragment Docking to Proteins with the Multi-copy Simultaneous Search Methodology

Collin M. Stultz and Martin Karplus

7.1
Introduction

The design of compounds (leads) which bind pre-specified targets is typically one of the first steps in the design of therapeutic compounds [1]. Thus, methods which simplify the discovery of such compounds can accelerate the rate at which new drugs are discovered. The finding of ligands for binding sites with known structure can be approached efficiently by using a series of steps that include: (1) determining optimal positions of small chemical fragments in the binding site, (2) linking the fragments together to form molecules that are complementary to the target, and (3) estimating the binding affinity of the resulting molecules. The multi-copy simultaneous search (MCSS) methodology [2], which addresses the first step in the above approach, determines energetically favorable positions of different functional groups in a binding site of interest. The method provides functionality maps of the binding site. Functional groups in these energetically favorable positions can then be linked together to construct ligands *de novo* which are complementary to the binding site of the target. Alternatively, these positions can be used to modify known ligands to improve their binding affinity. The method is quite general and can be used to construct functionality maps for proteins, DNA, and RNA. Moreover, in principle, protein flexibility can be incorporated in the process in a straightforward manner. In this chapter, we describe the main aspects of the methodology and demonstrate how the information arising from the functionality maps can be used to design ligands.

7.2
The MCSS Method

Here we review major aspects of the MCSS methodology used for calculating functionality maps.

Fragment-based Approaches in Drug Discovery. Edited by W. Jahnke and D. A. Erlanson
Copyright © 2006 WILEY-VCH Verlag GmbH & Co. KGaA, Weinheim
ISBN: 3-527-31291-9

7.2.1
MCSS Minimizations

The method begins by randomly distributing many replicas (e.g., between 1000 and 10 000 copies) of a small functional group in the binding site of a target molecule. The positions of the different functional group copies are then simultaneously minimized in the binding site, using cycles of conjugant gradient minimization. Functional group replicas which are converging to a common minimum energy position are removed during the minimization process to save time. In most applications, replicas are said to be converging to the same minimum energy position if their root mean square (rms) deviation is less than 0.2 Å. Use of such a small cutoff ensures that a variety of different functional group positions and orientations will be kept.

A central component of the method is the time-dependent Hartree (TDH) approximation, which enables a number of molecular trajectories to be simultaneously determined in the field of a target molecule [3, 4]. Making use of this concept, different functional group replicas do not see one another, but do see the full field of the target molecule. Typically, the functional group replicas are fully flexible and therefore sample different conformations during the minimization procedure, while the target molecule is fixed. When the target is rigid, the trajectories generated with the TDH approximation are exact; i.e., the MCSS minimum energy positions are identical to minima that would have been obtained if each randomly placed replica were individually minimized in the field of the fixed protein. By contrast, if the protein is flexible, the resulting MCSS minima may differ from positions obtained from individual minimizations [5, 6]. As such, many of the initial applications of MCSS focused on obtaining functionality maps to rigid targets. However, protein flexibility can be incorporated into MCSS minimizations and the resulting data can, in principle, be post-processed to determine the minima which are most promising. In subsequent sections of this chapter, we deal with the issue of including protein flexibility in MCSS calculations. In the remaining portions of this section we focus on MCSS minimzations with a rigid target.

7.2.2
Choice of Functional Groups

Chemical fragments used for MCSS functionality maps are typically small functionalities that span the range of different chemical moieties that could be used to build larger, more realistic, molecules, although in some applications essentially complete ligand candidates have been used. When MCSS was first introduced, calculations were performed with functional groups that typically contained less than eight atoms. Popular functional groups used in prior applications include N-methylacetamide, methanol, methyl ammonium, acetate, propane, benzene, phenol, cyclohexane, and water [e.g., 2, 7, 8, 9]. These groups collectively model polar, charged, and hydrophobic moieties of different sizes. N-Methylacetamide is chosen to model the peptide backbone and is of particular interest if the goal is to design peptide based ligands that are complementary to a given target.

As the speed of modern computers has improved, the size of functional groups used in routine MCSS calculations has significantly increased. It is now routine to perform MCSS calculations on rather complex chemical fragments that contain 20–30 atoms [e.g., 10]. Functional groups currently available with the standard version of MCSS span a range of possible chemical fragments (Fig. 7.1). A number of fragments were obtained from an analysis of the structures found in a database of known drug molecules (Figs. 7.2, 7.3) [11, 12]. Furthermore, adding novel and sometimes complex functional groups to the MCSS library is relatively straightforward and only involves creating the necessary topology, parameter, and symmetry files. Just as in the case of the standard CHARMM topology and parameter files [13], the MCSS topology file lists the partial charges, atom types, and connectivity of the atoms. The group parameter file contains the additional parameters that are needed for the energy calculations. MCSS runs can be performed with both a polar hydrogen parameter set [14] and an all-atom parameter set [15] or using a hybrid approach where the protein is represented with one particular parameter set (e.g., all-atom) and the MCSS functionalities are represented with the another parameter set (e.g., polar).

It is important to note that, if the chosen functional group has several rotatable bonds, care must be taken to ensure that enough randomly placed functional group replicas are included in the initial stages of the simulation to sample different conformations of the chemical fragment. Undersampling of functional group conformations makes it difficult to find functional group minima with a variety of different conformations.

7.2.3
Evaluating MCSS Minima

Although MCSS minimizations begin with several thousand randomly placed replicas, typically only a few hundred remain after repeated cycles of energy minimization. The resulting interaction energy of each MCSS minimum is an important part of the MCSS output. These interaction energies represent a sum of the intragroup energy and the group–protein interaction energy. In current versions of MCSS, a reference vacuum energy is subtracted from the MCSS interaction energy of each group. This correction enables one to compare interaction energies arising from different models of the same functional group [16, 17]. For instance, the *in vacuo* energy of a polar-hydrogen representation of benzene is lower than the energy of an all-atom representation of benzene in the same conformation because the all-atom representation contains explicit hydrogens which have overlapping vdw radii [17]. Hence, to correct for this, the *in vacuo* CHARMM energy of a minimized conformation of benzene is subtracted from each MCSS minimum.

MCSS interaction energies are typically computed using a vacuum potential. However, in principle, solvent effects can be accounted for at the outset of MCSS calculations by minimizing the randomly placed functional group replicas, using

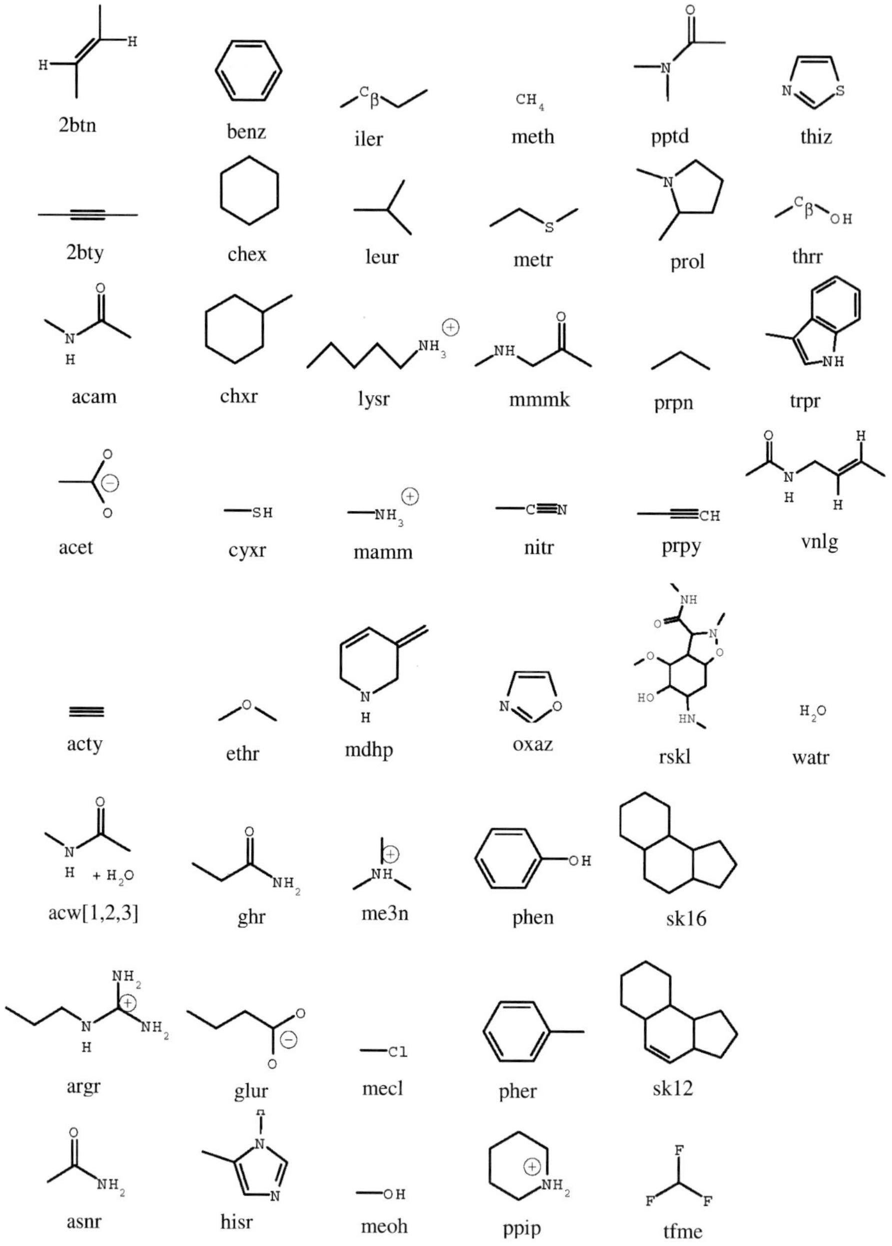

Fig. 7.1
Functional groups included in the standard MCSS distribution.

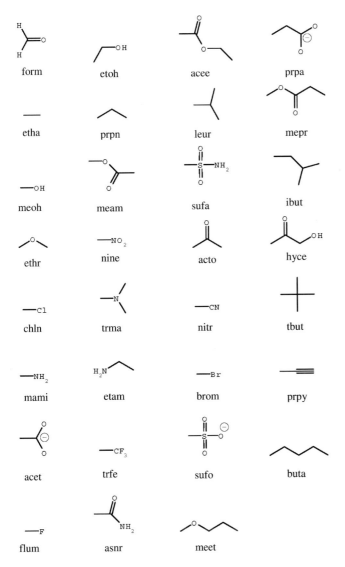

Fig. 7.2
Acyclic atoms in known drug databases which are included in MCSS.

a potential that incorporates an implicit model of solvent. A straightforward implicit solvent model to implement is based on a distance-dependent dielectric constant. We note that it has been argued that the use of a distance-dependent dielectric constant during MCSS minimizations increases the number of minima that one finds relative to protocols that employ a vacuum potential [18]. However, in previous applications on endothiapepsin, minima obtained with a distance-dependent dielectric constant were very similar to minima obtained with $\varepsilon=1$ [5]. Differ-

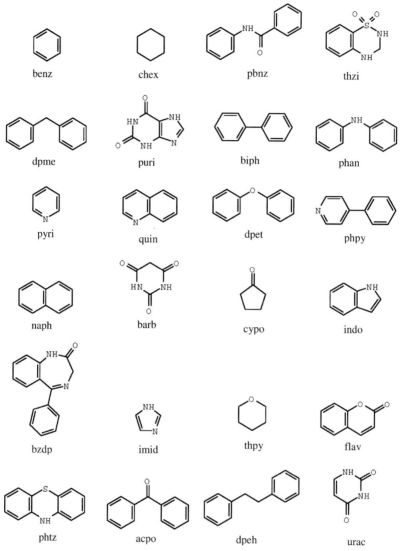

Fig. 7.3
Cyclic functional groups included with MCSS.

ences between the minima with implicit solvation and the minima on the vacuum potential surface are likely to be most significant for charged groups.

A number of other implicit solvent models have been developed recently that could be used for MCSS minimizations [19]. However, MCSS minimizations that employ such models do require additional computational time to optimize a system consisting of several thousand functional groups and a protein composed of several thousand atoms. An alternate approach is to perform MCSS minimizations with a vacuum potential and then post-process the resulting minimum energy positions in a manner that accounts for solvent.

The most extensively used method for post-processing MCSS results is to remove minima with interaction energies above some cutoff value that is based on the solvation enthalpy of the functional group in question [2]. This provides an approximate way to account for the fact that functional groups within the binding site are, at least, partially desolvated. Typically, only groups with interaction energies below the solvation energy are kept for further analysis. As the solvation energy of charged ions can be approximated by one-half the solvation enthalpy, in practice, the latter value is used as a cutoff [20]. The main advantage of using an interaction energy cutoff to account for solvent is that it does not require additional lengthy calculations. In addition, while the method may significantly reduce the number of functional groups that are kept, it usually leaves a considerable number of minima in the binding site for further analysis.

Alternate, more computationally intensive approaches for post-processing MCSS minima have been developed. In one study, Caflisch [18] ranked minima arising from an MCSS calculation on α-thrombin by estimating the binding free energy of each minimum with an approximate free energy formula given by:

$$\Delta G_{binding} \equiv \Delta E_{intra} + \Delta E_{inter,vdw} + \Delta G^{inter}_{elec} + \Delta G^{p}_{desolv} + k\Delta G^{m}_{desolv} + \Delta G_{np} \quad (1)$$

where ΔE_{intra} denotes the intramolecular energy of the MCSS chemical fragment and is the sum of the bonded, vdw, and electrostatic contributions. The intermolecular vdw interaction energy between the protein and the functional group is included in an additional term denoted by $\Delta E_{inter,vdw}$. The electrostatic interaction energy between the protein and functional group is denoted by ΔG^{inter}_{elec}. The energy required to desolvate the binding site of the protein is given by the energy required to desolvated the surface of the functional group that interacts with the protein and is denoted by ΔG^{m}_{desolv}. In this implementation, the desolvation energy of the functional group is multiplied by an additional factor, $k<1$, which serves as an approximate way to account for the fact that the desolvation contribution for a functional group which is part of a larger ligand may be smaller. This is likely to be true for the majority of small functional groups that are considered by MCSS, because in the unbound state small functional groups are surrounded by other atoms in the ligand which prevent the group from being completely solvated. In the work on α-thrombin, a value of $k = 0.4$ was used. This value was derived from a comparison of the electrostatic desolvation energy upon binding of NMA and a glycine containing dipeptide to a macromolecular target [18].

The electrostatic free energy terms were computed with a continuum dielectric model based on the linearized Poisson–Boltzmann equation and numerical solutions were obtained with the program UHBD [21]. The nonpolar contribution to the binding free energy, ΔG_{np}, is assumed to be linearly related to the loss in solvent-accessible surface area – a common approach for estimating the hydrophobic contribution to binding [22, 23].

The approximate binding free energy formula ignores some contributions which are, in general, important for ligand binding. Most notably, the loss of configurational entropy upon binding is not included, even in an approximate way. This contribution is likely to be negligible for the small, relatively rigid, functional groups considered in the initial study. However, if more flexible functional groups are used, then additional terms are needed to approximate the change in configurational entropy upon binding.

For α-thrombin, application of Eq. (1) to the MCSS minima leads to considerable reordering of the minima relative to the ordering obtained with a vacuum potential [18]. Differences between the vacuum energy rankings and the rankings obtained with Eq. (1) are primarily due to the desolvation terms. In many cases, apolar functional groups with favorable vacuum interaction energies have unfavorable binding free energies because they partially desolvate charged groups on the protein. Similarly, a number of charged groups with favorable vacuum energies have unfavorable binding free energies, primarily because of the large desolvation penalty associated with the binding of charged moieties. The inclusion of desolvation terms has the greatest effect in determining the relative affinities of MCSS minima for different functional groups. However, any potential benefit from such calculations must be weighed against the increase in computational time that is incurred with such an analysis. We note that an alternate approach would be to use the MCSS minima, and their vacuum interaction energies, to create potential ligands and then evaluate the binding energy of the resulting ligands with more rigorous approaches.

7.3
MCSS in Practice: Functionality Maps of Endothiapepsin

The MCSS method has been used to obtain functionality maps for the fungal protease, endothiapepsin [24]. The structure of endothiapepsin bound to the peptide-based inhibitor H-261 reveals that the binding site of the enzyme is relatively open and contains several hydrophobic pockets which form binding sites for hydrophobic moieties in the inhibitor [25] (Fig. 7.4A). MCSS minimizations were conducted on the binding site of endothiapepsin with the functional group probes N-methyl acetamide (NMA), methyl ammonium, methanol, and propane. Polar minima such as NMA and methanol were found throughout the binding site and make hydrogen bonds to polar moieties in the protein (Fig. 7.4B, C). Apolar functionalities are found throughout the binding site within apolar pockets (Fig. 7.4D). Charged groups such as methylammonium make salt bridges with charged side-chains in the binding site and therefore they occur in smaller numbers (Fig. 7.4E). In addition, as charged groups tend to occur near the surface of the protein, charged minima may be located on the surface in regions relatively distant from the binding site.

Comparison of MCSS functionality maps with the positions of corresponding functionalities in the inhibitor demonstrates that MCSS can reproduce the posi-

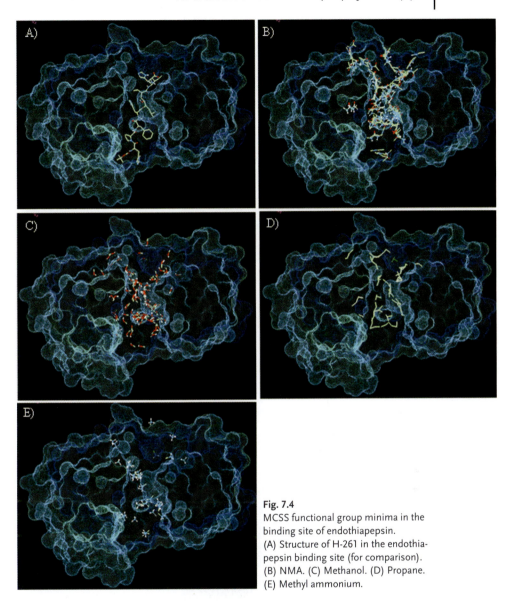

Fig. 7.4
MCSS functional group minima in the binding site of endothiapepsin.
(A) Structure of H-261 in the endothiapepsin binding site (for comparison).
(B) NMA. (C) Methanol. (D) Propane.
(E) Methyl ammonium.

tions and orientations of chemical fragments found in inhibitors. Propane minima are found in hydrophobic pockets occupied by phenylalanine, leucine and valine side chains (Fig. 7.5A, B). Methanol minima are also found near the lone hydroxyl moiety in H-261 that forms hydrogen bonds with two catalytic aspartates in the binding site (Fig. 7.6). Similarly, most peptide bonds within H-261 are located near an NMA minimum (e.g., Fig. 7.7A, B). Some peptide bonds, however, within the inhibitor are not located near an MCSS minimum.

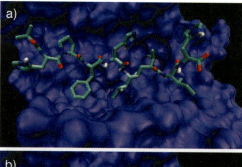
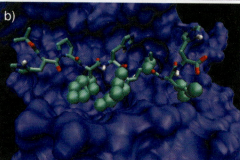

Fig. 7.5
(A) Structure of inhibitor H-261 in the binding site of endothiapepsin. A loop within the protein which overlaps the binding site has been removed to reveal the inhibitor. (B) MCSS propane minima within each of the hydrophobic pockets occupied by hydrophobic side-chains in H-261. Propane molecules are shown as vdw spheres and H-261 is depicted in liquorice mode.

Figure 7.7C shows a peptide bond in the inhibitor and its closest NMA minimum. Although the orientations of the two peptide hydrogens are similar, the NMA minimum has a root mean square deviation (msd) that is more than 3 Å from the peptide bond shown. Such data suggest that the peptide bond in H-261 is in an orientation that does not correspond to a minimum energy position.

As MCSS determines positions for small functionalities and not complete inhibitors, discrepancies between the functional group positions in co-crystal struc-

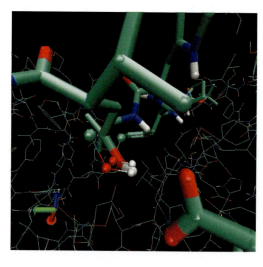

Fig. 7.6
Positions of MCSS methanol minima superimposed on the backbone of H-261. MCSS minima are represented as a ball and stick diagram, whereas the inhibitor is shown as a solid liquorice figure.

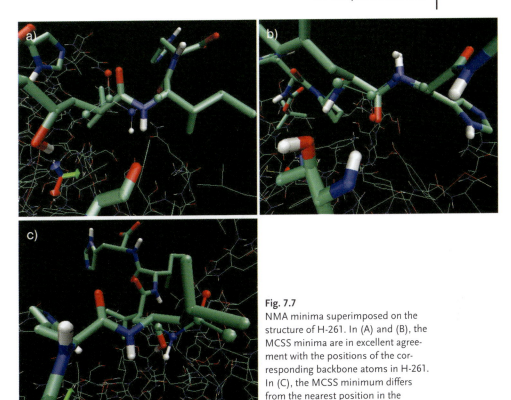

Fig. 7.7
NMA minima superimposed on the structure of H-261. In (A) and (B), the MCSS minima are in excellent agreement with the positions of the corresponding backbone atoms in H-261. In (C), the MCSS minimum differs from the nearest position in the inhibitor.

tures of bound inhibitors and the positions of MCSS minima are not unexpected. What is more interesting is the fact that MCSS typically finds a number of new minima that were not observed in known crystallographic structures, suggesting that the affinity of these inhibitors may be improved if low energy functionalities are incorporated into the inhibitor structures.

7.4
Comparison with GRID

One of the earliest algorithms used to determine functional group binding sites in a target molecule is the program GRID [26]. In its initial implementation, functional groups were represented as spherical probes and the interaction energy of the probe with the target was evaluated at discrete points on a grid. The interaction energy is evaluated using a simple potential energy function that is the sum of electrostatic, van der Waals, and a hydrogen bond contribution [26, 27]. The method has evolved since its initial stages and it now includes some multi-atom probes in its library, where the interaction energies for such probes are evaluated as the sum of interaction between single sphere probes and the protein. GRID uses these calculations to produce a contour map of the binding site of interest that can be used to identify energetically favorable positions. By contrast, MCSS represents functional groups as an atomic model for chemical fragments and the CHARMM potential energy function is used to evaluate the interaction energy. As such, MCSS not only provides optimal positions for functional groups in the binding site, but it also provides information regarding the orientation of these functional group minima. In addition, since MCSS functional groups may be inherently flexible, the interaction energy consists of both an intragroup energy and a group–protein interaction energy. This is an important difference, as GRID does not account for any internal strain that may be introduced within the fragment upon binding. Therefore, MCSS may be a more appropriate method for studying flexible chemical fragments.

In a comparison study, MCSS and GRID functionality maps were obtained for the hydrophobic pocket of poliovirus capsid protein and the src SH3 domain, using a number of different functional group probes [27]. GRID and MCSS agree very well when methane is used as a molecular probe. This is not unexpected, given that the CHARMM polar hydrogen representation for methane is a molecular sphere – a representation similar to that in the GRID library. However, agreement between MCSS and GRID is only qualitative when water is used as a functional group probe. While energetically favorable positions identified by GRID are typically found in the vicinity of MCSS minima, MCSS finds additional minima that are not identified by low-energy GRID contours. More precisely, contours at energies near the solvation free energy of water miss many minima found by MCSS.

As GRID models water as a neutral sphere with implicit hydrogens, energetically favorable positions for water are primarily determined by the GRID hydrogen bonding potential; and orientations of water molecules in the binding site are inferred based on an allowed set of geometric preferences. MCSS, by contrast, uses the CHARMM potential which no longer incorporates an explicit hydrogen bonding term; i.e., the hydrogen bond contribution in CHARMM arises from the electrostatic term in the potential function and the angular dependence is largely determined by the van der Waals term. As a result, some MCSS water minima within the poliovirus capsid protein have different hydrogen bonding patterns relative to corresponding GRID minima. More importantly, small changes in the positions of some GRID minima lead to significant changes in the hydrogen bonding patterns and the corresponding hydrogen bonding energies. This highlights one shortcoming of the GRID potential function and suggests that additional interactions that are not involved in direct hydrogen bonds should be included in the GRID energy function.

As GRID only computes interaction energies for spherical probes at pre-defined grid points in a binding site, it can compute contour maps with considerable efficiency; i.e., contour maps for large binding surfaces can be computed faster with GRID than with MCSS. Given this, one hybrid approach which may prove useful for obtaining functionality maps of large binding sites is to first obtain GRID contour maps and use them to determine the positions that warrant additional, more detailed, functional group mapping with MCSS.

7.5
Comparison with Experiment

In the initial test of the MCSS method, a small subset of functional groups consisting of acetate, methanol, methyl ammonium, methane, and water was used to map the binding site of the influenza coat protein, hemagglutinin [2]. It was demonstrated that MCSS could reproduce the positions of the corresponding functionalities in a sialic acid bound to hemagglutinin. In a subsequent study, the binding site of HIV-1 protease was mapped with the functional groups N-methyl acetamide, acetamide, methanol, acetate, methyl ammonium, and ethyl guanidium. The resulting functional group minima are all within 2.4 Å of the corresponding functionalities in the known HIV-1 inhibitor MVT-101, suggesting that one could approximate functional group positions within known inhibitors with MCSS [28]. Subsequent comparisons of MCSS minima to the structures of known inhibitors have been used to validate the MCSS approach in a number of applications, e.g., [7, 29, 30].

More recently, an experimental approach has been developed for finding binding sites of small organic molecules within a target molecule of known three-dimensional structure [31] (see Chapter 4). In the multiple solvent crystal structure (MSCS) method, the crystal structure of a given protein is determined in the presence of different organic solvents. The resulting three-dimensional structures reveal positions of organic solvent molecules on the surface of the protein. As such, the positions arising from the method can be directly compared to the MCSS minimum energy positions obtained when the same organic solvent is used as a functional group probe.

It should be noted that experiments with organic solvent may introduce significant alterations in the structure of the protein. In an application of MSCS, positions for acetonitrile molecules in the binding site of porcine pancreatic elastase were obtained [31]. To ensure that protein would be stable in the organic solvent, elastase was first cross-linked using a solution containing glutaraldehyde, prior to being exposed to a 100% acetonitrile solution. Crystal structures of cross-linked elastase in acetonitrile differed from the native, uncross-linked structure by almost 0.5 Å. Unfortunately, the original report does not detail the precise differences within the binding site; and it is just these differences that would have the largest impact on the positions of MCSS minima. MCSS minima were obtained by minimizing randomly placed chemical fragments, using a vacuum potential and post-processing the data to account for the affects of aqueous solvent. There was qualitative agreement between the MSCS solvent positions and the minima identified by MCSS. However, it is important to note that the observed binding sites of organic molecules in the presence of nonaqueous solvent will likely differ as the dielectric constant of neat organic solvents will differ from the dielectric constant of water. These considerations suggest that comparisons between the results of MSCS experiments and MCSS calculations should be interpreted with care.

In another application, MSCS was applied to thermolysin and the results of those experiments were compared to results obtained with MCSS [32]. In that work, the structure of thermolysin was determined in the presence of the organic solvents acetone, acetronile, and phenol. All experimentally determined positions were within 3.5 Å of a MCSS minimum energy position and most MSCS positions were within 2.8 Å of a MCSS minimum. These data again suggest that MCSS can be used to effectively map binding positions of small functional groups.

NMR has also been used to experimentally determine binding sites for small organic compounds within a given target [33] (see Chapter 8). Using an ^{15}N-labeled target, chemical shifts in pre-specified regions of the target protein are observed when a ligand binds to the region of interest. This information is used to both identify molecules from a database of compounds which bind the target in different areas of the binding site. Recently, Sirockin et al. [34], identified binding sites on FKBP12 for three functional groups [(2S)1-acetylprolinemethylester, 1-formylpiperidine, 1-piperidinecarboxamide] and compared these data to the corresponding MCSS functionality maps. For each ligand, the MCSS functionality maps contained at least one minimum energy position that satisfied the NOE restraints arising from NMR data. Moreover, using an approximate binding free energy function similar to that described by Caflisch et al. [18] and Eq. (1), Sirockin et al. demonstrate that each minimum that satisfies the NOE restraints is among the ten best minima in the corresponding functionality map. These data support the use of MCSS for discovering the binding sites of small molecules as well as for fragments of larger compounds.

7.6
Ligand Design with MCSS

7.6.1
Designing Peptide-based Ligands to Ras

In a recent application, Zeng et al. used MCSS to design peptides that would inhibit the association of the Ras-Raf complex [35, 36]. As the Ras-Raf-GTP complex may play a role in promoting cell growth and oncogenesis, the process of inhibiting complex formation is an area of considerable interest [37].

Zheng et al. used a procedure that is similar, in spirit, to the approach described by Caflisch et al. for constructing peptide-based ligands to HIV-1 protease [4]. They constructed peptide-based inhibitors which prevent the formation of Ras-Raf complex in a stepwise manner: (1) they obtained energetically favorable positions for NMA, benzene, phenol, methanol, acetate, methylammonium, methylguanidium, and propane with MCSS, (2) a peptide backbone was constructed from the

NMA minimum energy positions, (3) the functional groups were connected to the constructed backbone, and (4) the peptides arising from this procedure were aligned to determine the probabilities of different side-chains occupying each Cα position. Calculations began with an average structure obtained from a 2ns simulation of Ras bound to the Ras-Binding-Helical (RBH) domain within Raf [35]. The basic idea underlying the design strategy was to extend the RBH to obtain a peptide that would bind at the Ras-Raf interface and prevent formation of the complex [36]. MCSS calculations on the Ras-RBH structure identified three potential binding sites that mimic regions observed in crystal structures of a homologous protein, Rap, bound to Raf [35]. More importantly, the locations of MCSS minima identified novel interactions that could be utilized in the design of a peptide-based inhibitor.

To construct peptides from the MCSS minima, a Cα trace was created from the positions of the various NMA minima on the surface of the protein in the region of the RBH domain. Side-chains were then added to the Cα atoms of polyglycine to construct peptide ligands using a simple distance criterion to determine which functional groups could be attached to each Cα backbone atom in the polyglycine model. The designed peptides were then merged to the RBH domain, yielding a total of 104 peptides which were then aligned to facilitate the creation of a consensus sequence. Interestingly, the consensus sequence was similar to known effector sequences which modulate the binding of Ras to Raf. Additional sequences were constructed using residues that occurred in the alignment with probability >0.3. Each of these potential leads were further modified to improve its solubility and to take advantage of known sequence motifs that are thought to bind Ras. Their resulting peptides were able to inhibit the Ras-Raf interaction in vitro by 23–39% using an Elisa-based assay [36].

Although the details of the peptide construction and design are similar to previously described MCSS-based peptide design methods, the approach of Zeng et al. has some important and interesting differences. The data arising from MCSS is used to obtain probabilities for different side-chains occupying various positions in the bound structure rather than to design a single peptide that bound Ras with high affinity. In this manner, Zheng et al. exploit the wealth of information contained in many different functional group minima, instead of just a few that can link up with the peptide backbone. In addition, Zeng et al. explicitly account for water-mediated hydrogen bonds between charged functional groups and the protein by associating a small number of water molecules (i.e., 2–4) with each replica. A corresponding approach was used by Joseph-MCarthy et al. [38]. Each replica and its associated water molecules are treated as one functional group copy; i.e., waters see their associated replica and the protein, but waters associated with different functional group copies do not see one another. As a result, many of the functional group minima are engaged in water-mediated charged bridges. One drawback to this approach is that, given the significant conformational flexibility associated with multi-atom functional groups and their associated water molecules, a considerable number of functional group replicas are needed to adequately sample the different conformations that the functional groups and their waters can adopt. Zheng et al. only include 400–1000 functional group replicas at the outset of their MCSS calculations. Therefore, it is likely that the number of possible functional group minima is undersampled and minima are missed. Nevertheless, they obtain interesting results with the subset of minima that they do find. Overall, this method presents an interesting example of how information from MCSS and experiment can be combined to obtain peptides which inhibit the association of two proteins.

7.6.2
Designing Non-peptide Based Ligands to Cytochrome P450

In another application, MCSS was used to design nonpeptide-based inhibitors to a member of the cytochrome P450 family, lanosterol 14α-demethylase (CYP51) [39]. Such molecules have proven to be a useful source of antifungal agents.

MCSS calculations were conducted on the substrate binding pocket of CYP51, using the functional groups benzene, propane, butane, cyclohexane, phenol, methanol, ethanol, ether, and water. The calculations began with structural models of CYP51 from *Mycobacterium* and *Candida albicans* and each MCSS minimization began with 5000 randomly placed functional groups in each binding pocket. The minimizations yielded a large number of functional group minima for both protein models. However, unlike the work of Zheng et al., for each functional group considered, the lowest energy minimum in each subsite was chosen for further analysis; i.e., of the more than 1000 functional group minima found, only 13 minima were used to construct potential leads. The 13 different functional group minima were connected manually to obtain putative leads. Of note, the only constraints used during the process of connecting the various minima were that no internal strain was introduced into the structure and that the resulting molecules were synthetically accessible.

Four promising compounds arose from this analysis that were subsequently shown to have IC_{50} values in the micromolar range against reconstituted CYP51 from *Candida albicans*; and the calculated CHARMM interaction energies of the different lead compounds correlated quite well with the measured IC_{50} values. An analysis of the potential leads within the CYP51 binding site suggested additional modifications that led to an improved IC_{50} for some forms CYP51. This example illustrates how MCSS can be used to design lead compounds *de novo*. Moreover, this work represents the first example of *de novo* design of a ligand for any member of the cytochrome P450 system.

7.6.3
Designing Targeted Libraries with MCSS

MCSS generates maps of energetically favorable positions in the binding site of a target molecule. Such maps typically contain many functional group positions. Given the number of potential compounds that can be designed from such data, it is natural to use this information to construct combinatorial libraries that can efficiently explore a wide range of potential ligands. One advantage of using MCSS to design combinatorial libraries, as opposed to designing individual molecules, is that the scoring function used to select the most promising compounds from a combinatorial library need not be very accurate. More precisely, the molecules within the library are typically screened with an efficient experimental assay. In this setting, MCSS can be used to direct the construction of relatively small libraries that contain molecules with a high probability of being successful.

In recent work, MCSS was used to create a small targeted library from functionality maps to picornavirus [17, 38]. MCSS calculations on the hydrophobic binding pocket of P3/Sabin poliovirus was obtained using NMA, methanol, water, acetic acid, methylammonium, magnesium ion, magnesium ion+water, 3-methuloindole, 5-methylimidazole, phenol, benzene, toluene, cyclohexane, propane, isobutene, and methylbenzimidazole (both protonated and unprotonated forms) [17, 38]. A comparison of the resulting MCSS minima with compounds that are known to bind within the largely hydrophobic binding pocket showed good agreement. More importantly, this comparison suggested that the bound compounds did not take advantage of all the available low energy configurations of its associated functional groups.

The functional group minima detail the positions and orientations of energetically favorable positions within the hydrophobic binding pocket. An analysis of the minima reveals that they form three distinct clusters within the hydrophobic pocket. These data were used to construct libraries of compounds that contained three functionalities connected by different linker moieties [38]. Although the first library contained at most 75 compounds (the exact number was not determined), eight of these compounds bound P1/Mahoney as determined with a noncell-based assay. Four of the eight compounds were identified and used to construct six additional molecules which were subsequently shown, using an immunoprecipitation assay, to prevent viral transformation from the native particle to its infectious intermediate. A crystallographic structure of one of the compounds bound to the P1/Mahoney structure verified that the functionalities in the designed compound agreed well with the positions and orientations of the corresponding MCSS functional groups. These results offer strong validation for the MCSS methodology and demonstrate that MCSS functionality maps can be used to design very small libraries that contain compounds which are complementary to the target molecule.

7.7
Protein Flexibility and MCSS

It is well established that protein flexibility can have a significant effect on ligand binding [40]. Many potential ligands are likely to be missed if only rigid targets are considered in the design process. It is desirable, therefore, to develop methods that can incorporate protein flexibility in the early stages of a drug design protocol.

One straightforward approach for including protein flexibility in MCSS calculations is to perform MCSS minimizations on different X-ray crystallographic structures of a given protein target. As different crystallographic structures may sample different side-chain conformations, such data provide insight into the different functionality maps that one can obtain from different structures. Similarly, molecular dynamics simulations could be used to position side-chains in the target in different conformations and these structures can be used to produce functionality maps for each protein conformation. In this manner, one obtains several different functionality maps for a given functional group, where each map is constructed for a different target. While such approaches are potentially quite useful, it is not clear how to combine the data from functionality maps that arise from the different structures.

A comparison has been made of functionality maps for HIV-1 protease, generated using both a rigid protein and a flexible protein [41]. To investigate how different protocols for including flexibility in the target structure affect the positions of MCSS minima, the functionality maps obtained for HIV-1 protease using standard MCSS with different protein structures were compared to functionality maps obtained with quenched molecular dynamics. Functionality maps using the probes methanol and methyl ammonium were obtained, as these functionalities are representative of the polar and charged chemical fragments that are utilized in standard MCSS runs.

All quenched molecular dynamics simulations that include functional group replicas were performed with the locally enhanced sampling (LES) methodology [4]. As mentioned above, in this approximation, each functional group copy interacts with the full field of the protein, but the functional groups do not interact with each other and the protein interacts with the mean field of all of the functional group replicas. Unlike the rigid protein case, when the protein is flexible, the minima that arise from the molecular simulations are not exact; i.e., functional group minima arising from individual minimizations of the protein with one functional group replica may differ from the minima arising from quenched simulations that using the LES method [5, 6]. As such, the minima arising from these simulations are approximations to the exact result; and different metrics have been developed to determine how far each of these "approximate" minima are from the "true" minima that would have been obtained with repeated minimizations of a single copy of the functional group [6].

Functionality maps with an alternate crystal structure yield minima with interaction energies that are comparable to the first crystallographic structure. However, protocols that employ quenched molecular dynamics yield minima with less favorable interaction energies relative to the minima obtained with standard MCSS and a crystallographic structure. In quenched molecular dynamics simulations with the LES method, protein atoms in the binding site see the full field of other residues in the protein, but see only the mean field of the functional group replicas. As there are 1000 replicas, the field arising from any one copy is quite small relative to the field arising from the protein atoms. Therefore, protein residues preferentially optimize their interactions with other residues and not with the functional group replicas themselves. The result is that the functional group minima are less favorable relative to the minima obtained with the original crystallographic structure.

Quenched dynamics involving specific functional group minima in a localized region of the binding site was shown to find minima with significantly lower (more favorable) interaction energies [41]. In light of this, one approach for incorporating protein flexibility in MCSS calculations is to perform standard MCSS minimizations with different crystallographic structures of a given protein followed by local optimizations of specific minima which are of particular importance. Such highly optimized chemical fragments could serve as the starting point for designing molecules that bind their targets with high affinity.

7.8
Conclusion

MCSS has proved to be a valuable tool in the arsenal of methods used for drug discovery. In this chapter, we have focused on the application of MCSS to proteins, but it should be noted that the method has been applied to both DNA and RNA and fruitful results have been obtained [42, 43].

As MCSS finds energetically favorable positions of chemical fragments in a binding site of interest, additional methods are needed to construct ligands from MCSS data. A variety of different approaches have been developed to construct more complex compounds from MCSS minima. Some methods link NMA minima together to form a peptide backbone which serves as the starting point for the design of peptides that bind given targets [e.g., 7]. Other methods focus on the design of small molecules that are not peptides. Such approaches can be divided into those that find compounds through a database search and those that build, or modify, compounds without the need for a database search. The algorithm HOOK, for example, searches predefined small molecule databases to discover molecules that bind a target such that chemical fragments in the potential ligand occupy positions corresponding to MCSS minima [44]. By contrast, the dynamic ligand design (DLD) method links existing MCSS functional group minima together to form molecules *de novo* which are complementary to the target molecule [45, 46].

The MCSS method is a flexible strategy for finding energetically favorable positions in a binding site of interest. The data arising from this method can be used to design individual molecules that are complementary to the given target, or to design libraries that have a high probability of containing molecules which may be suitable leads. Given the ease of use of the method and the wealth of data contained in a functionality map, the approach is well suited for the design and optimization of lead compounds.

Acknowledgments

We would like to thank Vincent Zoete for help with preparing Figs. 7.1–7.3, and Diane Joseph-McCarthy for helpful comments on the manuscript. This work was supported, in part, by a grant from the NIH. C.M.S. is a recipient of a Burroughs Wellcome Fund Career Award in the Biomedical Sciences.

References

1 Anderson, A. C. **2003**, The process of structure-based drug design. *Chem. Biol.* 10, 787–797.

2 Miranker A., M. Karplus **1991**, Functionality maps of binding sites: a multiple copy simultaneous search method. *Proteins* 11, 29–34.

3 Gerber, R. B., V. Buch, M. A. Ratner **1982**, Time-dependent self-consistent field approximation for intramolecular energy transfer. 1. Formulation and application to dissociation of van der Waals molecules. *J. Chem. Phys.* 77, 3022–3030.

4 Elber, R., M. Karplus **1990**, Enhanced sampling in molecular dynamics: use of the time-dependent hartree approximation for a simulation of carbon monoxide diffusion through myoglobin. *J. Am. Chem. Soc.* 112, 9161–9175.

5 Ulitsky A, R. Elber **1993**, The thermal equilibrium aspects of the time-dependent Hartree and the locally enhanced sampling approximations: formal properties, a correction, and computational examples for rare gas clusters. *J. Chem. Phys.* 98, 3380–3388.

6 Stultz, CM, M. Karplus **1998**, On the potential surface of the locally enhance sampling approximation. *J. Chem. Phys.* 109, 8809–8815.

7 Caflisch, A., A. Miranker, M. Karplus **1993**, Multiple copy simultaneous search and construction of ligands in binding sites: application to inhibitors of HIV-1 aspartic proteinase. *J. Med. Chem.* 36, 2142–2167.

8 Ji, H., W. Zhang, M. Zhang, M. Kudo, Y. Aoyama, Y. Yoshida, C. Sheng, Y. Song, S. Yang, Y. Zhou, J. Lu, J. Zhu **2003**, Structure-based de novo design, synthesis, and biological evaluation of nonazole inhibitors specific for lanosterol 14r-demethylase of fungi. *J. Med. Chem.* 46, 474–485.

9 Zeng, J., T. Nheu, A. Zorzet, B. Catimel, E. Nice, H. Maruta, A. W. Burgess, H. R. Treutlein **2001**, Design of inhibitors of Ras–Raf interaction using a computational combinatorial algorithm. *Protein Eng. Fold. Des.* 14, 39–45.

10 Leonardi, A., D. Barlocco, F. Montesano, G. Cignarella, G. Motta, R. Testa, E. Poggesi, M. Seeber, P. G. De Benedetti, F. Fanelli **2004**, Synthesis, screening, and molecular modeling of new potent and selective antagonists at the r1 d adrenergic receptor. *J. Med. Chem.* 47, 1900–1918.

11 Bemis, G. W., M. Murcko **1996**, Properties of known drugs. 1. Molecular frameworks. *J. Med. Chem.* 39, 2887–2893.

12 Bemis, G. W., M. Murcko **1999**, Properties of known drugs. 2. Side chains. *J. Med. Chem.* 42, 5095–5099.

13 Brooks, B., R. Bruccoleri, B. Olafson, D. States, S. Swaminathan, M. Karplus **1983**, Charmm: a program for macromolecular energy, minimization, and molecular dynamics calculations. *J. Comput. Chem.* 4, 187–217.

14 Neria, E., S. Fischer, M. Karplus **1997**, Simulation of activation free energies in molecular systems. *J. Chem. Phys.* 105, 1902–1921.

15 MacKerell A. Jr., et al. **1998**, All-atom empirical potential for molecular modeling and dynamics studies of protein. *J. Phys. Chem.* B102, 3586–3616.

16 Evensen, E., D. Joseph-McCarthy, M. Karplus **1997**, MCSS ver. 2.1, Harvard University, Cambridge, Mass.

17 Joseph-McCarthy, D., J. M. Hogle, M. Karplus **1997**, Use of the multiple copy simultaneous search (MCSS) method to design a new class of picorna-

virus capsid binding drugs. *Proteins Struct. Funct. Genet.* 29, 32–58.
18 Caflisch, A. **1996**, Computational combinatorial ligand design: application to human α-thrombin. *J. Comput. Aided Mol. Des.* 10, 372–396.
19 Roux, B., T. Simonson **1999**, Implicit solvent models. *Biophys. Chem.* 78, 1–20.
20 Roux, B., H. Yu, M. Karplus **1990**, Molecular basis for the Born model of ion solvation. *J. Phys. Chem.* 94, 4683–4688.
21 Davis, M. E., J. A. McCammon **1989**, Solving the finite difference linearized Poisson–Boltzmann equation: a comparison of relaxation and conjugate gradient methods. *J. Comput. Chem.* 10, 386–391.
22 Chothia, C. **1974**, Hydrophobic bonding and accessible surface area in proteins. *Nature* 248, 338–339.
23 Nicholls A., KA, Sharp, B. Honig **1991**, Protein folding and association: insights from the interfacial and thermodynamic properties of hydrocarbons. *Proteins Struct. Funct. Genet.* 11, 281–296.
24 Stultz, C.M., M. Karplus **2000**, Dynamic ligand design and combinatorial optimization: designing inhibitors to endothiapepsin. *Proteins Struct. Funct. Genet.* 40, 258–289.
25 Veerpandian, B., J. B. Cooper, A. Sali, T. L. Blundell **1990**, X-ray analyses of aspartic proteinases: the three dimensional structure of endothiapepsin complexed with a transition-state isotere inhibitor of rennin at 1.6Å resolution. *J. Mol. Biol.* 216, 1017–1029.
26 Goodford, P. J. **1985**, A computational procedure for determining energetically favorable binding sites on biologically important macromolecules. *J. Med. Chem.* 28, 849–857.
27 Bitetti-Putzer, R., D. Joseph-McCarthy, J. M. Hogle, M. Karplus **2001**, Functional group placement in protein binding sites: a comparison of GRID and MCSS. *J. Comput. Aided Mol. Des.* 15, 935–960.
28 Caflisch, A. C., A. Miraniker, M. Karplus **1993**, Multiple copy simultaneous search and construction of ligands in binding sites: application to inhibitors of HIV-1 aspartic proteinase. *J. Med. Chem.* 36, 2142–2167.

29 Joseph-McCarthy, D., S. K. Tsang, D. J. Filman, J. M. Hogle, M. Karplus, Use of MCSS to design small targeted libraries: application to picornavirus ligands. *J. Am. Chem. Soc.* 123, 12758–12769.
30 Joseph-McCarthy, D., J. C. Alvarez **2003**, Automated generation of MCSS-derived pharmacophoric DOCK site points for searching multiconformation databases. *Proteins Struct. , Funct. Genet.* 51, 189–202.
31 Allen, K. N., C. R. Bellamacina, X. Ding, C. J. Jeffery, C. Mattos, G. A. Petsko, D. Ringe **1996**, An experimental approach to mapping the binding surfaces of crystalline proteins. *J. Phys. Chem.* 100, 2605–2611.
32 English, A. C., C. R. Groom, R. E. Hubbard **2001**, Experimental and computational mapping of the binding surface of a crystalline protein. *Protein Eng.* 14, 47–59.
33 Shuker, S. B., P. J. Hajduk, R. P. Meadows, S. W. Fesik **1996**, Discovering high-affinity ligands for proteins: SAR by NMR, *Science* 274, 1531–1534.
34 Sirockin, F., C. Sich, S. Improta, M. Schaefer, V. Saudek, N. Froloff, M. Karplus, A. Dejaegere **2002**, Structure activity relationship by NMR and by computer: a comparative study. *J. Am. Chem. Soc.* 124, 11073–11084.
35 Zeng, J., H. R. Treutlein **1999**, A method for computational combinatorial peptide design of inhibitors of Ras protein. *Protein Eng.* 12, 457–468.
36 Zeng, J., T. Nheu, A. Zorzet, B. Catimel, E. Nice, H. Maruta, A. W. Burgess, H. R. Treutlein **2001**, Design of inhibitors of Ras–Raf interaction using a computational combinatorial algorithm. *Protein Eng.* 14, 39–45.
37 Barbacid, M. **1987**, *ras* genes, *Annu. Rev. Biochem.* 56, 779–827.
38 Joseph-McCarthy, D., S. K. Tsang, D. J. Filman, James M. Hogle, M. Karplus **2001**, Use of MCSS to design small targeted libraries: application to picornavirus ligands. *J. Am. Chem. Soc.* 123, 12758–12769.
39 Haitao J., W. Zhang, M. Zhang, M. Kudo, Y. Aoyama, Y. Yoshida, C. Sheng, Y. Song, S. Yang, Y. Zhou, J. Lu, J. Zhu **2003**,

Structure-based de novo design, synthesis, and biological evaluation of non-azole inhibitors specific for lanosterol 14r-demethylase of fungi. *J. Med. Chem.* 46, 474–485.

40 Teague, S.J. **2003**, Implications of protein flexibility for drug discovery. *Nat. Rev. Drug Discov.* 2, 527–541.

41 Stultz, C. M., M. Karplus **1999**, MCSS functionality maps for a flexible protein. *Proteins Struct. Funct. Genet.* 37, 512–529.

42 Leclerc, F., M. Karplus **1999**, MCSS-based predictions of RNA binding sites. *Theor. Chem. Acc.* 101, 131–137.

43 Schechner, M., F. Sirockin, R. H. Stote, A. P. Dejaegere **2004**, Functionality maps of the ATP binding site of DNA gyrase B: generation of a consensus model of ligand binding. *J. Med. Chem.* 47, 4373–4390.

44 Eisen, M. B., D. C. Wiley, M. Karplus, R.E. Hubbard **1994**, HOOK: a program for finding novel molecular architectures that satisfy the chemical and steric requirements of a macromolecule binding site. *Proteins Struct. Funct. Genet.* 19, 199–221.

45 Miranker, A., M. Karplus **1995**, Dynamic ligand design. *Proteins Struct. Funct. Genet.* 23, 472–490.

46 Stultz, C. M., M. Karplus **2000**, Dynamic ligand design and combinatorial optimization: designing inhibitors to endothiapepsin. *Proteins Struct. Funct. Genet.* 40, 258–289.

Part 3: Experimental Techniques and Applications

8
NMR-guided Fragment Assembly
Daniel S. Sem

8.1
Historical Developments Leading to NMR-based Fragment Assembly

NMR-guided fragment assembly was first implemented in the Fesik laboratory, using a method termed "SAR by NMR" [1]. This and related approaches are now widely used in many different variations, but all with the goal of NMR-guided fragment assembly to generate tight-binding inhibitors [2]. Before elaborating on these techniques and their application, which is the topic of this chapter, a brief background will be given on the scientific foundations upon which SAR by NMR and related methods were built.

The notion that two weak binding ligands (now commonly referred to as "fragments") would achieve higher affinity for a protein target upon chemical linkage dates back to seminal work by Page and Jencks describing the chelate effect in chemistry and enzymology [3, 4]. Their work on model compounds demonstrated an advantage on the order of 45 entropy units for chemically equivalent uni- versus bi-molecular reactions, effectively explaining the entropic advantage of binding two fragments versus one to a protein target. Later experimental work in the 1970s demonstrated that linking two weak-binding substrates can produce affinity increases of around 10^5-fold for enzymes like elastase [5] and myosin ATPase [6]. More relevant to the field of drug discovery is work by Abeles and Nakamura in 1985 on HMG CoA reductase [7], the target of cholesterol-lowering statin drugs. But, up until the early 1990s, chemical linkage of ligands was based either on a knowledge of the biochemical reaction catalyzed by an enzyme (e.g. between PLP and an amino acid), or on a crystal structure with a proximal pair of tight-binding ligands bound in a ternary complex. It was not until the introduction of the SAR by NMR method in 1996 [1] that it became possible to routinely link very weak-binding fragments, which might bear no resemblance to known substrates. Since then, other methods have been developed to link fragments [8], including a disulfide-based functional assay [9], mass spectrometry ("SAR by MS") [10], and X-ray crystallography [11, 12, 13]. Each of these methods has its own advantages and disadvantages. Ideally, they are best used synergistically in combination, but this review focuses only on NMR-based fragment assembly. Furthermore, many reviews of NMR-based fragment as-

---

Fragment-based Approaches in Drug Discovery. Edited by W. Jahnke and D. A. Erlanson
Copyright © 2006 WILEY-VCH Verlag GmbH & Co. KGaA, Weinheim
ISBN: 3-527-31291-9

sembly are already available and the reader is referred to these articles for additional perspectives on this expansive topic [2, 8, 14, 15]. This chapter attempts to provide a broad and applied coverage of the field, useful for those implementing fragment-based inhibitor and drug design. It also presents some of the author's personal experiences in fragment-based drug design efforts while at Triad Therapeutics, including some new and unpublished results. The goal is therefore to provide a self-contained and practical guide to NMR-based fragment assembly, while referring the reader to other sources for more detailed information on experimental setup, or examples of successful applications of the methods.

8.2
Theoretical Foundation for the Linking Effect

Page and Jencks [3, 4] provided a cogent description of the thermodynamic advantage of linking two weak-binding fragments (A and B), which results in higher affinity for the linked fragments (A–B) binding to an enzyme or receptor. The affinity or rate increase can be as large as 10^8-fold, due largely to entropic effects. Page and Jencks [3] noted that a bimolecular reaction is disfavored relative to a unimolecular reaction by 45 entropy units, because of the additional three degrees of rotational and translational entropy that are lost in forming the transition state. This "chelate effect" is a significant contributor to the rate enhancement that enzymes provide by proper placement of reactive groups, but it also explains the high affinity achieved when using fragment assembly to do inhibitor design. In a later analysis [4], Jencks discussed the specific example of two linked ligands (A–B) binding to an enzyme, noting that the overall affinity of A–B for the receptor has contributions from:
1. ΔG_A^i, the intrinsic binding energy of fragment A;
2. ΔG_B^i, the intrinsic binding energy of fragment B;
3. ΔG_s, the connection energy that represents the change in probability of binding due to linkage.

The first two terms simply represent the affinities of the two fragments [$-RT \ln(K_A K_B)$], while the third term is related to the three degrees of rotational and translation entropy that are absent for A–B relative to A and B. This latter term provides additional affinity for the linked fragments, beyond the $K_A K_B$ product, according to:

$$\Delta G_{AB}^\circ = \Delta G_A^{i\,\circ} + \Delta G_B^{i\,\circ} + \Delta G_s$$

But, Jencks notes that this full binding energy may not be realized for a number of reasons, including the fact that the individual fragments in A–B may not be oriented as optimally as in the isolated A and B fragments. Murray and Verdonk [16] (chapter 3, see also chapter 9) extended the analysis of Jencks in light of recent results from fragment assembly projects and concluded that loss of rigid body entropy corresponds to about three orders of magnitude in affinity and that bi-ligands (A–B) may often retain a significant amount of rigid body entropy while bound to an enzyme. In practice, it is unusual to achieve an affinity increase for linking two fragments that goes beyond the $K_A K_B$ product; and this may be due to linkers that do not optimally present fragments A and B in their binding sites, or retain a significant amount of rigid body entropy while bound. It should be noted, especially for fragments that are linked over a large distance, that there is an entropic penalty for binding A–(–)$_n$–B, due to the tethering of a long flexible linker (large n) upon binding to a receptor. This is analogous to the entropic penalty experienced when a disulfide bond is formed at the ends of a long linker, as described by Florey [17, 18] with the following equation:

$$\Delta S_{\text{linker}} = 0.75 R \ln(n + 3)$$

where R is the gas constant and n is the number of statistical segments between the tethering points, which are cysteine sulfurs in the case of disulfide bond formation, or ligands A and B in the case of a bi-ligand A–(–)$_n$–B binding to an enzyme. So, the linker cannot be so short as to strain and disrupt the proper presentation of A and B in their respective binding sites, but it cannot be so long so as to increase the entropic penalty of binding the bi-ligand. Indeed, the original NMR-based fragment assembly results of Shuker et al. [1] demonstrated the need for avoiding linkers that are either too long or too short (Fig. 8.1).

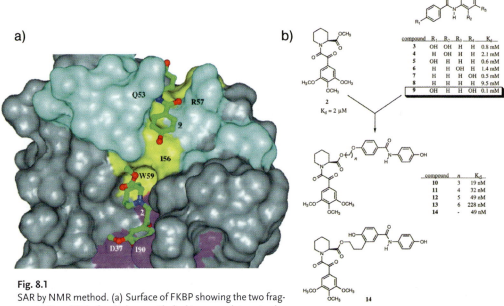

Fig. 8.1
SAR by NMR method. (a) Surface of FKBP showing the two fragments (2 and 9) before linkage, with docking pose determined based on chemical shift perturbations and protein-ligand NOEs. (b) K_i values for binding to FKBP for fragments before and after linkage. The bottom table summarizes the effect of linker length on affinity. Reproduced with permission from [1].

8.3
NMR-based Identification of Fragments that Bind Proteins

8.3.1
Fragment Library Design Considerations

All NMR-based fragment assembly strategies require a library of strategically selected or designed fragments, which are ultimately screened for binding to protein targets. Some practical considerations in designing such libraries are summarized here, with the reader referred to other sources for more detailed information [19, 20]. Fragment libraries are subject to many of the same design constraints that are applied to any combinatorial chemistry screening library. For example, combinatorial libraries must be highly pure, with well characterized compounds, and should contain no protein-reactive compounds. In the early days of combinatorial chemistry, this was not recognized and many hours were wasted characterizing false positives from screening campaigns. It is now recognized that careful attention to library quality upfront saves significant time in the long run. In this regard, Yan et al. [21] at ChemRx note that "the urge for compound number is gradually converted to the desire for compound quality"; and they recommend high-throughput purification and characterization of every compound in a library. They recognize the value of NMR-based characterization for such efforts, since many undesirable impurities are invisible to current QC screening methods, such as UV_{214}, UV_{254}, and evaporative light-scattering. But, they conclude that NMR characterization of every compound is not practical for large libraries, due to the time-consuming nature of spectral interpretation. Kenseth and Coldiron [22] also argue that compound purity needs to be more carefully addressed, but they advocate NMR-based characterization since NMR is unique in its ability to elucidate compound structure and current throughput with flow-probes permits a 96-well plate to be analyzed in just 4 h.

Libraries used in fragment assembly projects are unique relative to traditional combinatorial chemistry libraries in that they are small enough (<10 000) that every member can be fully characterized with NMR. In fact, libraries used in NMR fragment assembly projects are of unusually high purity and are especially well characterized relative to traditional combinatorial chemistry libraries [20]. This is because such initial characterization is required for the later stages of NMR fragment assembly projects, since compounds are identified from strategically designed pools (generally of no more than ten compounds) based on their ^1H chemical shifts. It is also important to pool compounds in such a way that there is minimal spectral overlap in a 1D ^1H NMR spectrum [20], thereby avoiding downstream problems deconvoluting hits in NMR-based screening.

In addition to the above QC/purity issues, libraries are usually screened to avoid structures that are reactive with proteins, such as Michael acceptors, anhydrides, epoxides, alkyl halides, acyl halides, imines, aldehydes, aliphatic ketones or esters, and other structures known to react with proteins [23]. Such prescreening avoids false positives from fragments that act by covalent modification of the protein target, which is usually undesirable because of the lack of specificity it implies (except for suicide substrates). Furthermore, it is also important to avoid combinations of compounds that might react with each other [24], such as electrophile/nucleophile combinations. An even more comprehensive filtering of over 200 functional groups can be achieved using the online REOS (Rapid Elimination Of Swill) tool [25], but Lepre cautions against the blind application of such a strong filter since it can reject potentially useful compounds (it rejected 73% of the compounds in the CMC database). An additional filtering strategy suggested by Hann et al. [24] is to avoid compounds containing any atom other than C, O, H, N, S, P, F, Cl, Br, I, or anything with <10 atoms, or anything that does not have at least one of the following bonds: C–N, C–O, or C–S, since most drug fragments would satisfy these criteria.

Since the goal of fragment assembly projects is to design an inhibitor that is to become a drug lead, an especially important consideration is that the fragments be "drug-like", thereby increasing the likelihood that they will have good ADMET properties (absorption, distribution, metabolism, excretion, and toxicology). Especially predictive for good oral bioavailability is the "Lipinski rule of five" [26], which when adjusted for fragments (assuming there are two fragments in a drug lead) translates into the rule of three [27], whereby all fragments should have:

1. molecular weight <300
2. ClogP ≤3
3. hydrogen bond donors ≤3
4. hydrogen bond acceptors ≤3
5. number of rotatable bonds ≤3
6. polar surface area ≤60.

It should be noted that the small size of these fragments does not preclude the possibility of finding compounds with reasonable affinity for a protein, since Kuntz has shown that nanomolar affinity can be achieved with compounds having as few as 10–20 atoms [28]. In order to assist researchers involved in fragment assembly projects, we have screened databases of commercially available compounds to identify those that satisfy the "rule of three"; and these compounds can be viewed at the Chemical Proteomics Facility at Marquette website (www.marquette.edu\cpfm) using a convenient web interface (Fig. 8.2), and sd files containing these structures can be downloaded for further filtering (see below).

Additional considerations in building a drug-like fragment library are elaborated below, but it is routine practice to make sure that commonly occurring substructures [29–32] (chapters 5 and 6) in drug databases also occur with high frequency in a fragment library. Furthermore, since fragments are ultimately going to be subjected to chemical linkage and medicinal chemistry optimization, they must be amenable to routine chemical modifications. This usually means that they must also be fairly hydrophilic and small, since most chemical modifications increase both size and hydrophobicity. Such fragments are referred to as "lead-like" [33]. A strategy to select both drug-like building blocks and to ensure chemical tractability is RECAP (retrosynthetic combinatorial analysis procedure), whereby building blocks are selected by computationally fragmenting compounds

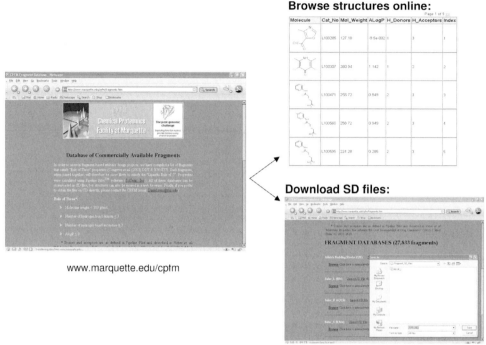

Fig. 8.2
Online fragment library available for browsing and download from the Chemical Proteomics Facility at Marquette (CPFM). Fragments have been filtered initially just by the "rule of three" [27] and for reactive groups [23]. Filtering was done using Pipeline Pilot software (SciTegic, San Diego, Calif.), version 3.0.6.

in the Derwent World Drug Index at 11 predefined bonds, which suggest the possibility of joining fragments using known chemistries [34]. Such linkages included amide C–N, ester C–O, amine C–N (3° and 4°) and ether C–O bonds.

8.3.2
The "SHAPES"" NMR Fragment Library

The SHAPES strategy was an early NMR-based fragment assembly approach to drug design. A unique strength of the method was the design strategy behind the fragment library which it employed, which has served as a model for current implementations of fragment-based NMR. The initial SHAPES library was designed based on an analysis of 5120 relevant drugs in the CMC database (Comprehensive medicinal chemistry database, version 94.1; MDL, San Leandro, Calif.) [35]. What Bemis and Murcko [36] found was that all of these compounds could be described by 1179 generic molecular shapes, which they refer to as frameworks. But a mere 32 of these frameworks could describe 50% of all drugs, with the top 10 shown in

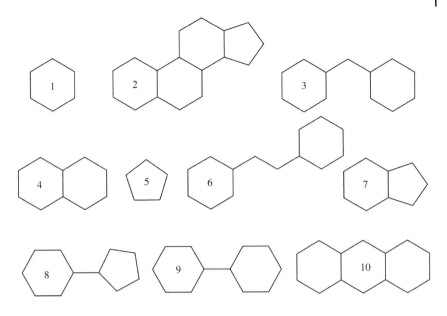

Fig. 8.3
The ten most commonly occurring generic "shapes" that describe the drugs present in the Comprehensive Medicinal Chemistry (CMC) database [36]. The relative frequency of occurrence of drugs with each graph framework is indicated.

Fig. 8.3. A framework is a generic description of a molecule's shape that does not specify atom types; and it is defined as the "union of ring systems and linkers", where linkers join the ring systems in a molecule. What is also intriguing, from the perspective of fragment assembly, is that the most commonly occurring ring system (six-membered ring) is joined by linkers varying in length from zero to seven atoms (Fig. 8.4 a, b), with an obvious preference for shorter linkers. But, the longer linkers still occur with reasonable frequency, suggesting that, although tethering with short linkers is preferred for entropic reasons, it may be acceptable to go as high as a seven-atom linker.

In a practical sense, this information on frameworks can be used to design a fragment library for NMR screening, by choosing actual rings/substructures that match the most frequently occurring frameworks and are themselves highly represented in existing drugs. After specifying atom types and bond orders, to convert these generic shapes to "complex" frameworks, it was found that 41 of these structures were contained in 25% of drugs, thereby providing a guide to preparing a small yet fairly universal screening library [15]. The actual SHAPES screening library was derived from the above analyses, then searching a database of over 10^6 commercially available compounds that pass the various filters. Further details of

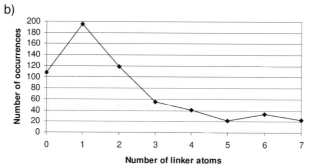

Fig. 8.4
(A) The graph framework that describes the commonly occurring diphenyl group, with linker lengths of zero to seven methylene carbons. (B) Number of occurrences of different linker lengths, indicating a preference for 0–3, but with reasonably high frequency of occurrence for linker lengths as long as seven atoms.

the library design process are described in many reviews and articles, to which the reader is referred [20, 23, 24, 29, 30, 34, 36–38]. The SHAPES NMR screening library is therefore composed of compounds that have these commonly occurring drug substructures, as defined in Fig. 8.5 [37]. In terms of deciding which substituents (called "side-chains") can be present on the fragments, the library is biased to be substituted in a way that mimics how drugs in the CMC are commonly substituted, based on an analysis of the most commonly occurring side-chains on drugs [39], summarized in Fig. 8.6. Compounds are also chosen based on having high aqueous solubility and minimum spectral overlap in 1D ^{1}H NMR spectra.

8.3.3
The "SAR by NMR" Fragment Library

The fragment library used in the SAR by NMR method was designed with some of the same considerations mentioned above. But, library design is the central and unique feature of the SHAPES approach to fragment assembly, whereas the screening and linkage strategy is the unique feature of the SAR by NMR method.

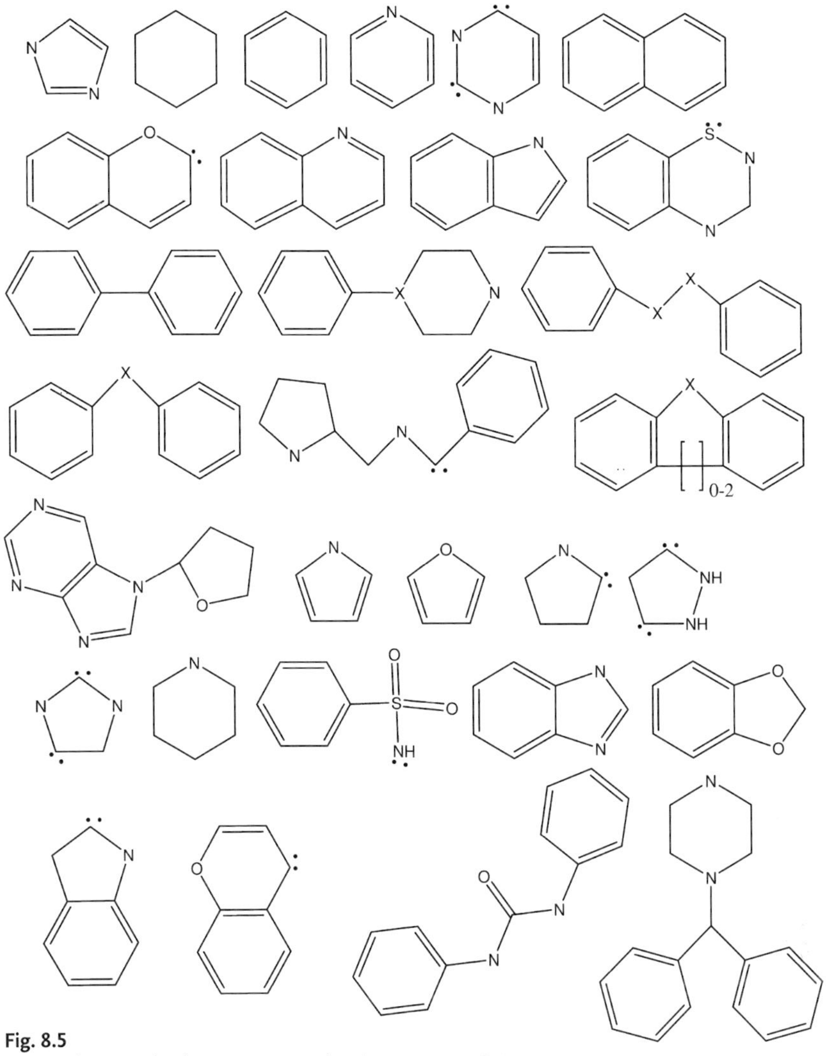

Fig. 8.5
Atomic frameworks that are contained within compounds in the SHAPES fragment library [37]. Attachment points are indicated with dots, and X indicates N, O, S or C.

As such, although the SAR by NMR method predates SHAPES, there are fewer details available on the design strategy behind the SAR by NMR fragment library; and available details appeared in the literature after the SHAPES publications [36, 37, 39]. In both cases (SAR by NMR and SHAPES), the full description of the library is not publicly available, although general design philosophy and composition have been published.

The SAR by NMR library has varied in size from 1000–10 000 compounds, with an average molecular weight of 200 g mol^{-1}. This library is comprised of relatively

C=O	C-CH$_3$	C-OH	C-O-CH$_3$	C-Cl	N-CH$_3$	C-NH$_2$	C-CO$_2^-$
3545 :	3125 :	2566 :	916 :	823 :	719 :	549 :	454

C-F	S=O	C-CH$_2$-OH	C-CH$_2$-CH$_3$	C-O-C(O)-CH$_3$	C-NO$_2$	C-N-(CH$_3$)$_2$
355 :	331 :	235 :	214 :	55 :	137 :	108

N-CH$_2$-CH$_3$	C-CF$_3$	C-C(O)-NH$_2$	C-C(O)-O-CH$_3$	C-C(O)-O-CH$_2$-CH$_3$
93 :	85 :	84 :	81 :	76

Fig. 8.6
The most commonly occurring side-chains [39] on drugs in the CMC. Side-chains are attached to frameworks, such as those in Fig. 8.3. Side-chains are specified in bold, and are sometimes categorized differently based on what atom they are attached to. Relative frequencies of occurrence in the CMC database are specified.

simple molecules with a small number of rotatable bonds and functional groups. The full (99%) library of 10 080 compounds contains the 104 fragments presented in Fig. 8.7 [40], with the frequency of occurrence specified for each substructure. This library was also compared against the 154 000 compounds in the Derwent World Drug Index and the Maccs Drug Data Report (MDL), as a measure of how "drug-like" the library is, again with frequency of occurrence specified for each substructure. It was also compared against the 177 000 compounds in the Available Chemical Directory (MDL), as a measure of availability of compounds for follow-up structure–activity studies. Early library screening handled mixtures of ten compounds at a total concentration of 10 mM (each compound at 1 mM), although high-throughput screening was proposed with 100-compound mixtures [41]. This library was screened against numerous targets at Abbott as part of SAR by NMR drug discovery efforts, but also as a way to classify and characterize protein targets, as discussed in the next section. Of special note is that the Abbott group [40] identified the most favored protein-binding substructure, besides the more trivial R–CO$_2^-$ group (a side-chain in the language of Bemis and Murcko), as the diphenyl group with a zero- or one-atom linker (Fig. 8.4).

8.3.4
Fragment-based Classification of Protein Targets

It has been shown previously by methods other than NMR that a useful way to characterize proteins is by the collections of ligands that they bind [42–47]; and this profiling of proteins based on binding sites has utility both in inhibitor/drug design and in functional genomics. With regard to the latter [48, 49], one simply

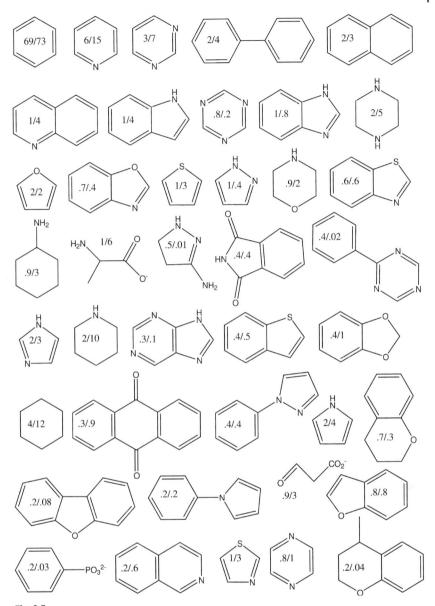

Fig. 8.7
Commonly occurring fragments or substructures contained within the Abbott SAR by NMR library. Within each structure is specified the fraction of compounds in the 10 080-member SAR by NMR library with this substructure, followed by the fraction of compounds in the 154 000 member Derwent World Drug Index and Maccs Drug Data Report (MDL) with this substructure. Fragments were computationally generated with RECAP [34].

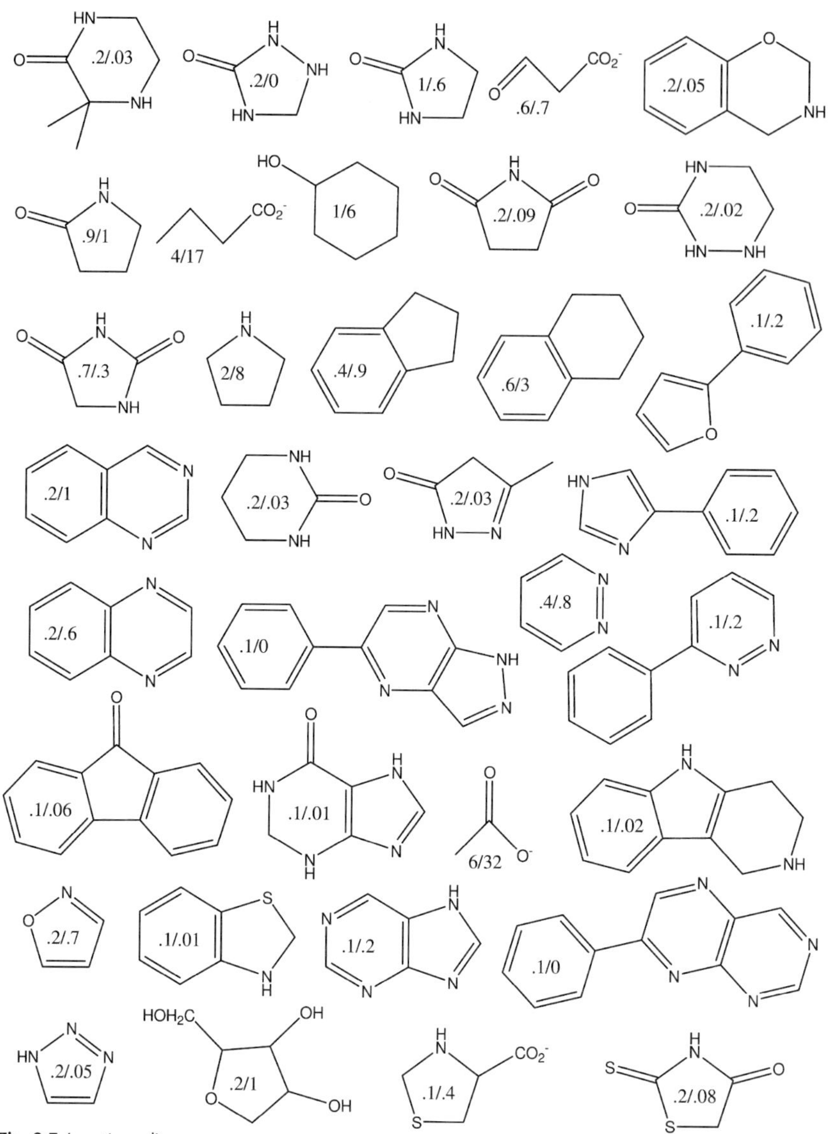

Fig. 8.7 (continued)

screens for binding of cofactors and cofactor fragments using methods described in Section 8.4.1 (usually STD or WATERLOGSY). Once cofactor or cofactor fragments are identified, this information is used to categorize the protein into an appropriate gene family, such as kinase, oxidoreductase, GTPase, etc. Identification of cofactor usage provides clues as to what type of reaction an enzyme may catalyze, or what type of regulatory ligand a protein may bind. In one application, it was shown that a protein previously thought to be a cGMP- or cAMP-dependant

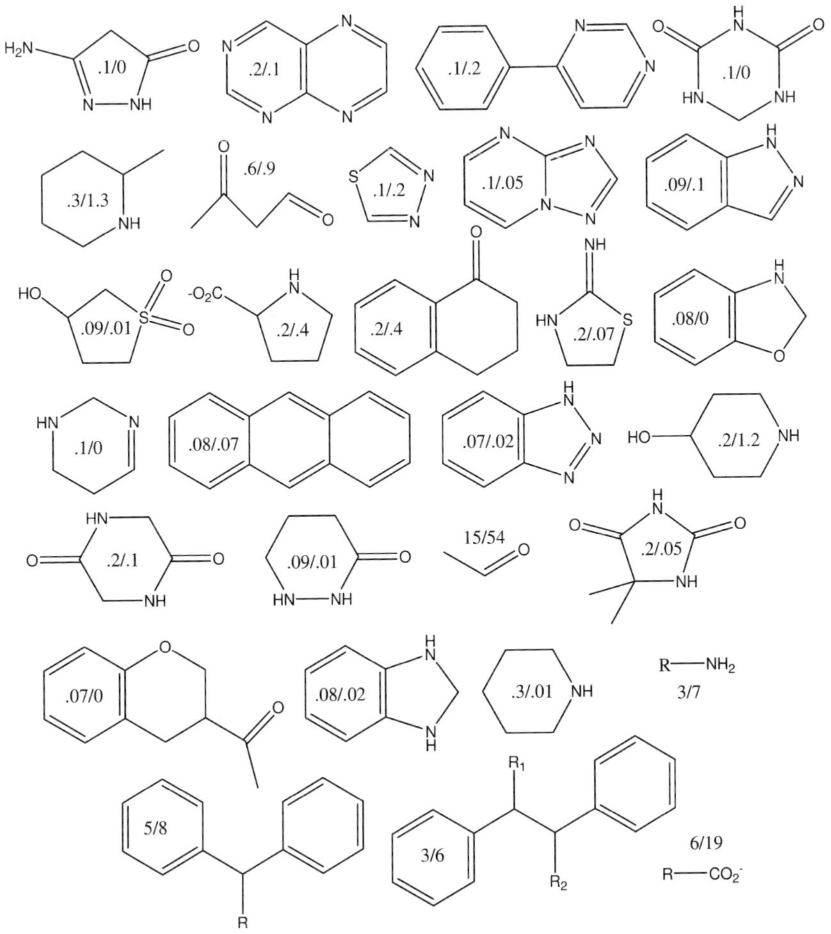

Fig. 8.7 (continued)

kinase, based on a bioinformatic analysis, actually bound cCMP with higher affinity [49]. Since cofactors are essentially privileged scaffolds, ligands for these binding sites can then be used in gene family-focused fragment assembly projects (see Sections 8.6.1, 8.6.2).

With regard to NMR-based ligand-profiling using a more generic fragment library, Hajduk et al. [40] have shown that, of 11 protein targets screened with their SAR by NMR fragment library, six bound substructures with the $-CO_2^-$ group, five with the diphenyl group (linker = 0 atoms) and three with the diphenyl group (linker = 1 atom), making these by far the most promiscuous of the substructures. In this sense, they are considered "privileged substructures", with utility as templates for designing inhibitors for multiple proteins (see Section 8.6). It is noteworthy that, although the diphenyl group is present in many known protein inhibitors, it is also possible to achieve high levels of specificity (>250-fold) for specific protein

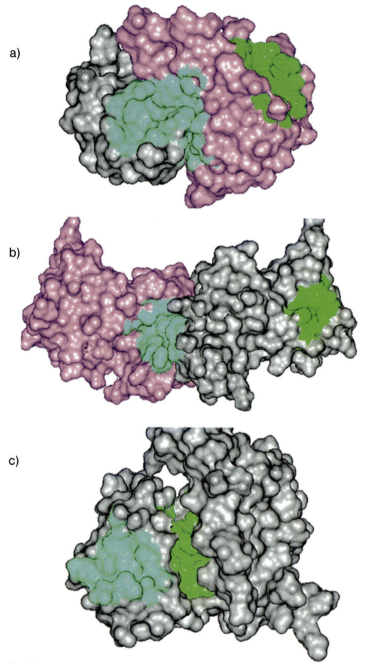

Fig. 8.8
Ancillary binding sites identified in the fragment-based screening of 23 targets at Abbott, based on chemical shift perturbations. Of the four newly discovered binding sites, only three were somewhat distal from the known binding site. These were sites present on: (A) the DNA binding domain of E2, (B) survivin and (C) CMPK. Reproduced with permission from [47].

targets, based on ring substituents. Indeed, they note that the diphenyl group is present in 4.3% of all drugs. Other substructures were found to bind to no more than one protein out of the 11 screened, but were still important diversity elements to be contained in the library. In later work [47], they profiled 23 protein drug targets with the SAR by NMR library and categorized proteins based on the inhibitors they bound as well as the binding site location. They found a correlation between a protein's ability to bind weak-binding fragments and the ultimate ability to design high affinity (< 200 nM) inhibitors. After analyzing certain structural features of these binding sites, they found these to be correlated with the ability to bind fragments as well. These parameters included pocket dimensions, surface complexity, and polar or apolar surface area. When these parameters were used to computationally screen protein structures outside of the training set (of 23 structures), they were able to identify proteins that are "druggable", defined as proteins which have been the subject of successful drug discovery efforts. In another observation of special relevance for fragment assembly projects, they found four previously undetected binding sites with no obvious role in binding native substrates. These ancillary sites could be pursued in drug discovery efforts as a source of additional affinity or specificity for a given target. In one of these proteins (FKBP), the extra binding site is close enough to the primary site where it could be used in a fragment assembly project [1]. In the other three, the ancillary binding site is more remote and might only be reached with very long linkers (Fig. 8.8).

8.4
NMR-based Screening for Fragment Binding

8.4.1
Ligand-based Methods

The first step of any fragment assembly project is to screen for weak-binding fragments using NMR. NMR-based screening methods can be broadly divided into those that detect protein and those that detect ligand. Both types of screening strategies have already been reviewed [14, 48, 50, 51] and are described only briefly here, with emphasis on implementing the most widely used techniques. Ligand-based methods have the advantage of not requiring isotopically labeled protein. These methods rely on fast exchange between the bound and free states, whereby some NMR-observable property of the ligand in the bound state is averaged with that for the free state. If ligand L binds to protein P, then the observed NMR signal (S_{obs}) is the weighted average for signal in the bound and free states, given by [48, 51]:

$$S_{obs} = S_{bound}([EL]/[L_{total}]) + S_{free}(1 - [EL]/[L_{total}]) \qquad (1)$$

where S_{obs} is the observed NMR signal. Commonly used NMR signals in a screening experiment include $1/T1$, $1/T2$ or diffusion rates (protein-based methods usually use chemical shift); and the reader is referred to other literature describing applications using each of these [52–54]. Perhaps the simplest to implement is T2, by monitoring changes in linewidths [1/(πT2)] in 1D ^{1}H NMR spectra of ligand with or without protein present. There are also ligand-based methods that rely on magnetization transfer from protein, and these include NOE pumping and reverse NOE pumping [55, 56], as well as STD (saturation transfer difference) [57–59]. A related method is WATERLOGSY [60], whereby magnetization is transferred from protein-bound water molecules to ligand. Perhaps the most widely used method at present is STD, whereby ligand is maintained in excess (10- to 20-fold) over protein and spectra are taken with irradiation off resonance (–20 ppm) and on resonance for a protein signal that does not overlap with ligand, such as 1.0 ppm (side-chain methyls). The two spectra are then subtracted from each other and a signal is only present for those ligands which come into contact with the protein. Irradiation is typically for several seconds, using a train of Gaussian-shaped pulses. This method is extremely sensitive and fast and permits screening pools of compounds (typically around ten), since it is possible to identify the ligand that binds based on its 1D ^{1}H NMR spectrum. Another widely used screening method is to look for the presence of transferred NOEs for bound ligand. Transferred NOEs were used for some of the SHAPES fragment screening at Vertex [37].

The main disadvantage of most ligand-based screening methods is that they do not detect tight-binding ligands. Slow-exchanging ligands do not give a weighted average NMR signal according to Eq. (1), and since ligand is in excess over protein, all one sees is the signal from free ligand. To address this deficiency, tight-binding ligands are detected based on their ability to displace a weaker-binding "spy" ligand, thereby decreasing its NMR signal [61–63]. An especially useful implementation of screening by competition has been with fluorinated weak–binding ligands, since ^{19}F sensitivity is excellent and there is no need to worry about overlapping signal from protein, solvent or other ligands [64–66].

Since it is common to find errors or unnecessarily complicated descriptions in the literature, with regard to the equations needed to measure K_d values, some time is spent in this chapter on this topic, in order to aid the reader in the practical application of fragment-based inhibitor design. Derivations and equations are provided for both ligand-based (here) and protein-based (Section 8.4.2) titrations. A direct STD-based titration can yield a K_d for a weak-binding ligand [48, 51]. The STD effect itself is given by:

$$STD = (I_{off} - I_{on})/I_{off} \tag{2}$$

which represents the percent change in signal intensity for a signal in the on-resonance relative to the off-resonance irradiated spectrum. Accordingly, a K_d can be obtained from the equation:

$$STD = - STD_{max}/(1 + [L]/K_d) + STD_{max} \tag{3}$$

where STD_{max} is the maximum STD effect (or the effect multiplied by an amplification factor [48]) and [L] is the ligand concentration. It will be true that [L] = [L]$_o$ if

[L]$_o$ > [P]$_o$, which is usually the case in STD experiments. But, in a competition experiment, one measures an IC$_{50}$, which represents the concentration of inhibitor (I) which decreases the STD effect for the weak-binding ligand (L) by 50%. If the K_d is already known for the weak-binding ligand ($K_{d,L}$), then the IC$_{50}$ can then be converted to a K_d for the tight-binding ligand ($K_{d,I}$) according to the following equation [64]:

$$K_{d,I} = IC_{50}/(1 + [L]/K_{d,L}) \tag{4}$$

8.4.2
Protein-based Methods

Unlike ligand-based screening methods, protein-based methods can provide valuable structural information. The main disadvantage of these methods is that they consume more protein and require isotopically labeled protein, which therefore precludes their use for some protein drug targets. But for proteins that *can* be isotopically labeled, it is possible to implement protein-based methods in a high-throughput screening campaign (discussed below), if adequate quantities of protein are available. Protein-based screening was central to the original SAR by NMR fragment assembly strategy, where there was a need to screen for ligand binding based on changes in the chemical shifts of protein atoms. If the protein has full chemical shift assignments and a structure has been calculated, then these chemical shift changes provide structural information that allows the proper placement of fragments in protein binding sites (Fig. 8.1 a). This is done by monitoring changes in ^{1}H, ^{13}C, and/or ^{15}N chemical shifts in a uniformly labeled protein, as protein is titrated with ligand. For proteins of higher molecular weight, protein-based screening is made possible by selectively labeling amino acids in methyl positions [67], or by selectively detecting loop (exchangeable) amides using SEA-TROSY [68]. In cases like these where protein is quite large, full chemical shift assignments are not available, but it is still possible to extract structural information to guide fragment assembly [67, 69–71]. An interesting variation of the protein-based screening method is RAMPED-UP (rapid analysis and multiplexing of experimentally discriminated uniquely labeled proteins) NMR [72]. In this method, three different proteins are screened in the same NMR tube by labeling each with a different amino acid type. In this way, it is possible to screen for fragments that are selective for one protein target over another, in the same NMR tube.

In order to determine K_d values in protein-based screening, chemical shift changes are quantitatively monitored in a titration experiment, with increasing ligand concentration [48, 51]. Chemical shift-based calculations of K_d rely on the fact that the changes in chemical shift for some protein atom, upon binding ligand L, reflect the fractional saturation of protein P according to the following equation:

$$(\Delta\delta_{obs})/(\Delta\delta_{max}) = [PL]/[L]_o \tag{5}$$

where $\Delta\delta_{obs} = \delta_{obs} - \delta_P$, the difference between the measured protein chemical shift (δ_{obs}) at some concentration of ligand L and that for protein in the complete absence of ligand (δ_P). $\Delta\delta_{max}$ is the change in chemical shift after protein is fully saturated with ligand, $[L]_o$ is the total concentration of ligand, and $[PL]$ is the concentration of the complex. This equation can only be used if binding is in fast exchange on the chemical-shift timescale $[k_{ex} > 1/(B_o \, \Delta\delta_{max})]$, so that δ_{obs} is the weighted average of bound and free chemical shifts, according to Eq. (1). Other equations needed to calculate a K_d value are [51]:

$$K_d = [P][L]/[PL] \qquad (6)$$

$$[L]_o = [L] + [PL] \qquad (7)$$

$$[P]_o = [P] + [PL] \qquad (8)$$

Obtaining a K_d value is simple if one of the components can be kept in large excess over the other. For example, if $[L]_o \gg [P]_o$, then the equations simplify, since $[L] = [L]_o$ at all concentrations of protein. In this case, one commonly used plot to obtain a K_d value is given by the Scatchard equation:

$$\Delta\delta_{obs}/[L]_o = -\Delta\delta_{obs}/K_d + \Delta\delta_{max}/K_d \qquad (9)$$

where a plot of $\Delta\delta_{obs}/[L]_o$ versus $\Delta\delta_{obs}$ gives a slope of $-1/K_d$. A second commonly used analysis is with the Benesi–Hildebrand equation:

$$1/\Delta\delta_{obs} = (K_d/\Delta\delta_{max})(1/[L]_o) + 1/\Delta\delta_{max} \qquad (10)$$

where a plot of $1/\Delta\delta_{obs}$ versus $1/[L]_o$ gives a slope of $K_d/\Delta\delta_{max}$ and a y-intercept of $1/\Delta\delta_{max}$. Although these graphical methods are convenient, it is best to get K_d from a non-linear least squares fit to the original equation for a rectangular hyperbola:

$$\Delta\delta_{obs} = \Delta\delta_{max}[L]_o/(K_d + [L]_o) \qquad (11)$$

The above equations bear an obvious resemblance to those commonly used for steady-state enzyme kinetic analyses.

Now if it is not possible to keep one of the components in excess over the other, then there is no choice but to solve the full quadratic equation and then perform a non-linear least squares fit to:

$$\Delta\delta_{obs} = \frac{\Delta\delta_{max}\left[[(K_d + [L]_o + [P]_o) - \sqrt{(K_d + [L]_o + [P]_o)^2 - 4[L]_o[P]_o}\,\right]}{2[P]_o} \qquad (12)$$

to get K_d. In this equation, there is the luxury of being able to use total concentrations of protein and ligand, without worrying about whether one is always in excess over the other. The reader is referred to two recent review articles for additional details on fitting NMR data to obtain K_d values [48, 51].

8.4.3
High-throughput Screening: Traditional and TINS

NMR is now routinely used in reasonably high-throughput screening efforts. It is most practical for such screening to be done with ligand-based methods such as STD or WATERLOGSY, since they require smaller amounts of protein. It is common to screen with 10–20 µM protein, using a 10- to 20-fold excess of ligand in STD screening. Experiments are on the order of 5 min per sample, can be easily performed on pooled samples (typically ten), and can be fully automated. Capillary NMR flow probes and/or cryoprobe technology can improve sensitivity and throughput as well. Still, it was recently estimated that to screen 10 000 compounds in pools of ten would take 3 months and consume 50 mg of protein. Protein consumption for these methods is therefore still a significant concern for targets that are hard to express. For this reason, TINS (target immobilized NMR screening) was recently developed [73], whereby a target protein is immobilized on agarose or sepharose media and ligands are screened for T2 effects (line broadening). Immobilized protein was reused for screening 2000 compounds and found to retain full binding activity, suggesting that this approach could be used for screening fragment libraries when protein target is only available in limited quantities.

Protein-based methods have also been used in high-throughput screening campaigns. Hajduk et al. [41] reported the acquisition of 2D ^{1}H-^{15}N HSQC spectra in 10 min on 50 µM protein samples using a cryoprobe on a 500-MHz instrument. They proposed that pools of 100 compounds at a time could be screened and that 200 000 compounds could be screened in 1 month. In later work, they describe the value of integrating NMR-based screening methods with other HTS technologies [74]. NMR is especially helpful in identifying false positives from other assays, thereby saving significant wasted effort pursuing undesirable compounds. Ultimately, the Abbott group screened 23 protein targets with most of its ~10 000-member SAR by NMR library, thereby validating the feasibility of doing moderately high-throughput screening with protein-based methods [47].

8.5
NMR-guided Fragment Assembly

8.5.1
SAR by NMR

SAR by NMR was reported in 1996 by Shuker, Hajduk, Meadows, and Fesik [1] as a fragment assembly approach to inhibitor design, using NMR as a structural guide. It is essentially a five-step method [75] that involves screening for weak-binding fragments in two binding sites. In step 1, NMR is used to screen for a weak binding ligand in a first site. In step 2, the ligand is optimized for higher affinity, using NMR structural assays to ensure that binding mode is retained. In step 3, the protein is saturated with the first ligand and is then used to screen for a "second-site" ligand. In step 4, the second ligand is optimized in the same manner as in step 2 for the first ligand. In step 5, the two ligands are joined based on structural information obtained from NMR. The screening steps are typically done using labeled protein and monitoring chemical shift changes. The initial implementation of SAR by NMR was with ^{15}N-labeled protein, although later work was done using ^{13}C-labeled protein [76]. For larger proteins, it is helpful to use selective labeling of side-chain methyl groups, or to improve resolution in ^{1}H-^{15}N spectra with TROSY [77]. The most important step in the SAR by NMR process is step 5, the linking step, which is enabled by determining a complete structure of the bound ternary complex between protein, fragment 1, and fragment 2. In the initial SAR by NMR application, a 19 nM inhibitor of FKBP was identified based on linkage of 2 µM and 100 µM inhibitors (Fig. 8.1b). The advantage of this linked-fragment strategy over previous SAR approaches is that fewer compounds need to by synthesized, since there is a combinatorial advantage that is equal to the size of the fragment library squared. So applying SAR by NMR to a 10 000-member library is similar to screening 10^8 compounds. But, in some sense, it even provides greater diversity since the method samples all possible ways of orienting two fragments, whereas a single chemical linkage would sample only one position and linker length. Over the years, SAR by NMR has been applied to many targets, including 23 proteins at Abbott. In the screening of these proteins, hit rates ranged from 0–0.9% for various samplings of the ~10 000-member fragment library. For ten of these proteins, a potent (<300 nM) inhibitor was eventually identified. In recently reported implementations of SAR by NMR ([78], summarized in [75]), previously identified potent inhibitors were fragmented and these fragments were used along with new fragments, to design new core structures. This was usually done to improve bioavailability or general pharmacokinetic behavior of drug leads. New core structures were identified with this strategy for adenosine kinase (IC_{50} = 10 nM), urokinase (IC_{50} = 2.8 µM), stromelysin (62 nM), and LAF-1 (leukocyte function associated antigen 1; IC_{50} = 20 nM).

The SAR by NMR approach to assembling fragments is conceptually similar to some of the commonly used *in silico* strategies for *de novo* drug design, such as the MCSS method [79]. Therefore, Sirockin et al. [80] recently compared MCSS (using the CHARMM forcefield [81]) with experimental SAR by NMR studies, to see whether *in silico* methods could accurately predict binding modes for fragments. The computational strategy employed the MCSS library of fragments, which was supplemented with relevant SAR by NMR fragments. The *in silico* method was able to rank binding modes consistent with NOE data in the top 5%, suggesting that it may be a useful complement to experimental SAR by NMR, serving possibly as an initial screen. Noteworthy is that this high level of predictability was possible only because of the inclusion of desolvation effects in the scoring function for binding energy [80]. Finally, efforts at docking fragments for SAR by

NMR projects should be aided by a knowledge-based scoring function developed for this purpose [82].

8.5.2
SHAPES

The SHAPES strategy involves screening the SHAPES library (Section 8.3.2) to identify which fragments bind to protein targets, using ligand-based detection methods such as transferred NOE, WATERLOGSY, and STD, which are discussed in greater depth in Section 8.4.1. The early [37] and more recent [15, 38] applications of the SHAPES strategy in ligand design and drug discovery are summarized in several articles and reviews, to which the reader is referred for more detail. An especially exciting successful application involved identification of inhibitors for Jnk3 (c-Jun N-terminal kinase-3), a MAP kinase that is a target for stroke and Parkinson's disease. In this case, three classes of inhibitors were identified (thiazole, uracil, and isoxazole), with inhibition as low as 800 nM for the isoxazole class [15, 38]. The latter compound was further optimized with medicinal chemistry to yield a < 10 nM inhibitor. The SHAPES strategy was also used for fragment assembly of inhibitors for adipocyte lipid-binding protein, a possible target for type II diabetes. In this case, the most potent inhibitors had K_i values as low as 300 nM [15, 38]. Finally, SHAPES is even being used to identify tight-binding ligands for structured RNA [38], which is increasingly being pursued as a drug target. In this case, it was found that WATERLOGSY was more sensitive than STD for screening the SHAPES library, although STD provided useful information on binding mode based on epitope mapping (discussed further in Sections 8.4.1 and 8.5.4).

8.5.3
Second-site Binding Using Paramagnetic Probes

In steps 3 and 4 of SAR by NMR, it was necessary to saturate the first binding site and then screen for "second-site" ligands. While the traditional SAR by NMR strategy does this using chemical shift perturbations, one can also use paramagnetic probes [83, 84]. In this strategy, a first ligand is labeled with an organic nitroxide radical probe like TEMPO (2,2,6,6-tetramethyl-1-piperidine-N-oxyl). The location of the second ligand, relative to the first, is established based on the distance-dependant relaxation effect of the probe on the second ligand. The significant advantage offered by this method is that protein does not need to be labeled; and sensitivity is quite high. The latter is true because the relaxation effect is dependant upon the gyromagnetic ratio of the electron, which is 658 times larger than that for protons. The relaxation rate enhancement is given by:

$$R2_{para} = 1/15\,[S(S+1)]\,[\gamma_H^2\,g^2\,\beta^2/r^6]\,[4\tau_c + 3\tau_c/(1+\omega_H^2\,\tau_c^2)] \tag{13}$$

where S is the electron spin on the probe, γ_H is the proton gyromagnetic ratio, g is the electronic g factor, τ_c is the correlation time of the vector connecting the electron and nuclear spin, β is the Bohr magneton, ω_H is the proton Larmor frequency and r is the distance between the probe and the proton on the second-site fragment. Depending on this distance, an estimated relaxation enhancement of around 50-fold is possible (relative to diamagnetic enhancement). Importantly, this has been shown to roughly correspond to the level to which protein concentration can be decreased with this approach to second-site screening. The main disadvantage of this approach is the need to label the first ligand, but this problem is partially alleviated if one labels protein instead, on amino acids such as lysine, tyrosine, histidine, methionine, or cysteine [83]. This method, termed SLAPSTIC (spin labels attached to protein side chains as tools to identify interacting compounds), was applied to second site screening in FKBP and the anti-apoptotic protein Bcl-xL [84]. In addition to not requiring isotopic labeling, the method avoids false positive second-site hits, since a relaxation effect can only occur if both the probe and the second-site ligand are bound simultaneously. It is important to note that, like all ligand-based screening methods, it only identifies ligands if they are binding in fast exchange because the actual $R2$ measured is a weighted average for the free and bound ligand. This limitation can be removed if screening is done in competition mode with a weak-binding ligand present [83, 84].

8.5.4
NMR-based Docking

In order to properly link two weak-binding fragments, one can determine a full structure of the complex with NMR, or simply find a way to orient the second ligand relative to the first, so that a linker can be designed. The latter approach is very pragmatic, and is enabled by methods like spin labeling (above), interligand NOE measurements [85], and some of the combinatorial methods described below (NMR SOLVE and NMR ACE). But, beyond linking fragments, structural data can be helpful for optimization of leads, such as engineering changes that improve the bioavailability, metabolism, or toxicological properties of drugs. In this sense, it is helpful to know which regions of a fragment are solvent-exposed and therefore available for modification. Such NMR-based docking can be implemented for very large proteins, by looking for magnetization transfer (NOE or STD) to selectively labeled side-chains, such as methyl groups, especially when the rest of the protein has been deuterated. This strategy has been implemented, without the need to assign protein residues, in dihydrodipicolinate reductase [67] and more recently in FKBP and MurA [86]. In addition to NOE measurements, one can also use calculated chemical shifts to aid in docking calculations [87]. Use of chemical shift changes to guide docking was applied to a series of compounds [88] and has been suggested as a means to score the poses that one obtains from *in silico* docking efforts [89]. For proteins that are larger or cannot be isotopically labeled, it is possible to determine which portions of the protein are solvent-exposed and which are buried, based on epitope mapping studies using STD [58, 90]. In these studies, protons on the ligand that show the largest STD effects are thought to be buried, while those with little or no STD effect are most likely solvent-exposed. Since relaxation effects can lead to significant changes in STD values [91], thereby suggesting inappropriate docking modes, it is best to do a full CORCEMA analysis [92, 93] to generate a docked structure, if possible.

8.6
Combinatorial NMR-based Fragment Assembly

8.6.1
NMR SOLVE

Since privileged scaffolds bind to multiple members of a gene family, they are excellent templates for combinatorial libraries tailored to that gene family, as long as diversity elements are directed into adjacent binding pockets that are likely to provide specificity for one target over another. This is especially true in cofactor-dependant enzymes, since the cofactor site is always adjacent to a substrate site (Fig. 8.9a, b). Because cofactor and substrate always react with each other, the relative relationship of the two sites is highly conserved within a gene family. Again, the central problem that must be addressed after identifying a privileged scaffold is where to build the library. That is, one must determine what position should be substituted with variable fragments (Fig. 8.9a), so that the library directs the attached fragments into a specificity pocket (Fig. 8.9b). NMR SOLVE (structurally-oriented library valency engineering) was developed to provide this information [67, 69–71, 94]. It relies on the binding of a privileged scaffold to a protein and measuring NOEs to that scaffold from atoms that are known to be located at the interface between the two binding sites. In this way, the best combinatorial library expansion point is identified on the privileged scaffold, since this position is adjacent to the specificity site. In order to accomplish this, one can assign interface atoms by mapping the binding site with some reference ligand, like the cofactor that normally binds the conserved site. This approach requires labeled protein, so although it provides rich structural information, it is limited to proteins that can be isotopically labeled. If the interface atom is on the other substrate, then it is not necessary to label the protein. A combination of NOEs from protein interface atoms (^{13}C-labeled threonines) and substrate interface atoms (2,6-pyridine-dicarboxylate) were used to design a dehydrogenase-focused library using the enzyme dihydrodipicolinate reductase (DHPR). In this case, the privileged scaffold bound weakly to three different dehydrogenases (25–100 µM). NMR SOLVE data obtained with DHPR (dehydrogenase-1 in Fig. 8.9b) suggested where to build the focused combinatorial library and ~300 low-molecular-weight fragments were attached at the identified site to create a focused combinatorial library for dehydrogenases (Fig. 8.9b). This library yielded inhibitors for each of the three dehydrogenases screened, with K_i values in the 40–200 nM range (Table 8.1). It should be noted that in the case of the dichlorophenyl fragment on the LDH inhibitor (Table 8.1), affinity for the isolated fragment was so low that it could not be detected. This suggests that this inhibitor could only have been identified in this

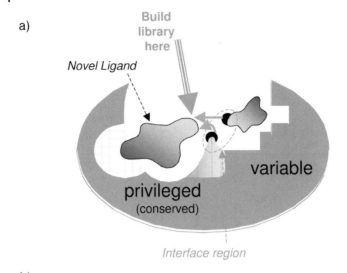

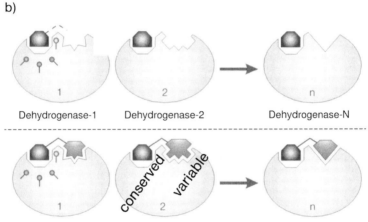

Fig. 8.9
Description of the NMR SOLVE method. (A) Cartoon representation, whereby a library expansion point is chosen based on NOEs from interface atoms on either the protein or an adjacent substrate. (B) Description of how NMR SOLVE studies on one representative of a protein family can be used to focus a combinatorial library that can produce inhibitors for multiple members of that gene family. In this case, the proteins are dehydrogenases which all share a common site for the NAD(P)H cofactor, but the concept applies to any family of proteins which bind a privileged scaffold. Modified from [14].

combinatorial fashion and not by traditional NMR-based fragment assembly approaches. Since the potential increase in potency from fragment linkage includes an entropic contribution, in addition to the product of K_d values for the two ligands, affinity gains can be quite large (Section 8.2). Therefore, it should not be surprising that individual fragments from bi-ligands may sometimes bind too weakly for detection.

Table 8.1 Affinity and specificity achieved with combinatorial fragment assembly guided by NMR SOLVE. Numbers are K_{is} values obtained in a steady-state enzyme kinetic assay, except for DOXPR where an IC_{50} was measured. As such, the actual K_{is} value for DOXPR will be lower. LDH = lactate dehydrogenase, DHPR = dihydrodipicolinate reductase, DOXPR = 1-deox-D-xylulose-5-phosphate reductoisomerase.

Structure	LDH	DHPR	DOXPR
CLM-1	55 μM	26 μM	>50 μM
	42 nM	>50 μM	10 μM
	12 μM	>25 μM	202 nM
	620 nM	100 nM	7.9 μM

8.6.2
NMR ACE

Combinatorial fragment assembly for some proteins must be pursued without the luxury of isotopic labeling. This is because many proteins, especially protein kinases, cannot be expressed in *E. coli* at levels high enough for labeling to be practical. For this reason, the NMR ACE (assembly of chemical entities) strategy was developed [95]. In this approach, weak-binding ligands are initially screened for binding to the target, which in the initial ACE project at Triad Therapeutics was p38α MAP kinase. Competitive binding studies with a privileged scaffold/ligand known to bind multiple members of the gene family are then used to categorize (Fig. 8.10a) weak-binding fragments as either privileged scaffolds or as potential specificity ligands to be attached to the privileged scaffold. Ideally, the "specificity ligands" would be directed to a subsite known to be unique to the desired target, if structural information is available. This approach enables the construction of a small focused combinatorial library directed to all proteins that bind the privileged scaffold, by combining the fragments in Fig. 8.10a. NMR ACE was used to design a potent inhibitor of p38α (Fig. 8.10b). An initial IC_{50} of around 326 µM was found for the privileged scaffold (competitive with SB203580). Fragment assembly using various linkers (four) led to an inhibitor with an IC_{50} of 1.7 µM, corresponding to a K_i of ~300 nM. In all, 23 fragments were identified or made that bound outside the ATP site, but had interactions (NOEs) with a fragment in the ATP site. Many of these also showed NOEs to an ATP-site reference compound (SB203580), such as the fragment shown in Fig. 8.10b which had its

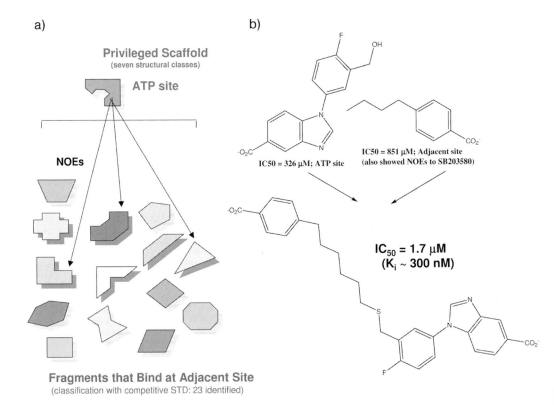

Fig. 8.10
NMR ACE strategy for designing bi-ligands and focused bi-ligand libraries, using unlabeled protein. (A) Cartoon description of how a privileged scaffold is joined to various specificity ligands, based on NOEs that are measured between them. In the initial p38α MAP kinase application at Triad Therapeutics (Huang, Jung, Kelly, Pellecchia, and Sem, unpublished data), seven structural classes of privileged scaffolds were identified, as were 23 different specificity ligands. (B) Sample application of NMR ACE to p38á MAP kinase, whereby a privileged scaffold that binds in the ATP site was joined to a fragment that binds outside the ATP site. Linkage yielded a >100-fold increase in affinity.

alkyl chain located within 3.0±0.5 Å of the pyridine ring of SB203580. Since the goal of this project was to specifically target p38α, it was desirable to have second-site fragments that bind in a region of the protein structure that is unique to p38α. To assess this, homology models were constructed for various MAP kinases and variability in sequence was assessed at each position and mapped onto the structure of p38α (Fig. 8.11a). It can be seen that there is a region adjacent to the ATP (SB203580) site that is highly variable; and this region is also outside of the region bound by peptide substrates (Fig. 8.11a, b). Therefore, any fragment that binds here would not be displaced by either SB203580 or peptide; and this competition strategy was used to characterize potential specificity ligands/fragments to be added to the privileged scaffold.

Since seemingly useful specificity ligands were often identified that showed no NOEs to the privileged scaffold, long alkyl chains were added to the scaffold, which were sometimes terminated methyl-ether groups in order to create an iso-

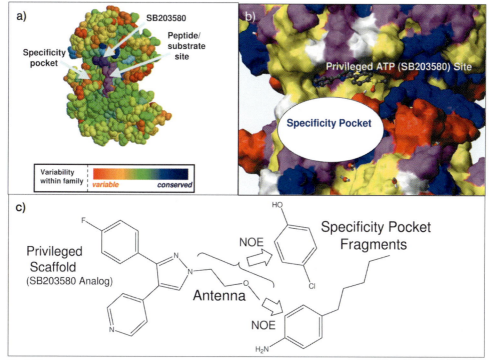

Fig. 8.11
Probing p38α binding pockets with NMR ACE antennas (Kelly, Dong, Lee, Jung, and Sem, unpublished data). (A) The crystal structure of p38α, with relative variability compared to other MAP kinases mapped onto the structure, in order to suggest possible pockets to probe for specificity. A potentially useful pocket is identified, which sits adjacent to both the SB203580 and peptide pockets. (B) The specificity pocket location just outside the SB203580 (ATP) site is shown in a close-up view. (C) Sample application of the use of an NMR ACE antenna to probe more distal binding sites, relative to the conserved ATP/SB203580 site.

lated methyl spin system. These were then used to probe for fragments in more distal binding sites, using NOE experiments. These "antennas" serve not only as structural probes of distal specificity sites, but also as potential linkers for fragment assembly. The antenna-variation of NMR ACE [95] has proved very useful for probing more distal sites for specificity fragments, such as that in Fig. 8.11c. A combination of the above methods, along with medicinal chemistry SAR studies, led to p38α inhibitors with IC_{50} values < 10 nM. Some of these showed very encouraging preclinical data in a rheumatoid arthritis project.

In a related project with Jnk2α2 kinase at Triad Therapeutics (Yan, Jung, and Lee, unpublished data), NMR ACE-based fragment assembly and subsequent SAR studies yielded a 20 nM inhibitor (TRD-97941), containing a novel core structure. Weak-binding fragments used to create this new core were first sub-classified, based on competitive STD studies using the ATP site inhibitor, SB203580. Fragments in different categories (ATP and non-ATP site) were then linked, based on the observation of interligand NOEs, to produce a 100 µM inhibitor. The use of transferred NOE data combined with modeling then allowed optimization of this lead to a 0.91 µM inhibitor. Next, NMR-based docking combined with medicinal chemistry SAR workup led to a further increase in potency to 20 nM. In all, with the aid of NMR at every step, a 20 nM inhibitor was produced after synthesis of only 30 compounds. This example clearly emphasizes the value of NMR in accelerating lead discovery and optimization. More importantly, it emphasizes how crucial it is to use the whole toolbox of NMR methods in the drug discovery process and to use the NMR information to focus and guide the efforts of traditional medicinal chemistry.

8.7
Summary and Future Prospects

Although fragment assembly strategies for inhibitor design predate SAR by NMR, it was that method which launched a whole series of innovative NMR-based approaches to fragment assembly. All of these methods rely on the initial screening of a strategically designed fragment library. Although protein-based screening methods provide the most structural information (hence the term structural screening) and can be implemented in a high-throughput fashion, it is most common to screen for first-site ligands with ligand-based methods like STD or WATERLOGSY. This is because protein is often not available in large quantities and also cannot be isotopically labeled. Fortunately, there are two ligand-based screening methods that can provide structural information: (a) screening with paramagnetic probes and (b) screening with interligand NOEs or STDs. Paramagnetic probes are attractive since they provide structural information along with high sensitivity. Interligand NOEs have also been used to generate combinatorial libraries from privileged scaffolds, tailored to gene families of proteins like dehydrogenases and kinases. This strategy enables a parallel approach to inhibitor and drug design that couples well with chemo-genomics and proteomics [94].

It is anticipated that some of the most exciting future innovations in NMR-based fragment assembly will take advantage of interligand NOE and STD effects to design inhibitors for extremely large proteins, or membrane-bound systems [96, 97] which previously could not be targeted by experimental structure-based methods. With regard to protein-based methods, it should now be possible to do SAR by NMR studies within living cells [98]. Other exciting recent developments include the ability to profile proteins for likelihood of success in a drug discovery process [47], which is a significant concern in the pharmaceutical industry, given the high cost of this process. Related to this point, it will become increasingly important to be able to use NMR-based methods to optimize drug leads or building blocks to have better bioavailability, along with optimized metabolic and toxicological properties [75, 78, 99]. Finally, since it has recently been shown that NMR fragment-based screening can be used to identify previously unrecognized binding sites on proteins [47], NMR-based screening and fragment assembly may open the door to inhibiting proteins at previously unexplored target sites or "hot spots" (Fig. 8.8). These ancillary sites seem to be present with relatively high frequency (four of 28 binding sites identified on 23 proteins screened [47]) and might also be explored as part of a chemical proteomic approach [48, 100] to functional genomics. Characterizing these ancillary sites may ultimately help to create a more accurate "systems view" of protein target space, since they probably have biochemical and physiological functions, such as interacting with other proteins directly or through effectors, as part of signaling or regulatory networks.

References

1 S. B. Shuker, P. J. Hajduk, R. P. Meadows, S. W. Fesik **1996**, *Science* 274, 1531–1534.
2 J. R. Huth, C. Sun, D. R. Sauer, P. J. Hajduk **2005**, *Methods Enzymol.* 394, 549–571.
3 M. I. Page, W. P. Jencks **1971**, *Proc. Natl. Acad. Sci. USA* 68, 1678–1683.
4 W. P. Jencks **1981**, *Proc. Natl. Acad. Sci. USA* 78, 4046–4050.
5 R. L. Thompson **1974**, *Biochemistry* 13, 5495–5501.
6 D. R. Trentham, J. F. Eccleston, C. R. Bagshaw **1976**, *Q. Rev. Biophys.* 9, 217–281.
7 C. E. Nakamura, R. H. Abeles **1985**, *Biochemistry* 24, 1364–1376.
8 D. A. Erlanson, R. S. McDowell, T. O'Brien **2004**, *J. Med. Chem.* 47, 3463–3482.
9 D. A. Erlanson, A. C. Braisted, D. R. Raphael, M. Randal, R. M. Stroud, E. M. Gordon, J. A. Wells **2000**, *Proc. Natl. Acad. Sci. USA* 97, 9367–9372.
10 E. E. Swayze, E. A. Jefferson, K. A. Sannes-Lowery, L. B. Blyn, L. M. Risen, S. Arakawa, S. A. Osgood, S. A. Hofstadler, R. H. Griffey **2002**, *J. Med. Chem.* 45, 3816–3819.
11 V. L. Nienaber, P. L. Richardson, V. Klighofer, J. J. Bouska, V. L. Giranda, J. Greer **2000**, *Nat. Biotechnol.* 18, 1105–1108.
12 D. Lesuise, G. Lange, P. Deprez, D. Benard, B. Shoot, G. Delettre, J. P. Marquette, P. Broto, V. Jean-Baptiste, P. Bichet, E. Sarubbi, E. Mandine **2002**, *J. Med. Chem.* 45, 2379–2387.
13 M. S. Congreve, D. J. Davis, L. Devine, C. Granata, M. O'Reilly, P. G. Wyatt, H. Jhoti **2003**, *Angew. Chem. Int. Ed.* 42, 4479–4482.
14 M. Pellecchia, D. S. Sem, K. Wuthrich **2002**, *Nat. Rev. Drug Disc.* 11, 211–219.
15 C. A. Lepre, J. Peng, J. Fejzo, N. Abdul-Manan, J. Pocas, M. Jacobs, X. Xie, J. M.

Moore **2002**, *Comb. Chem. High Throughput Screen.* 5, 583–590.
16 C. W. Murray, M. L. Verdonk **2002**, *J. Comp. Aided Mol. Des.* 16, 741–753.
17 P. J. Florey **1956**, *J. Am. Chem. Soc.* 78, 5222–5235.
18 G. E. Schulz, R. H. Shirmer **1990**, *Principles of Protein Structure*, Springer-Verlag, New York.
19 E. Jacoby, J. Davies, M. J. J. Blommers **2003**, *Curr. Top. Med. Chem.* 3, 11–23.
20 C. A. Lepre **2001**, *DDT* 6, 133–140.
21 B. Yan, L. Fang, M. Irving, S. Zhang, A. M. Boldi, F. Woolard, C. R. Johnson, T. Kshirsagar, G. M. Figliozzi, C. A. Krueger, N. Collins **2003**, *J. Comb. Chem.* 5, 547–559.
22 J. R. Kenseth, S. J. Coldiron **2004**, *Curr. Opin. Chem. Biol.* 8, 418–423.
23 G. M. Rishton **1997**, *DDT* 2, 382–384.
24 M. Hann, B. Hudson, X. Lewell, R. Lifely, L. Miller, N. Ramsden **1999**, *J. Chem. Inf. Comput. Sci.* 39, 897–902.
25 W. P. Walters, M. T. Stahl, M. A. Murcko **1998**, *DDT* 3, 160–178.
26 C. A. Lipinski, F. Lombardo, B. W. Dominy, P. J. Feeney **1997**, *Adv. Drug Deliv. Rev.* 23, 3–25.
27 M. Congreve, R. Carr, C. Murray, H. Jhoti **2003**, *DDT* 8, 876–877.
28 I. D. Kuntz, K. Chen, K. A. Sharp, P. A. Kollman **1999**, *Proc. Natl. Acad. Sci. USA* 96, 9997–10002.
29 C. Merlot, D. Domine, D. J. Church **2002**, *Curr. Opin. Drug Disc. Dev.* 5, 391–399.
30 C. Merlot, D. Domine, C. Cleva, D. J. Church **2003**, *DDT* n; 8, 594–602.
31 A. K. Ghose, V. N. Viswanadhan, J. J. Wendoloski **1999**, *J. Comb. Chem.* 1, 55–68.
32 M. Vieth, M. G. Siegel, R. E. Higgs, I. A. Watson, D. H. Robertson, K. A. Savin, G. L. Durst, P. A. Hipskind **2004**, *J. Med. Chem.* 47, 224–232.
33 S. J. Teague **1999**, *Angew. Chem. Int. Ed.* 38, 3743–3747.
34 X. Q. Lewell, D. B. Judd, S. P. Watson, M. M. Hann **1998**, *J. Chem. Inf. Comput. Sci.* 38, 511–522.
35 C. Hansch, P. G. Sammes, J. B. Taylor **1990**, *Comprehensive Medicinal Chemistry*, Vol. 6, Pergamon Press, Oxford.
36 G. W. Bemis, M. A. Murcko **1996**, *J. Med. Chem.* 39, 2887–2893.
37 J. Fejzo, C. A. Lepre, J. W. Peng, G. W. Bemis, Ajay, M. A. Murcko, J. M. Moore **1999**, *Chem. Biol.* 6, 755–769.
38 J. Moore, N. Abdul-Manan, J. Fejzo, M. Jacobs, C. Lepre, J. Peng, X. Xie **2004**, *J. Synchotron Rad.* 11, 97–100.
39 G. W. Bemis, M. A. Murcko **1999**, *J. Med. Chem.* 42, 5095–5099.
40 P. J. Hajduk, M. Bures, J. Praestgaard, S. W. Fesik **2000**, *J. Med. Chem.* 43, 3443–3447.
41 P. J. Hajduk, T. Gerfin, J.-M. Boehlen, M. Haberli, D. Marek, S. W. Fesik **1999**, *J. Med. Chem.* 42, 2315–2317.
42 L. M. Kauvar, H. O. Villar, J. R. Sportsman, D. L. Higgins, D. E. Schmidt **1998**, *J. Chromatogr. Biomed. Sci. Appl.* 715, 93–102.
43 L. M. Kauvar, D. L. Higgins, H. O. Villar, J. R. Sportsman, A. Engqvist-Goldstein, R. Bukar, K. E. Bauer, H. Dilley, D. M. Rocke **1995**, *Chem. Biol.* 2, 107–118.
44 E. Liepinsh, G. Otting, *Nat. Biotechnol.* **1997**, 15, 264–268.
45 K. N. Allen, C. R. Bellamacina, X. Ding, C. J. Jeffrey, C. Mattos, et al. **1996**, *J. Phys. Chem.* 100, 2605–2611.
46 A. C. English, C. R. Groom, R. E. Hubbard **2001**, *Prot. Eng.* 14, 47–59.
47 P. J. Hajduk, J. R. Huth, S. W. Fesik **2005**, *J. Med. Chem.* 48, 2518–2525.
48 B. J. Stockman, C. Dalvit **2002**, *Prog. NMR Spectr.* 41, 187–231.
49 H. Yao, D. S. Sem **2005**, *FEBS Lett.* 579, 661–666.
50 D. S. Sem, M. Pellecchia **2001**, *Curr. Opin. Drug Disc. Dev.* 4, 479–493.
51 L. Fielding **2003**, *Curr. Top. Med. Chem.* 3, 39–53.
52 P. J. Hajduk, E. T. Olejniczak, S. W. Fesik **1997**, *J. Am. Chem. Soc.* 119, 12257–12261.
53 M. Lin, M. J. Shapiro **1996**, *J. Org. Chem.* 61, 7617–7619.
54 M. Lin, M. J. Shapiro, J. R. Wareing **1997**, *J. Am. Chem. Soc.* 119, 5249–5250.
55 A. Chen, M. J. Shapiro **1998**, *J. Am. Chem. Soc.* 120, 10258–10259.
56 A. Chen, M. J. Shapiro **2000**, *J. Am. Chem. Soc.* 122, 414–415.

57 S. Forsen, R. A. Hoffman **1963**, *J. Chem. Phys.* 39, 2892.
58 M. Mayer, B. Meyer **1999**, *Angew. Chem. Int. Ed.* 38, 1784–1788.
59 J. Klein, R. Meinecke, M. Mayer, B. Meyer **1999**, *J. Am. Chem. Soc.* 121, 5336–5337.
60 C. Dalvit, P. Pevarello, M. Tato, M. Veronesi, A. Vulpetti, M. Sundstrom **2000**, *J. Biomol. NMR* 18, 65–68.
61 C. Dalvit, M. Fasolini, M. Flocco, S. Knapp, P. Pevarello, M. Veronesi **2002**, *J. Med. Chem.* 45, 2610–2614.
62 C. Dalvit, M. Flocco, S. Knapp, M. Mostardini, R. Perego, B. J. Stockman, M. Veronesi, M. Varasi **2002**, *J. Am. Chem. Soc.* 124, 7702–7709.
63 Y.-S. Wang, D. Liu, D. F. Wyss **2004**, *Magn. Res. Chem.* 42, 485–489.
64 C. Dalvit, E. Ardini, M. Flocco, G. P. Fogliatto, N. Mongelli, M. Veronesi **2003**, *J. Am. Chem. Soc.* 125, 14620–14625.
65 C. Dalvit, P. E. Fagerness, D. T. A. Hadden, R. W. Sarver, B. J. Stockman **2003**, *J. Am. Chem. Soc.* 125, 7696–7703.
66 J. W. Peng **2001**, *J. Magn. Res.* 153, 32–47.
67 M. Pellecchia, D. Meininger, Q. Dong, E. Chang, R. Jack, D. S. Sem **2002**, *J. Biomol. NMR* 22, 165–173.
68 M. Pellecchia, D. Meininger, A. L. Shen, R. Jack, C. B. Kasper, D. S. Sem **2001**, *J. Am. Chem. Soc.* 123, 4633–4634.
69 D. S. Sem, L. Yu, S. M. Coutts, R. Jack **2001**, *J. Cell. Biochem.* 37, S99–S105.
70 D. S. Sem **2001**, *NMR-SOLVE Method for Rapid Identification of Bi-ligand Drug Candidates*, US patent 6,333,149.
71 D. S. Sem, M. Pellecchia, A. Tempczyk-Russell **2001**, *NMR-SOLVE Method for Rapid Identification of Bi-ligand Drug Candidates*, US patents 6,620,589 and 6,797,460.
72 E. R. Zartler, J. Hanson, B. E. Jones, A. D. Kline, G. Martin, H. Mo, M. J. Shapiro, R. Wang, H. Wu, J., Yan **2003**, *J. Am. Chem. Soc.* 125, 10941–10946.
73 S. Vanwetswinkel, R. J. Heetebrij, J. Van Duynhoven, J. G. Hollander, D. V. Fillipov, P. J. Hajduk, G. Siegal **2005**, *Chem. Biol.* 12, 207–216.
74 P. J. Hajduk, D. J. Burns **2002**, *Comb. Chem. High Throughput Screen.* 5, 613–621.
75 J. R. Huth, C. Sun **2002**, *Comb. Chem. High Throughput Screen.* 5, 631–643.
76 P. J. Hajduk, D. J. Augeri, J. Mack, R. Mendoza, J. G. Yang, S. F. Betz, S. W. Fesik **2000**, *J. Am. Chem. Soc.* 122, 7898–7904.
77 K. Pervushin, R. Riek, G. Wider, K. Wuthrich **1997**, *Proc. Natl. Acad. Sci. USA* 94, 12366–12371.
78 P. J. Hajduk, A. Gomtsyan, S. Didomenico, M. Cowart, E. K. Bayburt, L. Solomon, J. Severin, R. Smith, K. Walter, T. F. Holzman, A. Stewart, S. McGaraughty, M. F. Jarvis, E. A. Kowaluk, S. W. Fesik **2000**, *J. Med. Chem.* 43, 4781–4786.
79 I. Halperin, B. Ma, H. Wolfson, R. Nussinov **2002**, *Proteins* 47, 409–443.
80 F. Sirockin, C. Sich, S. Improta, M. Schaefer, V. Saudek, N. Froloff, M. Karplus, A. Dejaegere **2002**, *J. Am. Chem. Soc.* 124, 11073–11084.
81 B. R. Brooks, R. E. Bruccoleri, B. D. Olafson, D. J. States, S. Swaminathan, M. Karplus **1983**, *J. Comput. Chem.* 4, 187–217.
82 I. Muegge, Y. C. Martin, P. J. Hajduk, S. W. Fesik **1999**, *J. Med. Chem.* 42, 2498–2503.
83 W. Jahnke **2002**, *ChemBioChem* 3, 167–173.
84 W. Jahnke, A. Florsheimer, M. J. J. Blommers, C. G. Paris, J. Heim, C. M. Nalin, L. B. Perez **2003**, *Curr. Top. Med. Chem.* 3, 69–80.
85 D. Li, L. A. Levy, S. A. Gabel, M. S. Lebetkin, E. F. DeRose, M. J. Wall, E. E. Howell, R. E. London **2001**, *Biochemistry* 40, 4242–4252.
86 P. J. Hajduk, J. C. Mack, E. T. Olejniczak, C. Park, P. J. Dandliker, B. A. Beutel **2004**, *J. Am. Chem. Soc.* 126, 2390–2398.
87 M. A. McCoy, D. F. Wyss **2000**, *J. Biomol. NMR* 8, 189–198.
88 A. Medek, P. J. Hajduk, J. Mack, S. W. Fesik **2000**, *J. Am. Chem. Soc.* 122, 1241–1242.

89 B. Wang, K. Raha, K. M. Merz **2004**, *J. Am. Chem. Soc.* 126, 11430–11431.

90 M. A. Johnson, B. M. Pinto **2004**, *Bioorg. Med. Chem.* 12, 295–300.

91 J. Yan, A. D. Kline, H. Mo, M. J. Shapiro, E. R. Zartlet **2003**, *J. Magn. Reson.* 163, 270–276.

92 V. Jayalakshmi, N. R. Krishna **2004**, *J. Magn. Reson.* 168, 36–45.

93 V. Jayalakshmi, N. R. Krishna **2002**, *J. Magn. Reson.* 155, 106–118.

94 D. S. Sem, B. Bertolaet, B. Baker, E. Chang, A. Costache, S. Coutts, Q. Dong, M. Hansen, V. Hong, X. Huang, R. M. Jack, R. Kho, H. Lang, D. Meininger, M. Pellecchia, F. Pierre, H. Villar, L. Yu **2004**, *Chem. Biol.* 11, 185–194.

95 D. S. Sem, M. Pellecchia, Q. Dong, M. Kelly, M. S. Lee **2004**, *Nuclear Magnetic Resonance Assembly of Chemical Entities*, US patent 2003/0112751 A1.

96 R. Meinecke, B. Meyer **2001**, *J. Med. Chem.* 44, 3059–3065.

97 B. Claasen, M. Axmann, R. Meinecke, B. Meyer **2005**, *J. Am. Chem. Soc.* 127, 916–919.

98 Z. Serber, V. Dotsch **2001**, *Biochemistry* 48, 14317–14323

99 H. Yao, A. D. Costache, D. S. Sem **2004**, *J. Chem. Inf. Comp. Sci.* 44, 1456–1465.

100 D. S. Sem **2004**, *Exp. Rev. Prot.* 1, 165–178.

9
SAR by NMR: An Analysis of Potency Gains Realized Through Fragment-linking and Fragment-elaboration Strategies for Lead Generation

Philip J. Hajduk, Jeffrey R. Huth, and Chaohong Sun

9.1
Introduction

Fragment-based screening has become a powerful complement to traditional high-throughput screening (HTS) technologies to identify lead compounds that have high potential for further optimization. Unlike conventional HTS, most fragment-based screening applications focus on the identification of low-molecular-weight, low-affinity compounds from which high-affinity drug candidates can be constructed. Such methodologies include NMR-based screening (both target-based and ligand-based) [1, 2] (chapter 8), high-throughput X-ray crystallography [3–5] (chapters 10 and 11), mass spectroscopy [6] (chapter 13), surface plasmon resonance [7], fragment tethering [8] (chapter 14), and dynamic combinatorial chemistry [9] (chapter 16). All of these fragment-based methods exploit the reduced complexity of utilizing low molecular weight (typically 100–300 Da) compounds, leading to an increased probability [10] of identifying highly efficient [11], chemically tractable ligands in well designed compound libraries of only 10^3–10^4 compounds [5, 12, 13]. However, while attractive starting points for design, fragment leads tend to bind with extremely low affinity to the target (e.g., K_D values of 100 μM to >1000 μM). Thus, while the compounds may still bind "efficiently" with respect to their size [11], there are typically at least two or three log unit gains in potency that need to be achieved before these compounds can become truly "useful" leads in drug discovery. Several strategies have been proposed for rapidly increasing the potency of fragment leads into ranges where lead optimization can begin (typically with IC_{50} values less than 1 μM). As shown in Fig. 9.1, two of the more common strategies are fragment-linking, in which two fragments that bind to proximal sites are tethered, and fragment-elaboration, in which libraries of compounds are prepared around a fragment lead in order to increase potency. This article explores the potency gains that have been reported or internally discovered for fragment-based ligand design using these approaches. In addition, we attempt to define some general principles that can aid the scientist in the optimal use of fragment leads in drug discovery.

Fragment-based Approaches in Drug Discovery. Edited by W. Jahnke and D. A. Erlanson
Copyright © 2006 WILEY-VCH Verlag GmbH & Co. KGaA, Weinheim
ISBN: 3-527-31291-9

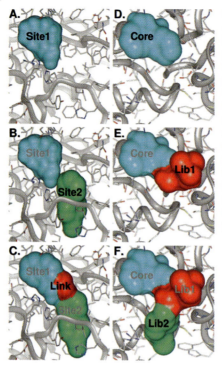

Fig. 9.1
Illustration of the linked-fragment (A–C) and iterative fragment elaboration (D–F) approaches. In the linked fragment approach, first-site (A, cyan surface) and second-site (B, green surface) ligands are identified that bind to proximal sites and linked (C, red surface) to yield high-affinity ligands. In iterative fragment elaboration, a first-site ligand (D, cyan surface), also called the "core," is elaborated using structure-based design or high-throughput parallel synthesis to identify substituents that improve affinity (E, red surface). These improved compounds are then utilized in a subsequent round of synthesis to identify additional substituents (F, green surface) that yield further gains in potency.

9.2
SAR by NMR

Since the description of SAR by NMR in 1996 [14], numerous linked-fragment approaches and applications have been described, as recently reviewed by Erlanson and co-workers [15]. As shown in Fig. 9.1, the premise of the SAR by NMR approach is the linking of two fragments that can simultaneously bind to proximal sites on a protein to yield high-affinity ligands. In our laboratory, we have made extensive use of two-dimensional heteronuclear correlation spectra in order to detect and characterize these weakly binding fragments. The power of using two-dimensional NMR is illustrated in Fig. 9.2, where distinct chemical shift perturbation patterns (or "fingerprints") can be observed with compounds that bind to two distinct sites on the protein. Thus, using this approach, first- and second-site ligands can be identified and characterized, based upon their unique patterns of induced shifts. Eight examples of the SAR by NMR approach are schematically given in Fig. 9.3. In each case, receptor-based NMR screening was used to identify and characterize the binding of the first- and second-site ligands; and linked molecules were designed on the basis of available structural information on the ternary complex.

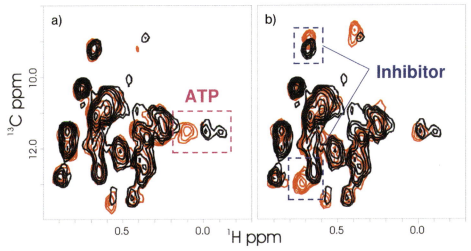

Fig. 9.2
Characterizing site-specific binding using $^1H/^{13}C$ correlation spectra on the bacterial protein murF [27]. Shown are expanded regions of the spectra containing only the isoleucine-δ_1 methyl resonances in the absence (black) and presence (red) of a test compound. Distinct chemical shift fingerprints can be observed for (A) ATP and (B) a novel inhibitor that does not bind to the ATP binding site [27]. Representative cross-peaks that exhibit differential effects with the two compounds are indicated with dashed boxes.

9.3
Energetic Analysis of Fragment Linking Strategies

One of the initial postulates of the SAR by NMR approach is that high-affinity (K_D <1 µM) leads could be rapidly derived from weakly binding (K_D ~1 mM) fragments, due to the fact that the binding energies of each fragment would be additive in a properly linked compound [16]. Thus, linking a second-site ligand with a K_D value of 1 mM to a first-site lead could, in principle, lead to a potency gain of three orders of magnitude, neglecting any gains in potency due to entropy or the linker itself. Table 9.1 details 17 applications of the linked-fragment approach to 11 different target proteins, including the reported potencies for the fragment leads and the linked compounds. Given these data, we can begin to test this additivity postulate with some statistical confidence and examine the energetics of linking two fragment leads. For simplicity, all of the potencies are represented as pK_D values (defined here as the negative base-10 log of the reported potency) and no distinction is made between K_D, K_I, or IC_{50} values. In this scenario, the theoretical gain is simply the pK_D value for the second site ligand, while the actual gain is the pK_D value for the linked compound minus the pK_D value for the first site ligand. As illustrated in Fig. 9.4, four of the linked compounds exhibited potency gains within 1.0 log unit of the expected gain (the boundaries being designated by gray dashed lines in Fig. 9.4), while four linked compounds actually exceeded the theoretical gain by more than an order of magnitude. The remaining nine exam-

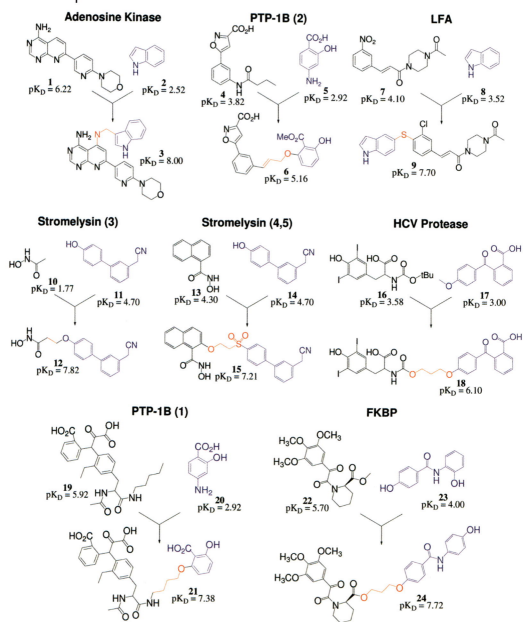

Fig. 9.3
Schematic views of eight examples of the SAR by NMR approach. In each case, the first-site ligand is colored black, the second-site ligand is colored blue, and the linking element is colored red. Values for pK_D were derived as described in the text. Examples are given for adenosine kinase [28], PTP-1B [29, 30], LFA [25], stromelysin [17,19], HCV protease [31], and FKBP [14].

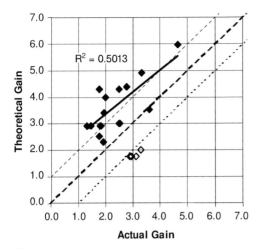

Fig. 9.4
Plot of the actual and theoretical potency gains realized through the linked fragment approach on 17 systems (see Table 9.1). The actual gain in potency (in base-10 log units) is defined as the pK_D of the linked compound minus the pK_D of the first site ligand, while the theoretical gain is defined as the pK_D of the second site ligand. The dashed black line corresponds to the ideal, where the actual gain is equal to the theoretical gain, while the dotted gray lines represent 1.0 log unit deviations in the actual and theoretical values. The observed correlation (solid line) excludes those examples for which the actual gain exceeded the theoretical gain (open black diamonds), most likely due to adventitious interactions with the linker (see text).

ples exhibited potency gains that were at least an order of magnitude less than the theoretical gain. On average, the actual potency gain was approximately 2.6 log units (corresponding to a 400-fold increase in potency), while the theoretical gain was approximately 3.3 log units.

Three of the four compounds that exceeded the theoretical potency gain by more than an order of magnitude were for the matrix metalloproteinase stromelysin (MMP-3, entries 11–13 in Table 9.1) [17]. In these cases, an acetohydroxamate group was linked to a biaryl using a simple alkyl linker; and structural studies confirmed an additional hydrophobic interaction between the linker and the protein [17, 18]. On the other extreme, an alternate series for MMP-3 in which a naphthylhydroxamate moiety was linked to a biaryl lead (entry 14 in Table 9.1), exhibited a gain in potency that was more than two orders of magnitude less than what could be expected based on additive energies [19]. While the naphthylhydroxamate group exhibited greater intrinsic potency than the acetohydroxamate (K_D value of 20 μM vs 17 mM), it was much more difficult to link this group to the biaryl and maintain the preferred orientation of both groups. Thus, significant potency losses were incurred upon linking (relative to theoretical gains) and extensive modification of the linker was required to produce compounds with IC_{50} values less than 100 nM (e.g., entry 15 in Table 9.1) [19]. It is noteworthy that more than half of the examples in Table 9.1 show gains in potency that are at least

Table 9.1 Potency gains realized through the linked-fragment approach [a].

Entry	Target	Technique	pK_D (1)[b]	pK_D (2)[c]	pK_D (link)[d]	Actual gain[e]	Δ[f]	Compound number[g]
1	AchE	ISFL[i]	7.74	5.96	12.40	4.66	−1.31	–
2	AK	NMR	6.22	2.52	8.00	1.78	−0.74	3
3	Bcl-2[h]	NMR	4.70	3.40	6.64	1.94	−1.46	–
4	Bcl-xL[h]	NMR	3.52	2.22	5.85	2.33	0.11	–
5	c-Src	CTGLA[j]	4.40	4.39	7.19	2.80	−1.59	–
6	FKBP	NMR	5.70	4.00	7.72	2.02	−1.98	24
7	GP	XRC	4.90	4.90	8.22	3.32	−1.58	–
8	HCV pro	NMR	3.58	3.00	6.10	2.52	−0.48	18
9	LFA (1)	NMR	4.10	3.52	7.70	3.60	0.08	9
10	LFA (2)	NMR	4.10	2.00	7.40	3.30	1.30	–
11	MMP-3 (1)	NMR	3.55	1.77	6.51	2.96	1.19	–
12	MMP-3 (2)	NMR	4.70	1.77	7.60	2.90	1.13	–
13	MMP-3 (3)	NMR	4.70	1.77	7.82	3.12	1.36	12
14	MMP-3 (4)	NMR	4.70	4.30	6.47	1.77	−2.53	–
15	MMP-3 (5)	NMR	4.70	4.30	7.21	2.51	−1.79	15
16	PTP1B (1)	NMR	5.92	2.92	7.38	1.46	−1.46	21
17	PTP1B (2)	NMR	3.83	2.92	5.16	1.33	−1.59	6

a) Unless otherwise noted, all examples and potencies were taken from Erlanson and co-workers [15], excluding the examples that did not fit the canonical linked-fragment approach (e.g., MMP-13 and AdSS, which more closely resemble the merged fragment approach) and those for which absolute potencies were not reported for both fragments (e.g., Bcl-xL, U1061A RNA, carbonic anhydrase, neuraminidase, CDK2, and caspace-3).
b) The negative log (base-10) of the experimental K_D (or IC_{50}) value for the first-site ligand.
c) The negative log (base-10) of the experimental K_D (or IC_{50}) value for the second-site ligand.
d) The negative log (base-10) of the experimental K_D (or IC_{50}) value for the linked ligand.
e) The actual gain is defined as the pK_D of the linked compound minus the pK_D of the first-site ligand.
f) Difference between the actual and theoretical gain.
g) The reference number for the linked compound (if shown) in Fig. 9.3.
h) Internal discovery [32].
i) *In situ* fragment linking.
j) Combinatorial target-guided ligand assembly.

10-fold less than the theoretical gains. Despite the remarkable absolute gains achieved with these systems, these losses highlight the difficulty in linking two fragments in the precise manner required to achieve the full energetic benefits.

Excluding the four cases where the actual gain in potency exceeded the theoretical gain by more than an order of magnitude, there is a reasonable correlation ($R^2 = 0.56$) between the actual and theoretical gains for these systems (see Fig. 9.4). Thus, these data appear to support the general postulate of additivity of binding energies when applying the linked-fragment approach. As illustrated in Fig. 9.4, the correlation between the actual and theoretical gains falls near the line for 1 log unit losses with respect to the theoretical gain. This guideline can be useful in evaluating whether linking two compounds will in fact yield a lead with the necessary potency for advancement. For example, if an inhibitor with sub-micromolar potency is an absolute requirement for further evaluation, then attempting to link two leads with only millimolar affinities will likely not produce the desired results. In this example, the *expected* potency could be calculated as:

$$pK_D \text{(expected)} = pK_D(1) + pK_D(2) - 1.0,$$

which results in an expected pK_D value of 5.0 (3.0 + 3.0 – 1.0), or an expected K_D value of 10.0 µM. While gains beyond the simple sum of the energies are possible (as illustrated for four examples in Table 9.1), this would appear to be a less likely occurrence and one should instead expect some energetic losses in the initial linking phase. Optimization of a linked compound beyond this potency can then be performed using traditional medicinal chemistry and structure-based drug design principles.

9.4
Fragment Elaboration

As described above, when two fragment leads can be identified that bind to proximal sites on a protein surface, remarkable gains in potency can be achieved by appropriately linking these two fragments together. It is usually straightforward and relatively easy to identify first-site fragment leads for protein targets using NMR and other methods. However, it is not always possible to identify two fragments that can simultaneously occupy adjacent pockets for a number of reasons. First, the binding affinities of these second-site leads tend to be at least 10-fold weaker than the first-site leads – requiring even more sensitive detection of these low affinity ligands. Second, it is not always straightforward to rule out either competitive binding with the first-site ligand or nonspecific association between the first- and second-site ligands that can lead to false positive results [20]. In these cases, fragment-linking is not recommended, and fragment elaboration strategies can instead be employed to increase the affinity of the first-site ligand into a useful range. As shown in Fig. 9.1, fragment elaboration involves systematically increasing the size (and hopefully the potency) of the first-site ligand. This can be done with small numbers of specific, tailored compounds, or high-throughput parallel synthesis approaches can be employed to produce large libraries of compounds around an appropriately functionalized core. The latter strategy allows for a much more extensive search of compound space and can give rapid feedback on the pragmatic utility of a given first-site lead. For example, with a library of 50–100 compounds, questions regarding the site of attachment, the observation of discernible structure–activity relationships, and the potential for large potency gains can be addressed with reasonable confidence. Given the availability of automated synthesis platforms and the ever-increasing diversity of chemical reactions amenable to high-throughput parallel synthesis [20], fragment elaboration through library generation is becoming a powerful tool for exploring and exploiting frag-

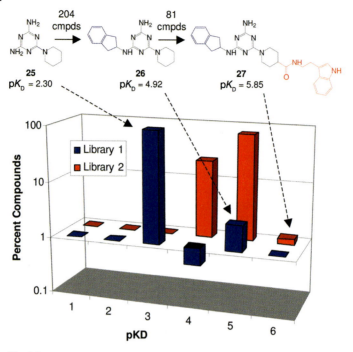

Fig. 9.5

The concept of iterative library design as applied to Erm-AM [21]. Starting from a weakly binding NMR lead (compound **25**), a library of 204 compounds were prepared in which an indan substituent (blue) was found to increase potency by >2.0 log units (compound **26**). A second-generation library of 81 compounds was then prepared around a core containing the indan group in order to exploit a different region of the binding site (red), resulting in an additional gain in potency of nearly 1.0 log unit (compound **27**).

ment leads in drug design. An example of this strategy is shown in Fig. 9.5 for Erm-AM [21]. In this case, a weakly binding NMR lead (compound **25**) was rapidly increased to single-digit micromolar potency (e.g. compound **27**) through the use of iterative library design.

9.5
Energetic Analysis of Fragment Elaboration Strategies

We are continuing to expand our use of high-throughput parallel synthesis in the development of fragment leads. Shown in Table 9.2 are statistics for 11 compound libraries designed against leads for six different protein targets. On average, these libraries yielded compounds with a 10-fold gain in potency over the parent. It is noteworthy, however, that gains in potency of more than two orders of magnitude can be achieved using this library approach, as illustrated for Erm-AM in Fig. 9.5. As with the linked-fragment approach above, rules can be derived from these data

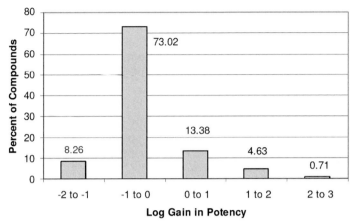

Fig. 9.6
Distribution of the potency gains observed for a set of 1405 compounds derived from libraries against six protein targets (see Table 9.2). The potency gain is defined as the pK_D of the enumerated compounds minus the pK_D of the core. The actual percentages for each bin are displayed.

to calculate the probability of achieving a desired potency gain from a given library of compounds designed around a core. A first step in this process is shown in Fig. 9.6, which gives the distribution of potency gains (relative to the parent compound) for all 1405 compounds contained in the libraries listed in Table 9.2.

Table 9.2 Potency gains realized through fragment elaboration strategies.

Entry	Target	Number of compounds	pK_D (core)[a]	Max. gain in pK_D[b]	Ref.
1	Bcl-xL (1)	16	3.44	2.01	32
2	Bcl-xL (2)	154	4.70	2.33	32
3	ERM (1)	204	2.30	2.62	21
4	ERM (2)	294	5.00	0.22	21
5	ERM (3)	81	4.92	0.93	21
6	HSP90	47	4.74	0.93	—[c]
7	MDM2 (1)	264	4.37	0.93	—[c]
8	MDM2 (2)	50	5.30	0.81	—[c]
9	PTP	75	5.92	1.02	26
10	Survivin (1)	182	5.91	1.45	—[c]
11	Survivin (2)	38	7.07	0.22	—[c]
	Average	128	4.9	1.2	

a) The negative log (base-10) of the experimental K_D (or IC_{50}) value for the core.
b) Maximum gain in potency expressed in log (base-10) units.
c) Internal discovery.

As can be appreciated from this figure, more than 80% of all compounds in these libraries exhibited potencies less than or equal to the parent and an additional 13% exhibited potency gains of 10-fold or less. However, approximately 5% of the library members exhibited at least a 10-fold gain in potency, with 0.7% exhibiting at least a 100-fold gain in potency. Thus, for these data, approximately one out of every 20 compounds showed at least a 10-fold gain in potency, while approximately one out of every 140 compounds exhibited a 100-fold gain. These data can fit an exponential decay that gives the probability [$P(N)$] of achieving an N-fold gain in potency from a given library:

$$P(N) = 27 \times \exp(-2 \times N)$$

According to this equation, the probability of achieving *at least* a 2 log unit gain in potency would be 0.5%, while achieving *at least* a 3 log unit gain in potency would be 0.07%. While the compound libraries and targets represented here are certainly limited in their diversity, these metrics can be useful for an initial evaluation and prioritization of fragment leads.

As illustrated with ErmAM (Fig. 9.5), a strategy of iterative library design was employed against four of the targets shown in Table 9.2, in which the best compound from the initial library was used as the "core" for a subsequent library (see Figs. 9.1, 9.5). In these scenarios, the potency gain realized relative to the initial core is actually the sum of the potency gains achieved for each of the individual libraries. This is a powerful approach to rapidly achieve large potency gains while minimizing the number of compounds that need to be synthesized. For example, according to the equation given above, the probability of achieving a 100-fold gain in potency ($N = 2$) from a single library of compounds is 0.5%, requiring a library size of approximately 200 compounds. However, the probability of achieving a 10-fold gain in potency ($N = 1$) is 3.7%, requiring a library of 27 compounds. Using an iterative library approach, a 100-fold gain in potency is just as likely to be achieved with two iterative libraries of 27 compounds (54 compounds total) as with a single library of 200 compounds. Thus, in evaluating the utility of different fragments, producing small compound libraries around multiple fragment leads is a preferred strategy, as those with even modest potency gains can then be further optimized in subsequent library design.

9.6
Summary

There is currently intense interest in the pharmaceutical community in utilizing and exploiting fragment-based design strategies for the development of novel therapeutics. In many ways, this is a reaction to the lack of productivity in producing high-quality clinical candidates in recent years, despite the tremendous resources that have been poured into high-throughput screening and combinatorial chemistry [22]. In another sense, this interest has resulted from a fundamental shift in our understanding and perception of chemical and receptor space. There is a boggling number of chemical entities that can possibly exist (with some estimates exceeding 10^{60} compounds [23]); and the costs associated with producing, storing, and distributing even a minute fraction of this number is tremendously high. Since fragment-based screening can cover a substantially larger portion of chemical space using relatively small compound libraries, these methods serve as an invaluable complement to current high-throughput screening approaches for lead identification. Given this interest, it is important to emphasize that several small molecule inhibitors designed using fragment-based approaches have gone beyond the pre-clinical setting into the clinic (e.g., matrix metalloproteinase inhibitors [24] and inhibitors of the LFA-ICAM interaction [25]). This is ample validation that fragment-based methods can be used to produce ligands not only with high affinity, but also with the appropriate drug-like properties for use in man.

References

1 Stockman, B. J., Dalvit, C. 2002, NMR Screening Techniques In Drug Discovery And Drug Design. *Prog. NMR Spectrosc.* 41, 187–231.

2 Meyer, B., Peters, T. 2003, NMR Spectroscopy Techniques for Screening and Identifying Ligand Binding to Receptors. *Angew. Chem. Int. Ed.* 42, 864–890.

3 Nienaber, V. L., Richardson, P. L., Klinghofer, V., Bouska, J. J., Giranda, V. L., et al. 2000, Discovering Novel Ligands For Macromolecules Using X-Ray Crystallographic Screening. *Nat. Biotechnol.* 18, 1105–1108.

4 Carr, R., Jhoti, H. 2002, Structure-Based Screening Of Low-Affinity Compounds. *Drug Discov. Today* 7, 522–527.

5 Hartshorn, M. J., Murray, C. W., Cleasby, A., Frederickson, M., Tickle, I. J., et al. 2004, Fragment-Based Lead Discovery Using X-ray Crystallography. *J. Med. Chem.* 48, 403–413.

6 Swayze, E. E., Jefferson, E. A., Sannes-Lowery, K. A., Blyn, L. B., Risen, L. M., et al. 2002, SAR by MS: A Ligand Based Technique for Drug Lead Discovery Against Structured RNA Targets. *J. Med. Chem.* 45, 3816–3819.

7 Boehm, H.-J., Boehringer, M., Bur, D., Gmuender, H., Huber, W., et al. 2000, Novel Inhibitors of DNA Gyrase: 3D Structure Based Biased Needle Screening, Hit Validation by Biophysical Methods, and 3D Guided Optimization. A Promising Alternative to Random Screening. *J. Med. Chem.* 43, 2664–2674.

8 Erlanson, D. A., Braisted, A. C., Raphael, D. R., Randal, M., Stroud, R. M., et al. 2000, Site-Directed Ligand Discovery. *PNAS* 97, 9367–9372.

9 Ramstrom, O., Lehn, J.-M. 2002, Drug Discovery by Dynamic Combinatorial Chemistry. *Nature Reviews Drug Disc.* 1, 26–36.

10 Hann, M. M., Leach, A. R., Harper, G. 2001, Molecular Complexity and Its Impact on the Probability of Finding Leads for Drug Discovery. *J. Chem. Inf. Comput. Sci.* 41, 856–864.

11 Hopkins, L. A., Groom, C. R., Alex, A. 2004, Ligand Efficiency: A Useful Metric for Lead Selection. *Drug Discov. Today* 9, 430–431.

12 Fejzo, J., Lepre, C. A., Peng, J. W., Bemis, G. W., et al. 1999, The SHAPES Strategy: An NMR-Based Approach for Lead Generation in Drug Discovery. *Chem. Biol.* 6, 755–769.

13 Lepre, C. A. 2001, Library Design for NMR-Based Screening. *Drug Discovery Today* 6, 133–140.

14 Shuker, S. B., Hajduk, P. J., Meadows, R. P., Fesik, S. W. 1996, Discovering High-Affinity Ligands for Proteins: SAR by NMR. *Science* 274, 1531–1534.

15 Erlanson, D. A., McDowell, R. S., O'Brien, T. **2004**, Fragment-Based Drug Discovery. *J. Med. Chem.* 47, 1–20.

16 Hajduk, P. J., Meadows, R. P., Fesik, S. W. **1999**, NMR-Based Screening in Drug Discovery. *Q. Rev. Biophys.* 32, 211–240.

17 Hajduk, P. J., Sheppard, G., Nettesheim, D. G., Olejniczak, E. T., Shuker, S. B., et al. **1997**, Discovery of Potent Nonpeptide Inhibitors of Stromelysin Using SAR by NMR. *J. Am. Chem. Soc.* 119, 5818–5827.

18 Olejniczak, E. T., Hajduk, P. J., Marcotte, P. A., Nettesheim, D. G., Meadows, R. P., et al. **1997**, Stromelysin Inhibitors Designed From Weakly Bound Fragments: Effects Of Linking And Cooperativity. *J. Am. Chem. Soc.* 119, 5828–5832.

19 Hajduk, P. J., Shuker, S. B., Nettesheim, D. G., Xu, L., Augeri, D. J., et al. **2002**, NMR-Based Optimization Of MMP Inhibitors With Improved Bioavailability. *J. Med. Chem.* 45, 5628–5639.

20 Huth, J. R., Sun, C., Hajduk, P. J. **2005**, Utilization of NMR-Derived Fragment Leads in Drug Design. *Methods Enzymol.* 394, 549–572.

21 Hajduk, P. J., Dinges, J., Schkeryantz, J. M., Janowick, D., Kaminski, M., et al. **1999**, Novel Inhibitors of Erm Methyltransferases from NMR and Parallel Synthesis. *J. Med. Chem.* 42, 3852–3859.

22 Campbell, S. J. **2000**, Science, Art and Drug Discovery: A Personal Perspective. *Clin. Sci.* 99, 255–260.

23 Bohacek, R. S., McMartin, C., Guida, W. C. **1996**, The art and practice of structure-based drug design: a molecular modeling perspective. *Med. Res. Rev.* 16, 3–50.

24 Wada, C. K., Holms, J. H., Curtin, M. L., Dai, Y., Florjancic, A. S., et al. **2002**, Phenoxyphenyl Sulfone N-Formylhydroxylamines (Retrohydroxamates) as Potent, Selective, Orally Bioavailable Matrix Metalloproteinase Inhibitors. *J. Med. Chem.* 45, 219–232.

25 Liu, G., Huth, J. R., Olejniczak, E. T., Mendoza, R., DeVries, P., et al. **2001**, Novel p-Arylthio Cinnamides as Antagonists of Leukocyte Function-Associated Antigen-1/Intracellular Adhesion Molecule-1 Interaction. 2. Mechanism of Inhibition and Structure-Based Improvement of Pharmaceutical Properties. *J. Med. Chem.* 44, 1202–1210.

26 Xin, A., Oost, T. K., Abad-Zapatero, C., Hajduk, P. J., Pei, Z., et al. **2003**, Potent, Selective Inhibitors of Protein Tyrosine Phosphatase 1B. *Bioorg. Med. Chem. Lett.* 13, 1887–1890.

27 Gu, Y.-G., Florjancic, A. S., Clark, R. F., Zhang, T., Cooper, C. S., et al. **2004**, Structure-Activity Relationships of Novel Potent MurF Inhibitors. *Bioorg. Med. Chem. Lett.* 14, 267–270.

28 Hajduk, P. J., Gomtsyan, A., Didomenico, S., Cowart, M., Bayburt, E. K., et al. **2000**, Design of Adenosine Kinase Inhibitors from the NMR-Based Screening of Fragments. *J. Med. Chem.* 43, 4781–4786.

29 Szczepankiewicz, B. G., Liu, G., Hajduk, P. J., Abad-Zapatero, C., Pei, Z., et al. **2003**, Discovery of a Potent, Selective Protein Tyrosine Phosphatase 1B Inhibitor Using a Linked-Fragment Strategy. *J. Am. Chem. Soc.* 125, 4087–4096.

30 Liu, G., Xin, Z., Pei, Z., Zhao, H., Hajduk, P. J., et al. **2003**, Fragment Screening and Assembly: A Highly Efficient Approach to a Selective and Cell Active Protein Tyrosine Phosphatase 1B Inhibitor. *J. Med. Chem.* 46, 4232–4235.

31 Wyss, D. F., Arasappan, A., Senior, M. M., Wang, Y. S., Beyer, B. M., et al. **2004**, Non-peptidic small-molecule inhibitors of the single-chain hepatitis C virus NS3 protease/NS4A cofactor complex discovered by structure-based NMR screening. *J. Med. Chem.* 47, 2486–2498.

32 Petros, A. M., Dinges, J., Augeri, D. J., Baumeister, S. A., Betebenner, D. A., et al. **2006**, Discovery of a Potent Inhibitor of the Antiapoptotic Bcl-xL from NMR and Parallel Synthesis. *J. Med. Chem.* 49, 656–663.

10
Pyramid: An Integrated Platform for Fragment-based Drug Discovery

Thomas G. Davies, Rob L. M. van Montfort, Glyn Williams, and Harren Jhoti

10.1
Introduction

The past two decades have seen considerable interest in new approaches to small-molecule drug discovery, aimed at expediting the identification of novel therapeutic agents. Technologies such as high-throughput screening and combinatorial chemistry have been widely adopted by the pharmaceutical industry, but have failed to generate a significantly increased level of productivity, largely due to high compound attrition rates during drug development [1]. The reasons for compound failure are multi-factorial, but a key issue appears to lie in the nature of hits discovered in high-throughput screens. Typical compound collections screened in such campaigns have evolved to be "drug-like" in properties such as molecular weight – roughly 350–400 Da [2, 3]. However, the addition of further functionality during the lead optimization phase can then result in molecules with excessive molecular weight and lipophilicity, properties widely recognized to have detrimental effects on pharmacokinetic properties such as bioavailablity [4]. In contrast, by screening a library of significantly smaller compounds, or "fragments", these problems associated with excessive molecular weight can be alleviated. Fragments are defined as small molecules of less than approximately 250 Da and are typically composed of just a few functionalities [5–8].

Although such fragments often have low (100 μM to 10 mM) affinity, they frequently exhibit high ligand efficiency, i.e. high values for the ratio of free energy of binding to the number of heavy atoms [9]. This is largely an indication of the "high quality" of the small number of protein–ligand interactions formed, which must outweigh the significant entropic rotational and translational penalty of binding [10] (chapter 3). Thus, they represent very attractive start-points in inhibitor design, interacting with "hot-spots" within target active sites, and are therefore suitable for rapid optimization into potent leads. It is important, however, that when a fragment hit is identified, such optimization is performed with carefully designed iterations consistent with maintaining good ligand efficiency. In this way, the advantages gained from starting small can be preserved throughout the drug discovery process. This new approach to inhibitor development, often referred to as "fragment-based

Fragment-based Approaches in Drug Discovery. Edited by W. Jahnke and D. A. Erlanson
Copyright © 2006 WILEY-VCH Verlag GmbH & Co. KGaA, Weinheim
ISBN: 3-527-31291-9

drug discovery" [6–8, 11, 12] has a number of other advantages over conventional screening, including a more efficient sampling of chemical space and a higher hit-rate due to lower molecular complexity [13]. Both of these factors mean that the number of compounds screened can be considerably reduced.

A key challenge in fragment based drug discovery is the detection of hits, because weakly binding fragments are difficult to detect reliably using "conventional" bioassay-based screening methods. However, a variety of biophysical methods such as X-ray crystallography [12, 14] (chapter 11), protein NMR [8, 15] (chapters 8 and 9), surface plasmon resonance [16], and mass spectrometry [17] (chapter 13) have been successfully used to detect the binding of fragments to a variety of protein targets. The use of crystallography as a screening technique has the additional advantage of very high sensitivity (capable of detecting binding in the millimolar range, and therefore allowing the screening of even smaller fragments) as well as providing precise structural details of the interaction between hit and target. Thus, this technique not only provides an efficient means to detect weak binders, but also allows for the most rapid and efficient hit optimization by structure-based design techniques. In addition, crystallography is less likely to suffer from the problem of false positives, which are intrinsic to most other screening techniques.

In this review, we describe how X-ray crystallography forms the central part of our fragment-based screening platform, which we term Pyramid [18]. We present a discussion of the issues involved in using crystallography as a screening technique, the technology developed to address these, and selected case studies. We also discuss the evolution of the Pyramid approach in which high-throughput crystallography is coupled with other biophysical techniques, such as NMR, resulting in an integrated fragment-based discovery platform.

10.2
The Pyramid Process

10.2.1
Introduction

Protein crystallography has historically been viewed by the pharmaceutical industry as a "low-throughput" technique, and hence its use and impact has been limited to the lead optimization phase. The key to transforming it into a technique with the potential for screening was the decrease in time taken to generate structural information on protein–ligand complexes. The Pyramid screening approach relies on the development of high-quality *fragment libraries*, coupled with *automated protocols* for rapid X-ray data collection, processing, and structure solution [19]. A flow-chart for a typical crystallographic fragment-screening experiment is shown in Fig. 10.1. Briefly, it involves soaking the crystals with fragments of interest, followed by X-ray data collection and processing, placement of water molecules in electron density and refinement of the ligand-free complex to potentially reveal difference electron density associated with the bound ligand. The electron

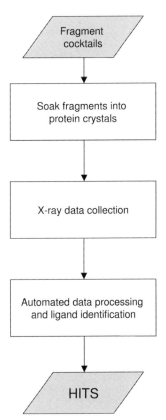

Fig. 10.1
Flow-chart for a crystallographic fragment-screening experiment.

density is then automatically interpreted, fitted and the complex further refined to give the final protein–ligand structure. The development and automation of the various steps in our Pyramid approach are explained in more detail below.

10.2.2
Fragment Libraries

10.2.2.1 Overview

The composition of the screening libraries themselves is critical to the success of any fragment-based method, and great care has to be taken in their design. We followed two complementary strategies in the assembly of our libraries. The first makes use of the assumption that "drug fragment space" can be represented by a relatively small number of compounds based on scaffolds and functional groups commonly found in known drugs. The second strategy is to design libraries targeted to particular proteins or protein classes, by exploiting knowledge of known ligands and key interactions with their protein targets. These targeted libraries can be further enriched by using virtual screening to select fragments that dock well into the active site of the target or protein family of interest.

In general, fragment selection is limited to compounds with molecular weights of 100–250 Da and with few functional groups, making them suitable for rapid synthetic optimization. Compounds with molecular weights higher than 250 Da are of limited value, because the increase in size and complexity generally reduces the chance of a hit [13]. However, for targeted libraries, the inclusion of fragments with a slightly higher molecular weight may sometimes be appropriate because their functionality addresses key interactions made between the target and known ligands.

10.2.2.2 Physico-chemical Properties of Library Members

When selecting compounds for screening libraries, it is important to consider their physico-chemical properties. There is a growing body of literature defining drug-like properties [2, 13]. For developing orally available drug candidates, Lipinski's "Rule of 5" [4], which limits the molecular weight, logP, and number of hydrogen bond donors and acceptors to values often found in oral drugs, provides a useful framework. To further define drug-likeness, these rules have been extended to include the number of rotatable bonds, number of rings, and heavy atoms [20]. Recently, the term "lead-like" was introduced for compounds suitable for further optimisation that have physico-chemical properties somewhat scaled down from the Lipinski values [3]. However, all these definitions are based on compounds identified through conventional bioassay screening of drug-size compound libraries and thus may not be appropriate descriptors for libraries of simple fragments.

Our own analysis of the calculated physico-chemical properties of a set of 40 fragments identified by crystallographic screening against three different targets (Table 10.1) showed that these hits generally have a molecular weight lower than 300 Da, with a number of hydrogen bond donors (HBD), hydrogen bond acceptors (HBA), and ClogP all equal to or smaller than three. In addition, the number of rotatable bonds (NROT) is equal to, or less than three and the polar surface area (PSA) less than 60 Å^2. Based on these results it appears that a "Rule of 3" would be a more useful guideline in the design of fragment libraries [18, 21]. The relevance of this "Rule of 3" for fragments was also supported by a recent paper

Table 10.1 Fragment screening hits – average calculated properties suggested a "rule of 3".

Target	N_{hits}	"Rule of 3" properties				Other properties	
		MWT	HBA	HBD	ClogP	NROT	PSA
Aspartic protease	13	228	1.1	2.9	2.7	3.5	44
Serine protease	13	202	1.7	3.1	1.8	2.9	56
Kinase	14	204	2.5	2.0	1.6	1.7	61
"Rule of 3" guidelines		<300	≤3.0	≤3.0	≤3.0		

describing the design of a set of four fragment libraries [22]. The authors point out that their complete fragment library consisting of 1315 compounds closely adheres to the Astex "Rule of 3" and claim that designing libraries of such hit-like fragments makes the screening effort more efficient.

10.2.2.3 Drug Fragment Library

This library was based on the idea that "drug fragment space" can be sampled with a relatively small number of compounds based on scaffolds and functional groups frequently occurring in known drugs. Several investigations have shown that only a small number of simple organic ring systems commonly occur in drug molecules [15, 23, 24]. Thus the first step in constructing our "Drug Fragment Library" was the identification of a set of low molecular weight ring systems (Fig. 10.2a) and additionally simple carbocyclic and heterocyclic rings often found in known drugs (Fig. 10.2b). Secondly, a set of desirable side-chains was chosen, which in addition to those frequently observed in drug molecules (Fig. 10.3a), included a set of lipophilic groups intended to pick up hydrophobic interactions in a protein binding site and a set of nitrogen substituents (Fig. 10.3b).

Based on this collection of ring systems and side-chains, a virtual library was generated by substituting side-chains onto each of the ring systems. Each ring carbon atom was substituted by the side-chains found in known drugs and by lipophilic side-chains, whereas ring nitrogens were only substituted by side-chains from the nitrogen-substituent group. With the exception of benzene and imidazole, which were allowed to be substituted at all positions with all side-chains, each ring system was substituted at only one position at a time. The resulting virtual library consisted of 4513 fragments, of which 401 were commercially available. Removal of insoluble compounds and known toxophores resulted in a final drug fragment library of 327 compounds [19].

10.2.2.4 Privileged Fragment Library

In addition to the fragmentation of known drugs we created a so-called "Privileged Fragment" library, which is based on fragments of good-quality lead molecules. A broad set of drug targets (39 enzymes and 25 receptors) guided the selection of moieties considered to be privileged from a medicinal chemistry perspective. The library was further enriched with some of the fragment hits from our Pyramid screening campaigns, because we considered that these compounds had shown some pedigree as valuable fragment hits. The application of stringent physico-chemical property criteria guaranteed that the average properties of the library complied with the "Rule of 3", resulting in a final library of 120 compounds.

10.2.2.5 Targeted Libraries and Virtual Screening

We designed several target specific fragment libraries exploiting the knowledge of key interactions between protein targets and known ligands. For example, a

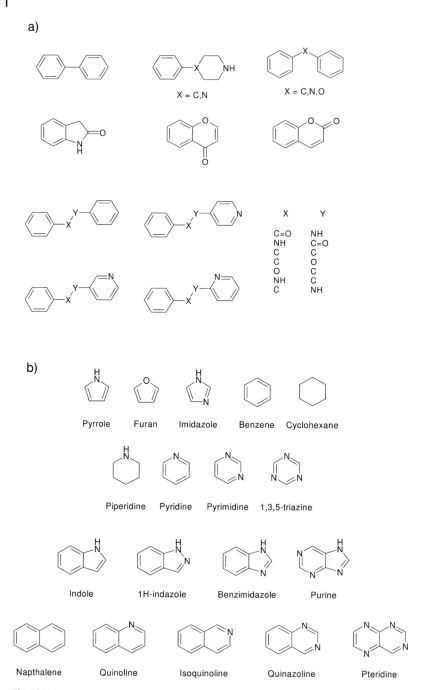

Fig. 10.2
(a) Low-molecular-weight ring systems and (b) simple carbocyclic and heterocyclic ring systems chosen for the drug fragment library.

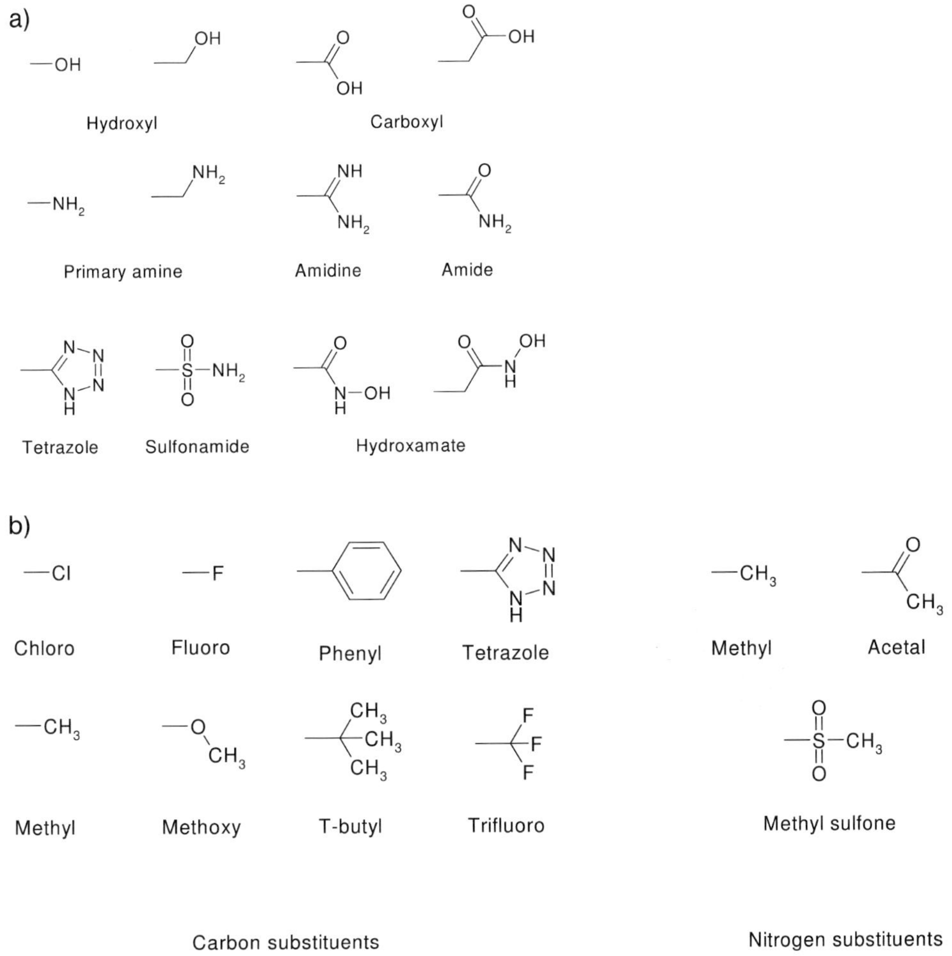

Fig. 10.3
(a) Side-chains well represented in known drug molecules and (b) lipophilic side-chains and nitrogen substituents chosen for the drug fragment library.

"Focused Kinase Library" was constructed containing motifs likely to bind in the ATP pocket of kinases. An initial set of scaffolds frequently found to bind in the ATP binding site of kinases and form hydrogen bonds with the backbone hinge region were identified through literature and patent analysis [25]. A virtual library of candidate fragments was generated by enumeration of these scaffolds with drug-like side-chains and commercially available compounds were then purchased. This library was further improved by the chemical synthesis of certain scaffolds and templates and by the addition of compounds suggested through virtual screening, as described in detail elsewhere [26]. In brief, a database of more than 3.6 million unique compounds called ATLAS (Astex Therapeutics Library of Available Substances) has been generated from various suppliers of chemicals and compound libraries [19]. ATLAS can be queried using substructure filters and physico-chemical property filters (such as molecular weight, ClogP, PSA, etc.) to produce libraries of commercially available compounds meeting specific user requirements. These compound libraries can then be automatically docked into the active site of the target of interest, using a proprietary version of GOLD [27] with a variety of scoring functions [28, 29]. The best scoring function for virtual screening of a particular target is generally selected based on test runs with a set of ligands for which the binding mode is known. The results from virtual screening runs can be queried using a web-based interface, which allows the user to select subsets of compounds for visualization, using a variety of filters such as molecular weight, scoring functions or components of scoring functions, predefined pharmacophores, steric or electrostatic clashes, the formation of specific hydrogen bonds, and 2D substructures. A flow-chart summarizing the approach to virtual screening is shown in Fig. 10.4. The version of the Focused Kinase Library used to obtain the results described in this review consisted of 116 compounds.

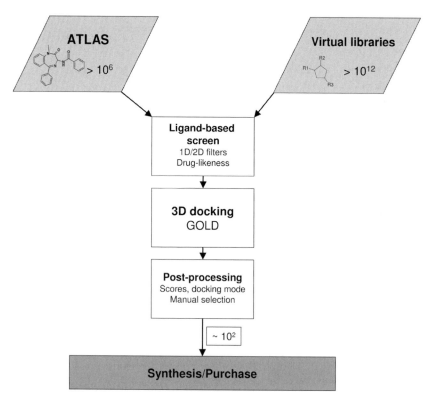

Fig. 10.4
Flow-chart of the virtual screening process.

In addition, a "Focused Phosphatase Library" was generated that was targeted against the phosphotyrosine binding pocket of protein tyrosine phosphatase 1B (PTP1B) [30]. In this case, the literature was searched for potential phosphotyrosine and carboxylic acid mimetics and commercial databases were searched for promising fragments. Where fragments of interest could not be obtained from commercial suppliers these were synthesized. To further improve the library and identify novel scaffolds, virtual screening runs were carried out against both the open and closed conformations of PTP1B. Finally, compounds that were expected to contain more than one negative charge at physiological pH were excluded from the library, resulting in a Focused Phosphatase Library of 264 compounds.

10.2.2.6 Quality Control of Libraries

Due to the relatively small size of our fragment libraries, it is particularly important that they are of the highest possible quality. In addition to compound selection based on calculated physico-chemical properties, we implemented a quality control process to ensure the final library members are suitable for Pyramid screening. Compounds that were less than 90% pure by LC/MS and ^{1}H NMR in DMSO, or were insoluble in DMSO at 100 mM were removed and replaced by more soluble/pure analogues. Furthermore, two 1D ^{1}H-NMR spectra were collected 24 h apart for each compound in each library to further address purity, solubility, and chemical stability over time in aqueous buffer. Compounds that failed one or more of these criteria were removed from the libraries. More recently, we expanded our quality control process by looking at aggregation of our fragments as single compound solutions and as part of a mixture of compounds (or cocktail), using NMR.

10.2.3
Fragment Screening

The most efficient method of obtaining structures of a protein–ligand complex is by soaking the ligand of interest into pre-formed protein crystals. This can often be achieved by placing a single crystal in a high concentration solution of ligand in mother liquor for a suitable length of time, allowing the ligand to diffuse through the solvent channels in the crystal and bind at energetically favorable sites. When screening for low affinity fragments, the high concentration (around 50 mM) of fragment in the soak solution reflects the requirement to drive the equilibrium towards near full occupancy of binding sites. For practical purposes, a ligand concentration 10-fold greater than the IC_{50} or K_d (giving a theoretical occupancy of approximately 90%) is usually sufficient. Ligand stocks are often formulated in DMSO, and therefore soaks generally contain 1–10% organic solvent, which can aid solubility.

An alternative procedure to obtain structures of protein ligand complexes is co-crystallization, in which the protein–ligand complex is prepared in aqueous phase and then crystallized with the ligand *in situ*. This method is less suitable for high-throughput fragment screening, because effectively a separate crystallization experiment is needed for each compound. This procedure can be further complicated if the presence of a ligand results in a change in the crystallization conditions. Moreover, co-crystallization is not optimal for the determination of weakly binding fragments, as the high concentration of ligand needed to fully occupy the binding site can interfere with the crystallization process itself. It should be noted, however, that some proteins will not crystallize without the presence of a ligand, which could be due to the ordering effect on mobile regions. In these cases, co-crystallization is the most likely alternative option. In addition, co-crystallization can be used in rare cases where fragment soaking causes crystals to crack, presumably by inducing conformational changes or binding at crystal contacts.

The efficiency of fragment screening can be increased dramatically by pooling or cocktailing the compounds in the library [12, 19, 31]. Identification of the bound fragment at the end of the X-ray experiment then becomes a case of determining the best fragment fit to the electron density. Excluding the trivial scenario of no binding, one can imagine three possible outcomes of a cocktail X-ray experiment. In the first scenario, only one fragment binds to the protein, its identity being unambiguously determined from the electron density. In a second scenario, removal of the initially identified fragment from the cocktail reveals the binding of secondary or even tertiary binders, and in this case the soaking is effectively a competition experiment. A third situation occurs when electron density can be explained by the simultaneous binding of two or more fragments of roughly similar affinity. In both latter cases, rounds of "deconvolution" are necessary to extract all relevant information, which at least partially negates the benefits of cocktailing.

For ease of deconvolution, cocktailing at Astex is usually performed in sets of four, and the selected components are chosen to be as chemically diverse as possible within a particular cocktail. This has the effect of reducing the number of hits per cocktail, as well as increasing the shape diversity, which expedites the automated interpretation of ligand electron density (see also Section 10.2.5, "Automation of Data Processing"). The initial partitioning of fragments into cocktails is achieved using a computational procedure, which minimizes chemical similarity [19].

10.2.4
X-ray Data Collection

High-throughput screening of fragments using crystallography requires fast and efficient X-ray data collection. Recent developments in hardware have been driven by the need to streamline and improve data collection at synchrotron beamlines where new third-generation sources, producing brighter and better collimated X-ray beams, allow higher quality data to be collected more rapidly [32]. Although largely driven by the high-throughput requirements of structural genomics consortia, fragment-based screening shares the need to collect good data in a short space of time. The rate-limiting step for third-generation synchrotrons is frequently the manual intervention required to change samples, where the time taken to mount and align crystals can easily exceed half that required to collect the data. Consequently, a significant number of synchrotrons have developed automatic sample changers and integrated them into their data collection systems. Sample changers such as ACTOR (Rigaku MSC), MARCSC (Marresearch) and BruNo (Bruker AXS) are increasingly becoming available to the commercial laboratory setup and have been a key step in the realization of high-throughput data collection in-house [33]. For example, at Astex we reported the collection of X-ray data from 53 crystals of PTP1B in approximately 80 h, using ACTOR [34, 35], and the system is in near-continual use on a range of projects.

Developments in technology have not been limited to sample changers. The latest generation of high-intensity X-ray generators (such as Rigaku's FR-E), coupled with steady improvements in X-ray optics have revolutionized in-house X-ray equipment to the point where the beam intensity is comparable to that obtainable at some synchrotrons. Parallel advances in X-ray detector design have resulted in a new generation of charged couple detectors (CCDs) such as the Quantum 315 Area Detector System Corporation (ADSC), which are larger, more sensitive, and have a faster readout. Coupled with stabilization of cost, their use has increased, and combined with brighter rotating anode generators, they are an important component of a high-throughput setup in a commercial laboratory. At Astex the high speed provided by two Jupiter CCDs is combined with two R-Axis HTCs (Rigaku) to give a flexible setup for routine high-throughput data collection.

Advances in the hardware involved in automating data collection demand a parallel development of software to control the system. The goal is to develop a "smart" system that can encompass control of crystal mounting and aligning, evaluation of experimental strategy based on initial images, data collection, and finally integration, scaling, and reduction of experimental intensities. An example of such a system that is currently being developed is the synchrotron software Blu-Ice [36]. At Astex this is achieved through the ACTOR-associated software Director (Rigaku MSC), coupled with the integration and scaling software d*TREK [37] as implemented in the CrystalClear package (Rigaku MSC).

10.2.5
Automation of Data Processing

Structure solution, refinement and analysis have traditionally been a major bottleneck in the rapid use of X-ray data. Automation of these steps combined with the full integration of the resulting information within an easily queried database environment has perhaps been the single most important factor in the application of crystallography as a primary screening technique. The different stages involved in our automated data processing procedure are shown in Fig. 10.5 and is briefly described below. Implicit in this approach is the availability of a suitable protein starting model for phasing, in the same space-group as the protein–ligand complex crystal.

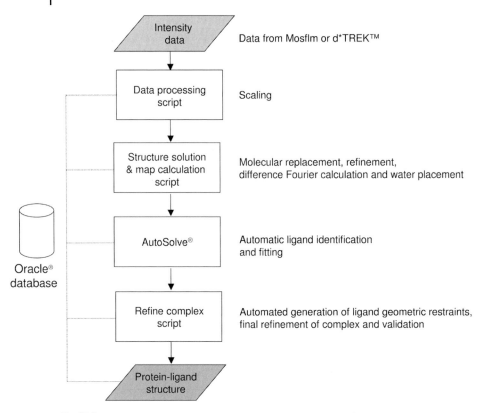

Fig. 10.5
Flow-chart of Pyramid automatic structure solution cascade. The cascade starts with reduction and scaling of the collected diffraction data, followed by a structure solution and map calculation step. In turn, this is followed by an automatic ligand identification and fitting step called AutoSolve. Once the correct ligand has been fitted, the resulting protein–ligand complex is subjected to further rounds of refinement followed by a structure validation step. All steps from scaling to ligand fitting are carried out within a company-wide Oracle database environment.

Automated data processing starts with the reduction and scaling of experimental intensities, followed by a combination of local 6D molecular replacement and rigid body refinement, using a suitable protein starting model. This effectively handles the small changes in isomorphism that can occur when protein crystals are soaked with small molecule ligands, whilst being significantly faster than a traditional "full" molecular replacement, as implemented in programs like Amore [38] or Molrep [39]. Molecular replacement is followed by cycles of restrained refinement interspersed with automated placement of water molecules into F_o-F_c electron density, except in a user-defined active site. The resulting F_o-F_c difference Fourier in the active site is then passed to AutoSolve [18] for ligand identification and fitting.

The process of electron density analysis and interpretation as implemented by AutoSolve has significantly accelerated crystallographic data analysis [5]. The initial step comprises the generation of ligand 3D geometries from SMILES strings using CORINA [40]. AutoSolve then identifies the bound ligand and passes the fitted ligand coordinates on for dictionary generation and subsequent protein–ligand refinement. In the majority of cases, AutoSolve is able to correctly identify the ligand in a cocktail. Ligand fitting using AutoSolve is driven via a web-based interface which is linked to a company-wide Oracle database. This allows for a fully automated process from data processing to examination of the final fitted solution and maps using AstexViewer [18, 41]. If required, the protein and ligand may then be further manipulated and refined manually.

The full integration of structural information with other experimental data (e.g. cloning, purification, bioassay, chemical synthesis) is of key importance for the most effective and timely use of data. Once identified as a "validated hit", the protein–ligand structure becomes viewable to computational and medicinal chemists within a number of in-house chemo- and bioinformatic platforms and allows further cycles of ligand design.

10.2.6
Hits and Diversity of Interactions

The Pyramid process has been applied to a range of protein targets at Astex, including kinases, proteases, and phosphatases. Three examples of typical fragment hits are presented below and demonstrate the diversity of interactions observed, including purely hydrophobic and those requiring conformational movement.

10.2.6.1   Example 1: Compound 1 Binding to CDK2

Cyclin-dependent kinase 2 (CDK2) is a key kinase involved in the regulation of the cell cycle, and inhibitors of CDK2 are believed to have potential as anti-cancer therapeutics [42–44]. Pyramid screening of Astex's Privileged Fragment Library identified the pyrazine based fragment, compound **1**, which has a measured affinity of 360 µM in an enzyme assay [19]. Compound **1** was clearly defined in $F_o - F_c$ electron density, and binds at the hinge region of the ATP binding site (Fig. 10.6a). It forms a single hydrogen bond with the protein, between the pyrazine nitrogen and the amide nitrogen of the hinge residue Leu83. Formation of a hydrogen bond with this residue is a conserved interaction in nearly all hinge-binding kinase inhibitors; and this hydrogen bond is probably a key determinant in driving binding in this case. In addition, it forms non-classical hydrogen bonds between the aromatic C–H's on the pyrazine, and the backbone carbonyl oxygens of Glu81 and Leu83. These "CHO" interactions, which are commonly observed in kinases between electron-deficient heterocycles and protein acceptors, exemplify the diverse range of interactions which are formed by fragments. In addition, compound **1** forms hydrophobic interactions with the side-chains of Ala31, Phe80, Phe82, and Leu134 which form the top and bottom of the lipophilic adenine site. Compound **1** represents an

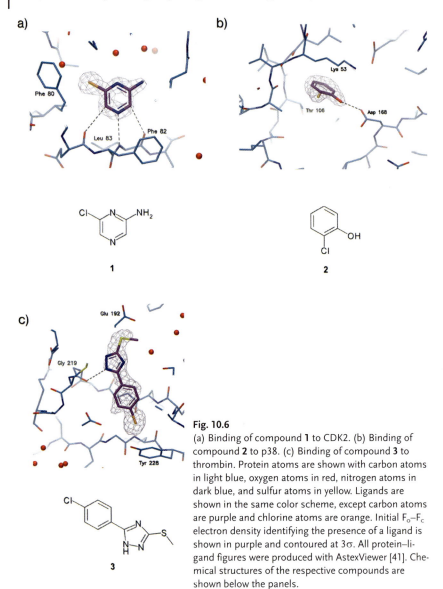

Fig. 10.6
(a) Binding of compound **1** to CDK2. (b) Binding of compound **2** to p38. (c) Binding of compound **3** to thrombin. Protein atoms are shown with carbon atoms in light blue, oxygen atoms in red, nitrogen atoms in dark blue, and sulfur atoms in yellow. Ligands are shown in the same color scheme, except carbon atoms are purple and chlorine atoms are orange. Initial F_o–F_c electron density identifying the presence of a ligand is shown in purple and contoured at 3σ. All protein–ligand figures were produced with AstexViewer [41]. Chemical structures of the respective compounds are shown below the panels.

attractive medicinal chemistry start-point for the design of novel lead compounds, with synthetically accessible and structurally useful vectors available for exploitation. In addition, it exhibits a high ligand efficiency of 0.59 kcal mol^{-1} heavy atom^{-1}, which compares favorably with the ligand efficiencies of 0.20–0.30 kcal mol^{-1} heavy atom^{-1} typically observed for hits discovered by conventional bioassay-based high-throughput screens. Elaboration and optimization of fragment hits against CDK2 has led to compounds which are now in the clinic.

10.2.6.2 Example 2: Compound 2 Binding to p38α

Pyramid screening of the enzyme p38 MAP kinase, a key modulator in the TNF pathway, and a potential target for the development of anti-inflammatory therapeutics [45] also yielded several novel fragment hits. For example, compound **2** from the Astex Drug Fragment Library, was clearly defined in F_o-F_c maps, despite an affinity of weaker than 1 mM in enzyme assays. This hit has since been optimized into a lead series with compounds exhibiting nanomolar affinity [46].

The chlorophenyl portion of compound **2** occupies a lipophilic pocket adjacent to the threonine gatekeeper, but does not form a hydrogen bond with the hinge region (Fig. 10.6 b). Targeting this pocket, which is more generally associated with tyrosine kinases, is believed to be the crucial driving force for inhibitor binding in this case. This result clearly indicates that it is possible to exploit key selectivity pockets adjacent to the ATP binding site, without using the conserved donor/acceptor motif of the hinge region. The phenolic hydroxyl group *ortho* to the chlorine atom forms a rarely observed hydrogen bond with the side-chain of Asp168, which adopts a conformation not observed in any public domain p38α ligand structure. This example illustrates how fragment-based screening can probe predominantly lipophilic interactions, as well as inducing conformation changes [19].

10.2.6.3 Example 3: Compound 3 Binding to Thrombin

Inhibitors of the serine protease thrombin, a key component of the blood-clotting cascade, are potentially useful as anticoagulants [47]. Compound **3** was identified during Pyramid screening of a "Targeted Thrombin Library" assembled using virtual screening approaches as outlined earlier; and it binds with an affinity of approximately 1 mM.

Compound **3** binds to thrombin with its chlorophenyl moiety buried deep within the S1 specificity pocket; and the chlorine atom interacting with the pi-electrons of the Tyr 228 side-chain, which forms the bottom of the pocket (Fig. 10.6 c). Molecular recognition and binding is dominated by lipophilic contacts in this pocket, although the triazole NH also forms a hydrogen bond with the backbone carbonyl of Gly 219. Detection of this uncharged S1 binder further validates Pyramid's ability to detect fragment binding largely driven by hydrophobic interactions.

10.3
Pyramid Evolution – Integration of Crystallography and NMR

The examples described above show that the Pyramid approach, based on crystallographic screening, is well suited for identifying low affinity fragments and has been highly successful in generating hits against a number of different targets. To further expand the potential of our fragment-based hit generation process, we have now broadened our crystallographic screening platform by complementing it with NMR spectroscopy, which is the primary alternative technique able to effectively detect weak binders. X-ray crystallography and NMR can be combined in a complementary manner, ranging from screening a target in parallel, to using NMR to reduce the number of compounds for crystallographic experiments. This development increases flexibility in our hit generation procedure, allowing tailoring to the target of interest and maximizing the structural information available for lead optimization.

There are a number of NMR methods used to identify ligand binding to proteins, which can be divided into target-detected methods such as SAR by NMR [8] and ligand-detected methods such as transverse relaxation NMR (T2) [48, 49], saturation transfer difference NMR (STD-NMR) [50], and water–ligand observed via gradient spectroscopy (Water-LOGSY) [51, 52]. In Water-LOGSY, the magnetisation of the bulk water is partially and selectively transferred to the free ligand via close proton–proton contacts in the hydrated protein–ligand complex. The resonances of bound ligands appear as positive peaks in the spectrum, whereas the peaks from the unbound hydrated ligands are negative and tend to be weaker [53]. In common with other ligand-detected methods, Water-LOGSY has the advantage that it does not require isotope labeling of the protein, uses relatively small amounts of protein (approximately 1–50 µM in a 0.5 ml sample) and can detect binding at ligand concentrations at or below the dissociation constant. Since it detects ligand resonances, it is possible to screen cocktails of fragments without the need for elaborate deconvolution which, together with an affinity range from 10^{-3} M to approximately 10^{-7} M, makes it an ideal method for fragment screening. However, a positive signal in a Water-LOGSY experiment is not always coupled with ligand binding, but can also be a sign of aggregation of the ligand [22]. Furthermore, the absence of a LOGSY signal can also be a sign of a compound binding with a slow off-rate, preventing the build-up of magnetization on the free ligand. Finally, the high concentrations of ligands typically used and the sensitivity of these experiments to protein binding may lead to signals from nonspecific binding to secondary sites on the studied target.

10.3.1
NMR Screening Using Water-LOGSY

We screened our Drug Fragment Library, Focused Kinase Library and Privileged Fragment Library against a variety of targets including different kinases and a serine protease, using the Water-LOGSY method. Where possible, we set up the experiments in a competition format in which an initial screen for fragment binding was followed by a displacement step with a known tight binder to discriminate fragments binding at the protein active site from nonspecific binders. In order to identify these binders in a cocktail, all cocktail spectra were compared with spectra of the single fragments in aqueous buffer obtained in the quality control process of the libraries. For example, the NMR screening of our Focused Kinase Library against the mitogen activated protein kinase p38 was carried out in cocktails of four fragments. Pyridinylimidazole SB203580, which binds in the p38 ATP binding site with an IC_{50} of 48 nM [54], was used to displace bound fragments in a subsequent competition step. Fig. 10.7a shows the NMR data for a typical cocktail

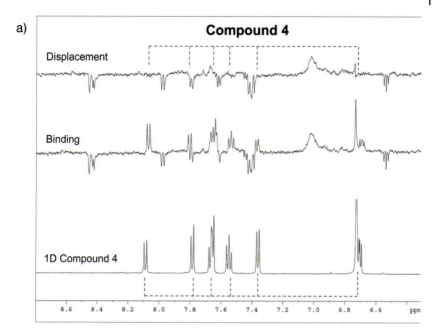

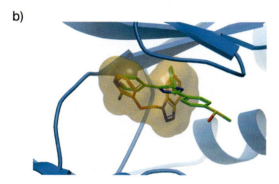

Fig. 10.7
(a) NMR spectra of the NMR screening of compound **4**. The bottom spectrum shows the 1D spectrum of 500 µM compound **4** dissolved in 50 mM phosphate/D$_2$O buffer at pH 7.4. The middle spectrum is the LOGSY spectrum of the p38 solution supplemented with the cocktail containing compound **4**. Binding of compound **4** is revealed by the positive peaks in the spectrum, which correspond to the peaks in its 1D spectrum. The top spectrum shows the LOGSY spectrum of the p38–cocktail solution after addition of SB203580. Displacement of compound **4** is shown by the absence of the positive peaks in the spectrum. (b) Crystallographic confirmation of active site binding of compound **4**. p38 is displayed as a ribbon diagram in blue. Compound **4** is shown with carbon atoms in orange, nitrogen atoms in dark blue, oxygen atoms in red, and its molecular surface in semitransparent orange. The superimposed SB203580 (PDB ID code 1a9u) is shown with carbon atoms in green, oxygen atoms in red, nitrogen atoms in dark blue, sulfur atoms in yellow, and fluorine atoms in purple.

from the Focused Kinase Library in which the middle trace shows the LOGSY spectrum of the protein–cocktail sample. The positive peaks in the spectrum are indicative of binding of one of the compounds which was identified from the 1D spectrum displayed at the bottom. Active site binding was confirmed by the upper LOGSY spectrum, which shows the disappearance of the positive compound peaks as a result of displacement with the tight active site binder. Detailed structural information, obtained by crystallographic screening, confirmed compound 4 as an active site binder which interacts with the hinge region and the gatekeeper pocket (Fig. 10.7b). Superposition of the p38 structure with SB203580 and compound, respectively, reveals an extensive overlap in their binding modes, illustrating the NMR displacement step in atomic detail.

10.3.2
Complementarity of X-ray and NMR Screening

The evolution of Pyramid to incorporate NMR not only provides extra flexibility in our fragment discovery capabilities, but also allows direct comparisons with crystallographic screening. Although in many cases the same hits are obtained through X-ray and NMR screening, we found that exploiting the strengths of each technique by their use in combination allows the detection of more hits than would be identified through either technique alone. For example, screening of our Focused Kinase Library against p38 using X-ray crystallography yielded a hit rate that was 4-fold lower than that for a typical kinase target. However, NMR screening of the Focused Kinase Library against p38 yielded a hit rate which was comparable to hit rates observed for NMR screening this library against other kinase targets. This may reflect some constraints of the p38 protein when in the crystal environment. In contrast, for the serine protease the number of fragment hits identified using crystallography was considerably higher than the number identified using NMR. This is probably due to the fact that the crystallographic screening is often able to reliably detect extremely weak binders, which may fall outside the potency limits for NMR experiments.

In addition to performing crystallographic and NMR fragment screening in parallel against a target, in some cases we have also used NMR as a pre-filter to select compounds for crystallographic analysis. For example, NMR screening of the Drug Fragment Library and Focused Kinase Library against one of our other kinase targets yielded 86 hits and 32 hits, respectively, and was completed in only 3 weeks of NMR time, including data processing and analysis. Of the 32 Focused Kinase Library NMR hits that were tested in X-ray crystallography, structures were obtained for 19 fragments. Of the 17 DFL NMR-hits that were progressed to crystallography, 13 protein–ligand structures were obtained. In most cases, those NMR hits that failed to generate X-ray structures were of lower potency or had poorer solubility than the average. The high number of progressed NMR-hits yielding X-ray structures in a short period of time further illustrates the advantages obtained by combining the speed of NMR screening with the highly detailed structural information obtained by crystallography. Furthermore, using NMR spectroscopy and X-ray crystallography in combination, we were able to identify secondary binding sites on the protein surface, which would have been difficult to confirm by using either technique on its own.

10.4
Conclusions

Despite increasing attempts to improve productivity in the pharmaceutical industry, the application of new technologies to drug discovery has frequently had limited impact. In contrast, fragment-based drug discovery has rapidly established itself as showing early promise. We have applied our integrated fragment-based approach, Pyramid, to a number of protein targets and have identified several chemically attractive hits. Hits from Pyramid have proved highly amenable to rapid optimization, with a low compound attrition rate. In addition, we have shown that fragment binding can be driven by the formation of hydrophobic as well as electrostatic interactions with the protein target, and can also induce conformational change.

The use of crystallography to screen fragment libraries brings a number of advantages, but has traditionally been impractical due to low throughput. We have approached this issue through the combined use of compound cocktailing and automated data processing and ligand fitting. These steps significantly reduce the time needed to screen fragment libraries and have transformed crystallography into a highly efficient technique, suitable for use as a primary screen. The Pyramid approach has recently evolved to include the use of NMR and other biophysical techniques. In this way, it allows the detection of fragment binding in systems not amenable to high-throughput crystallography and can provide additional hits which complement those discovered using crystallography alone. The integration of these techniques with medicinal chemistry, bioassays, and DMPK has produced a highly effective drug discovery engine which has produced clinical candidates within a short time-frame.

Acknowledgments

The Authors gratefully acknowledge the support of many people at Astex Therapeutics who have contributed to aspects of the work presented in this chapter. We also thank Chris Murray, Miles Congreve, Ian Tickle, and Tom Blundell for helpful discussions. T.G.D. and R.L.M.v.M. contributed equally to this work.

References

1 Campbell, S. F. **2000**, Science, art and drug discovery: a personal perpsective, *Clin.Sci.* 99, 255–260.
2 Oprea, T. I., Davis, A. M., Teague, S. J., Leeson, P. D. **2001**, Is there a difference between leads and drugs? A historical perspective, *J.Chem.Inf.Comput.Sci.* 41, 1308–1315.
3 Teague, S. J., Davis, A. M., Leeson, P. D., Oprea, T. **1999**, The design of leadlike combinatorial libraries, *Angew.Chem. Int.Ed.Engl.* 38, 3743–3748.
4 Lipinski, C. A., Lombardo, F., Dominy, B. W., Feeney, P. J. **2001**, Experimental and computational approaches to estimate solubility and permeability in drug discovery and development settings, *Adv.Drug Deliv.Rev.* 46, 3–26.
5 Blundell, T. L., Abell, C., Cleasby, A., Hartshorn, M. J., Tickle, I. J., Parasini, E., Jhoti, H. **2002**, High-throughput X-ray crystallography for drug discovery, in *Drug Design: Special Publication*, ed. Flower, D. R., Royal Society of Chemistry, Cambridge, pp. 53–59.
6 Erlanson, D. A., McDowell, R. S., O'Brien, T. **2004**, Fragment-based drug discovery, *J.Med.Chem.* 47, 3463–3482.
7 Rees, D. C., Congreve, M., Murray, C. W., Carr, R. **2004**, Fragment-based lead discovery, *Nat.Rev.Drug Discov.* 3, 660–672.
8 Shuker, S. B., Hajduk, P. J., Meadows, R. P., Fesik, S. W. **1996**, Discovering high-affinity ligands for proteins: SAR by NMR, *Science* 274, 1531–1534.
9 Hopkins, A. L., Groom, C. R., Alex, A. **2004**, Ligand efficiency: a useful metric for lead selection, *Drug Discov. Today* 9, 430–431.
10 Murray, C. W., Verdonk, M. L. **2002**, The consequences of translational and rotational entropy lost by small molecules on binding to proteins, *J.Comput.Aided Mol.Des.* 16, 741–753.
11 Carr, R., Jhoti, H. **2002**, Structure-based screening of low-affinity compounds, *Drug Discov.Today* 7, 522–527.
12 Nienaber, V. L., Richardson, P. L., Klighofer, V., Bouska, J. J., Giranda, V. L., Greer, J. **2000**, Discovering novel ligands for macromolecules using X-ray crystallographic screening, *Nat.Biotechnol.* 18, 1105–1108.
13 Hann, M. M., Leach, A. R., Harper, G. **2001**, Molecular complexity and its impact on the probability of finding leads for drug discovery, *J.Chem.Inf.Comput.Sci.* 41, 856–864.
14 Blundell, T. L., Jhoti, H., Abell, C. **2002**, High-throughput crystallography for lead discovery in drug design, *Nat. Rev. Drug Discov.* 1, 45–54.
15 Fejzo, J., Lepre, C. A., Peng, J. W., Bemis, G. W., Ajay, Murcko, M. A., Moore, J. M. **1999**, The SHAPES strategy: an NMR-based approach for lead generation in drug discovery, *Chem. Biol.* 6, 755–769.
16 Metz, G., Ottleben, H., Vetter, D. **2003**, Small molecule screening on chemical microarrays, in *Protein-Ligand Interactions from Molecular Recognition to Drug Design*, ed. Bohn, H. J., Schneider, G., Wiley–VCH, Weinheim, pp. 213–236.
17 Erlanson, D. A., Wells, J. A., Braisted, A. C. **2004**, Tethering: fragment-based drug discovery, *Annu. Rev. Biophys. Biomol. Struct.* 33, 199–223.
18 The following are trademarks of Astex Technology: Pyramid, AutoSolve, Astex Rule of Three, AstexViewer.
19 Hartshorn, M. J., Murray, C. W., Cleasby, A., Frederickson, M., Tickle, I. J., Jhoti, H. **2005**, Fragment-based lead discovery using X-ray crystallography, *J.Med.Chem.* 48, 403–413.
20 Veber, D. F., Johnson, S. R., Cheng, H. Y., Smith, B. R., Ward, K. W., Kopple, K. D. **2002**, Molecular properties that influence the oral bioavailability of drug candidates, *J.Med.Chem.* 45, 2615–2623.
21 Congreve, M., Carr, R., Murray, C., Jhoti, H. **2003** A "rule of three" for fragment-based lead discovery? *Drug Discov. Today* 8, 876–877.
22 Baurin, N., Aboul-Ela, F., Barril, X., Davis, B., Drysdale, M., Dymock, B., Finch, H., Fromont, C., Richardson, C., Simmonite, H., Hubbard, R. E. **2004**, Design and characterization of libraries of molecular fragments for use in NMR screening against protein targets, *J. Chem. Inf. Comput. Sci.* 44, 2157–2166.

23 Bemis, G. W. Murcko, M. A. **1996**, The properties of known drugs. 1. Molecular frameworks, *J. Med. Chem.* 39, 2887–2893.

24 Bemis, G. W., Murcko, M. A. **1999**, Properties of known drugs. 2. Side chains, *J. Med. Chem.* 42, 5095–5099.

25 Davies, T. G., Bentley, J., Arris, C. E., Boyle, F. T., Curtin, N. J., Endicott, J. A., Gibson, A. E., Golding, B. T., Griffin, R. J., Hardcastle, I. R., Jewsbury, P., Johnson, L. N., Mesguiche, V., Newell, D. R., Noble, M. E., Tucker, J. A., Wang, L., Whitfield, H. J. **2002**, Structure-based design of a potent purine-based cyclin-dependent kinase inhibitor, *Nat. Struct. Biol.* 9, 745–749.

26 Watson, P., Verdonk, M. L., Hartshorn, M. J. **2003**, A web-based platform for virtual screening, *J. Mol. Graph. Model.* 22, 71–82.

27 Verdonk, M. L., Cole, J. C., Hartshorn, M., Murray, C. W., Taylor, R. D. **2003**, Improved protein-ligand docking using GOLD, *Proteins* 52, 609–623.

28 Jones, G., Willett, P., Glen, R. C. **1995**, Molecular recognition of receptor sites using a genetic algorithm with a description of desolvation, *J. Mol. Biol.* 245, 43–53.

29 Baxter, C. A., Murray, C. W., Clark, D. E., Westhead, D. R., Eldridge, M. D. **1998**, Flexible docking using Tabu search and an empirical estimate of binding affinity, *Proteins* 33, 367–382.

30 Groves, M. R., Yao, Z. J., Roller, P. P., Burke, T. R. Jr., Barford, D. **1998**, Structural basis for inhibition of the protein tyrosine phosphatase 1B by phosphotyrosine peptide mimetics, *Biochemistry* 37, 17773–17783.

31 Verlinde, C. L. M. J., Kim, H., Bernstein, B. E., Mande, S. C., Hol, W. G. J. **1997**, Antitrypanosomiasis drug development based on structures of glycolitic enzymes, in *Structure-Based Drug Design*, ed. Veerapandian, P., Marcel Dekker, New York, pp. 365–394.

32 Blakely, M. P., Cianci, M., Helliwell, J. R., Rizkallah, P. J. **2004**, Synchrotron and neutron techniques in biological crystallography, *Chem. Soc. Rev.* 33, 548–557.

33 Muchmore, S. W., Olson, J., Jones, R., Pan, J., Blum, M., Greer, J., Merrick, S. M., Magdalinos, P., Nienaber, V. L. **2000**, Automated crystal mounting and data collection for protein crystallography, *Structure Fold. Des.* 8, R243–R246.

34 Sharff, A. J. **2004**, High throughput crystallography on an in-house source, using ACTOR, *Rigaku J.* 20, 10–12.

35 van Montfort, R. L., Congreve, M., Tisi, D., Carr, R., Jhoti, H. **2003**, Oxidation state of the active-site cysteine in protein tyrosine phosphatase 1B, *Nature* 423, 773–777.

36 McPhillips, T. M., McPhillips, S. E., Chiu, H. J., Cohen, A. E., Deacon, A. M., Ellis, P. J., Garman, E., Gonzalez, A., Sauter, N. K., Phizackerley, R. P., Soltis, S. M., Kuhn, P. **2002**, Blu-ice and the distributed control system: software for data acquisition and instrument control at macromolecular crystallography beamlines, *J. Synchrotron. Radiat.* 9, 401–406.

37 Pflugrath, J. W. **1999**, The finer things in X-ray diffraction data collection, *Acta Crystallogr.* D55, 1718–1725.

38 Navaza, J. **2004**, AMoRe: an automated package for molecular replacement, *Acta Crystallogr.* A50, 157–163.

39 Vagin, A., Teplyakov, A. **1997**, MOLREP: an automated program for molecular replacement, *J. Appl. Crystallogr.* 30, 1022–1025.

40 Gasteiger, J., Rudolph, C., Sadowski, J. **2004**, Automatic generation of 3D-atomic coordinates for organic molecules, *Tetrahedron Comput. Methodol.* 3, 537–547.

41 Hartshorn, M. J. **2002**, AstexViewer: a visualisation aid for structure-based drug design, *J. Comput. Aided Mol. Des.* 16, 871–881.

42 Fischer, P. M., Lane, D. P. **2000**, Inhibitors of cyclin-dependent kinases as anticancer therapeutics, *Curr. Med. Chem.* 7, 1213–1245.

43 Knockaert, M., Greengard, P., Meijer, L. **2002**, Pharmacological inhibitors of cyclin-dependent kinases, *Trends Pharmacol. Sci.* 23, 417–425.

44 Sausville, E. A., Zaharevitz, D., Gussio, R., Meijer, L., Louarn-Leost, M., Kunick, C., Schultz, R., Lahusen, T., Headlee, D., Stinson, S., Arbuck, S. G., Senderowicz,

A. **1999**, Cyclin-dependent kinases: initial approaches to exploit a novel therapeutic target, *Pharmacol.Ther.* 82, 285–292.

45 Salituro, F. G., Germann, U. A., Wilson, K. P., Bemis, G. W., Fox, T., Su, M. S. **1999**, Inhibitors of p38 MAP kinase: therapeutic intervention in cytokine-mediated diseases, *Curr.Med.Chem.* 6, 807–823.

46 Gill, A. L., Frederickson, M., Cleasby, A., Woodhead, S. J., Carr, M. G., Woodhead, A. J., Walker, M. T., Congreve, M. S., Devine, L. A., Tisi, D., O'Reilly, M., Seavers, L. C. A., Davis, D. J., Curry, J., Anthony, R., Padova, A., Murray, C. W., Carr, R. A. E., Jhoti, H. **2004**, Identification of novel p38α MAP kinase inhibitors using fragment-based lead generation, *J. Med. Chem.* 2004

47 Fenton, J. W., Ofosu, F. A., Moon, D. G., Maraganore, J. M. **1991**, Thrombin structure and function: why thrombin is the primary target for antithrombotics, *Blood Coagul. Fibrinolysis* 2, 69–75.

48 Jahnke, W., Rudisser, S., Zurini, M. **2001**, Spin label enhanced NMR screening, *J.Am.Chem.Soc.* 123, 3149–3150.

49 Sarazin, M., Chauvet-Derhodhile, M., Bourdeaux-Pontier, M., Briand, C. **1978**, NMR Study of the interaction between methotrexate and human serum albumin: the nature of the complexation site on the drug., *Proc. Eur. Conf. NMR Macromol.* 1978, 503–508.

50 Mayer, M., Meyer, B. **1999**, Characterization of ligand binding by saturation transfer difference NMR spectra, *Angew.-Chem.Int.Ed.Engl.* 38, 1784–1788.

51 Dalvit, C., Pevarello, P., Tato, M., Veronesi, M., Vulpetti, A., Sundstrom, M. **2000**, Identification of compounds with binding affinity to proteins via magnetization transfer from bulk water, *J.Biomol.NMR* 18, 65–68.

52 Dalvit, C., Fogliatto, G., Stewart, A., Veronesi, M., Stockman, B. **2001**, WaterLOGSY as a method for primary NMR screening: practical aspects and range of applicability, *J.Biomol.NMR* 21, 349–359.

53 Stockman, B. J., Dalvit, C. **2002**, NMR screening techniques in drug discovery and drug design, *Progr. Nucl. Magn. Res. Spect.* 41, 187–231.

54 Wang, Z., Canagarajah, B. J., Boehm, J. C., Kassisa, S., Cobb, M. H., Young, P. R., Abdel-Meguid, S., Adams, J. L., Goldsmith, E. J. **1998**, Structural basis of inhibitor selectivity in MAP kinases, *Structure* 6, 1117–1128.

11
Fragment-based Lead Discovery and Optimization Using X-Ray Crystallography, Computational Chemistry, and High-throughput Organic Synthesis

Jeff Blaney, Vicki Nienaber, and Stephen K. Burley

11.1
Introduction

Traditional drug discovery usually begins with a search for small molecule "hits" that demonstrate modest (IC$_{50}$ ≤ 10 μM) *in vitro* activity against the molecular target of interest. Such hits are subsequently optimized into preclinical drug candidates using iterative, trial-and-error methods and/or structure-directed design. The most commonly used approaches for finding hits include high-throughput screening (HTS) of large compound libraries (typically 100 000–2 000 000 compounds) or modification of substrate analogs and/or published active compounds. Although these methods have yielded a large number of successfully marketed drugs, optimization of HTS hits into clinical candidates remains a considerable challenge. Major shortcomings of traditional approaches include an inherent lack of chemical diversity for the initial hits and poor compliance of most hits with what we now recognize as advantageous lead-like properties (Chapter 1) [1–4]. Limitations in diversity of the screening library effectively biases sampling of potential starting points for drug discovery, and may, therefore, not yield the best lead series. Poor compliance with the requirement for lead-like properties often complicates and prolongs the lead optimization process, and certainly contributes to high failure rates seen in pharmaceutical discovery and development. New methods are, therefore, required if we are to improve process efficiency.

An emerging method in modern drug discovery utilizes screening for small fragments of drug molecules. Screening of fragments has many advantages, the most important of which is an increase in the diversity of lead-like starting points at a cost of sampling relatively small numbers of compounds. For example, a 1000-compound fragment library can give rise to over 160 × 10^6 readily accessible analogs, which significantly exceeds the size of a typical HTS screening collection. Moreover, small fragments have an increased probability of binding to a given target than do larger, more complex molecules [1]. Fragments are also likely to be more efficient ligands (i.e., more of their constituent atoms participate in interactions with the target protein). Because they contain fewer compounds, fragment

Fragment-based Approaches in Drug Discovery. Edited by W. Jahnke and D. A. Erlanson
Copyright © 2006 WILEY-VCH Verlag GmbH & Co. KGaA, Weinheim
ISBN: 3-527-31291-9

libraries may be custom-assembled to maximize fragment diversity and subsequently derived lead diversity.

In contrast, typical proprietary screening libraries are often biased towards certain structural classes, because these collections are composed of molecules synthesized for targets of historical importance, rather than molecules chosen to sample lead-like chemical space. The number of potential drug-like molecules is predicted to be $\sim 10^{60}$, which actually exceeds estimates for the total number of atoms comprising the universe [5]. Typical HTS libraries also consist of larger (i.e., less efficient) molecules that usually yield more potent starting points for synthetic chemistry than fragments (i.e., $IC_{50} < \sim 10$ µM vs $IC_{50} < \sim 1$ mM). Optimization of these larger molecules is, however, often complicated by the need to identify and remove functional groups to minimize molecular weight and hydrophobicity, while other functional groups must be simultaneously added or modified to increase activity. Thus, optimization of larger HTS hits into clinical candidates may require retrospective disassembly into smaller fragments. Therefore, starting lead optimization with a smaller, more efficient, albeit more weakly bound fragment often represents a more efficient approach to discovery of clinically viable lead candidates.

Fragment-based lead discovery has been underway for more than a decade. Initial reports described computational screening of fragments, using tools such as DOCK [6, 7] or MCSS [8] (Chapter 7). The concept of linking fragments together to create lead compounds was described as early as 1992 by Verlinde et al. [9]. Petsko, Ringe, and co-workers first reported an experimental approach, wherein small organic solvents were soaked into crystals to identify functional groups that could be combined into a lead compound [10] (Chapter 4). Another experimental technique for detecting fragment binding is the "SAR by NMR" technique, pioneered by Abbott [11] (Chapter 9). In this spectroscopic approach, fragments are detected by NMR screening and subsequently linked by methods similar to those described by Verlinde and co-workers [9].

Routine application of crystallography to detect and identify fragment hits using shape-diverse mixtures was also pioneered at Abbott [12]. These hits have been optimized by both traditional structure-based drug design and structure-directed parallel synthesis [13–16]. Crystallographic screening is ideally suited to fragment-based lead discovery, because the three-dimensional structure of the hit interacting with the target is obtained upon detection. Each hit can be validated immediately by establishing that it binds to the protein target in a well defined orientation that is compatible with synthetic optimization. Without three-dimensional structure validation, optimization of weakly binding fragments is extremely challenging, because of the high propensity for non-specific binding and false positives detected by biochemical assays. Hence, application of crystallographic screening and/or co-crystallization with fragment hits makes the fragment-based approach an eminently practical and highly successful means of discovering novel drug leads.

11.2
Overview of the SGX Structure-driven Fragment-based Lead Discovery Process

SGX Pharmaceuticals, Inc. (formerly Structural GenomiX, Inc.) has developed a lead discovery process that combines state-of-the-art structural biology tools, including a dedicated third-generation synchrotron beamline [SGX-CAT at the Advance Photon Source (APS)], parallel organic synthesis, and proprietary computational chemistry software. In addition to these tools, a diverse screening library of ~1000 lead-like compound fragments has been assembled. Each component of the library has built-in synthetic handles to aid in rapid elaboration of structurally validated fragment hits.

The SGX process (*FAST, fragments of active structures*) encompasses the following steps: (1) screening by crystallography using shape-diverse mixtures of fragments, (2) examination of crystallographic data to identify fragment "hits", (3) examination of hits *in situ* to identify structurally accessible synthetic handles, (4) virtual construction of "linear" chemical libraries that derivatize each handle, (5) computational analysis of virtual libraries, (6) visual examination of libraries with favorable calculated binding free energies, (7) library synthesis, characterization, and purification, (8) analysis of the results of library synthesis by crystallography and biochemical assays, (9) synthesis of combinatorial libraries that utilize the better fragment elaborations at each available synthetic handle, and (10) crystallographic and biochemical and cell-based assay characterization of the resulting lead series.

Approximately one-half of the compounds in the SGX fragment library contain one or more bromine atoms to facilitate routine synthetic elaboration of crystallographic screening hits. Furthermore, tuning the SGX-CAT beamline to the appropriate X-ray wavelength allows detection of anomalous dispersion signals unique to bromine, permitting unambiguous bromine atom identification in experimental electron density maps. As a result of incorporating bromine atoms in the crystallographic fragment screening library and collecting data at the appropriate wavelength, detection and validation of crystallographic screening hits from mixtures has become routine.

Screening can be completed in one to two days of X-ray beam time by dividing the ~1000-compound SGX fragment library into 100 ten-compound shape-diverse mixtures. Typically, 100–200 preformed target protein crystals are soaked with the 100 fragment mixtures (1–2 soaking attempts per mixture). The soaked crystals are then frozen by immersion in liquid nitrogen and shipped to the SGX-CAT beamline via courier. Diffraction data from each soaked crystal are automatically collected, reduced, and analyzed using custom software tools. Our hit identification software is flexible. In many cases, the program correctly picks out the bound fragment from the shape-diverse mixture and fits the individual ligand to the observed electron density feature. For more challenging cases, the software supports visual inspection of electron density maps and semi-automated ligand fitting. These processes suffice for identifying the crystallographic hit in most experiments. For infrequent ambiguous situations, single compound soaks are used to confirm hit identification.

The goal of *FAST* is to select fragment hits that can be modified at more than one site (chemical handle) and to prepare, in parallel, "linear" libraries with single site elaborations for all accessible chemical handles. Computational tools support both library design and synthetic prioritization. After the linear libraries are synthesized, the resulting compounds are evaluated by both biochemical assays and X-ray crystallography to provide a *bona fide* three-dimensional structure–activity relationship (SAR). Co-crystal structures of elaborated fragments bound to the target permit direct monitoring of the mode of binding and detection of protein conformational changes. Promising single site elaborations at each handle are then multiplexed to give a combinatorial library of optimized compounds where, for an ideal system, the binding energies of the single-site elaborations are additive. This phenomenon has been demonstrated in various SGX programs and reported by others [17]. Herein, we describe details of the SGX-*FAST* process drawing on our preclinical programs to exemplify each stage of our lead discovery/lead engineering process.

11.3
Fragment Library Design for Crystallographic Screening

11.3.1
Considerations for Selecting Fragments

Recent studies of hit-to-lead optimization proposed a general definition of "lead-like" properties (Chapter 1) that increase the probability of successful optimization of hits to clinical candidates and successful prosecution of clinical development. General conclusions from these studies provided guidance for the design of the SGX fragment library.

Drug- or Lead-likeness Statistical Studies
Lipinski's "rules" [18] describe properties of approved drugs: molecular weight (MW) < 500, calculated log P (ClogP) < 5, < 5 hydrogen-bond donors, and < 10 nitrogens + oxygens. However, these rules are not appropriate for either hits or leads [3, 4]. Hits usually increase in molecular weight, ClogP, and in number of rings and freely rotatable bonds during initial lead optimization and during subsequent clinical candidate optimization.

Screening hits and leads should, therefore, be smaller than the MW ranges embodied in Lipinski's rules. Teague et al. [4] initially proposed MW < 350 and ClogP < 3.0. Hann and Oprea [2] more recently proposed that lead-like molecules should have the following properties: MW $\leq$ 460, ClogP < 4.2, $\leq$ 10 freely rotatable bonds, $\leq$ 4 rings, $\leq$ 5 hydrogen-bond donors, and $\leq$ 9 hydrogen-bond acceptors. "Lead-like" properties were originally proposed for molecules with activities in the low µM range derived from classic HTS or combinatorial chemistry approaches.

Fragment hits have activities in the low µM to low mM range and, therefore, require selection criteria that focus on yet smaller, simpler molecules. Hann and Oprea [2] proposed a "reduced complexity" screening set, with the following properties: MW <350, ClogP ≤2.2, ≤6 freely rotatable bonds, ≤22 heavy atoms, ≤3 hydrogen-bond donors, and ≤8 hydrogen-bond acceptors. Congreve et al [19] proposed a similar "Rule of Three": MW <300, ClogP <3, <3 hydrogen-bond donors, and <3 freely rotatable bonds.

Retrospective Analyses of Clinical Development Outcomes
Clinical trials select for smaller molecules; the larger the molecule, the lower the chance it will give satisfactory results in Phase 3 studies and be approved by the FDA [20, 21]. Specifically, recently published studies have demonstrated that clinical candidates with MW <400 have a 50% greater probability of obtaining approval than those with MW ≥400.

Retrospective Analyses of HTS Outcomes
The probability of finding a screening hit is inversely proportional to compound complexity (i.e., the number of hydrogen bond donors and acceptors, rotatable bonds, rings, MW, etc.) [1]. Smaller, simpler compounds have higher hit rates, but lower target-binding affinities. Larger, more complex molecules have lower hit rates, but higher target-binding affinities.

Additional Considerations
We exploited two other important considerations when establishing our approach to fragment-based lead discovery and optimization. First, HTS and classic drug discovery hits are not infrequently incompatible with efficient follow-on syntheses, and may require substantial custom, labor-intensive chemistry for optimization. In practice, the probability of optimizing a hit increases with the synthetic amenability of the hit to follow-up elaboration. We therefore limited the SGX fragment screening library to compounds that support rapid, 48 or 96 at a time, automated parallel synthesis using well established synthetic routes. Second, aromatic bromine is a particularly useful substituent for an X-ray crystallographic approach to fragment discovery and optimization. The anomalous dispersion signal from one or more bromine atoms greatly assists in structural validation of fragment screening hits. In addition, it can be used to form carbon–carbon bonds via Suzuki coupling and related reactions.

11.3.2
SGX Fragment Screening Library Selection Criteria

Selection of compounds for the SGX fragment screening library was based on the following criteria:
- Minimizing MW, ClogP, and compound complexity.
- Maximizing synthetic accessibility by requiring fragments to include two or three synthetic handles.
- Including a substantial fraction of brominated compounds.
- Excluding groups that are incompatible with drug-like properties [22], except for a few specific synthetic handles (e.g., $ArNO_2$, $ArNH_2$).

- Maximizing diversity by ring system (both unique rings and rings with unique substitution patterns) and 4-point pharmacophore analysis [23].
- Including known "drug-like" ring systems represented in the MDDR [24].
- Ensuring high solubility compatible with crystallographic screening at fragment concentrations of ~10 mM.

11.3.3
SGX Fragment Screening Library Properties

The properties of the SGX fragment screening library of ~1000 compounds can be summarized as follows:
- 100% have ≤16 non-hydrogen atoms with an average MW of 174. (Bromine is treated as a methyl group for this MW calculation, because it will ultimately be replaced by a carbon during fragment elaboration.)
- 100% have ≥1 ring, with >200 unique ring systems; and ~30% of MDDR ring systems are present.
- 100% have ≥2 synthetic "handles", chosen from 25 substituents, each of which is compatible with parallel synthesis using at least several hundred commercially available reagents.
- 50% contain aromatic bromine.
- 90% have ≤3 hydrogen bond acceptors.
- 90% have ≤3 hydrogen bond donors.
- 94% have ≤3 freely rotatable bonds.
- 98% satisfy rules that exclude molecules and substituents that are not either drug-like or appropriate for screening, as judged by Hann and co-workers [22].
- 90% have clogD ≤2 (calculated at pH 7.4 [25]).
- 60% are highly soluble (≥500 µM).

11.3.4
SGX Fragment Screening Library Diversity: Theoretical and Experimental Analyses

Various considerations led to the selection of ~1000 compounds for the SGX fragment screening library.

Coverage of Lead-like Space
As discussed earlier, the estimated number of possible drug-like molecules [5] that could be included in a conventional HTS library is ~10^{60}. In contrast, the estimated number of possible lead-like molecules with MW <160 is only about 14×10^6 [26]. A fragment library containing a modest number of compounds (i.e., ~1000) can therefore be used to sample lead-like space much more efficiently than a conventional HTS library samples drug-like space. A typical HTS library, containing 10^5–10^6 compounds, encompasses ~10^{-55} of the total estimated drug-like space, whereas a fragment library of only 1000–10 000 lead-like compounds (MW <160) represents ~0.001–0.01% of the total estimated lead-like space. We can also expect that fragment screening hit rates will be higher than those for con-

ventional HTS [1]. Taken together, these arguments suggest that fragment libraries can be much smaller than HTS libraries, while providing much better sampling of the total lead-like chemical space.

Observed Hit Rates for Crystallographic Screening
Initial crystallographic fragment screening experiments at SGX conducted with a small, diverse pilot library of 80 compounds gave hit rates of ~1–5%, with biochemical activities in the low mM to high µM range. This outcome suggests that crystallographic screening of a modest library of ~1000 carefully selected compounds would give 10–50 hits for most targets, which should be sufficient to select 4–5 structurally diverse fragments for linear library elaboration. These results also suggest that it would be neither necessary nor desirable to bias compound selection for our fragment screening library towards any particular class of target in order to obtain acceptable hit rates. We therefore elected to limit our fragment screening library to small, structurally diverse, lead-like molecules that are amenable to rapid synthetic elaboration.

Potential Chemical Diversity
The number of commercial reagents available for each type of chemical handle represented in our fragment screening library ranges from a minimum of ~400 to a maximum of ~40 000. A cursory analysis of the potential diversity realizable on synthetic elaboration of each one of the ~1000 SGX fragment screening library compounds documents ready access to an enormous number of elaborated compounds. In the worst case scenario (i.e., utilization of only two chemical handles with only 400 possible modifications at each handle), ~1000 fragments can be elaborated into ~160×10^6 distinct compounds. In the best case scenario (i.e., utilization of three chemical handles with 40 000 possible modifications at each handle), ~1000 fragments can be elaborated into ~64×10^{15} distinct compounds, which is comparable to the age of the universe in minutes. Crystallographic screening of fragment libraries identifies starting points for lead optimization that permit strategic, cost-effective access to enormous potential chemical diversity, without the need to synthesize excessive numbers of compounds.

11.4
Crystallographic Screening of the SGX Fragment Library

Historically, targets have been screened for fragment binding by various methods including NMR [11] (Chapter 9), X-ray crystallography [12], and mass spectrometry [27]. In some cases, these methods have complementary strengths and weaknesses and may be employed together to improve the probability of discovering a favorable lead candidate. At SGX, each target is enabled for both crystallographic screening and complementary evaluation by one of the following methods: biochemical activity assays and gel filtration–mass spectrometry and surface plasmon resonance detection of target-fragment complexes.

11.4.1
Overview of Crystallographic Screening

To conduct crystallographic screening, SGX has adapted a method [12] in which the ~1000-compound screening library is divided into 100 mixtures of ten shape-diverse compounds. In addition to differentiating mixture components by shape, compounds are further differentiated by the presence of bromine atoms for approximately 50% of library components (i.e., ~five brominated compounds per mixture). Using the proprietary SGX-CAT synchrotron beamline at the APS, bromine-containing compounds can be readily differentiated from non-bromine-containing compounds by conducting diffraction experiments at an X-ray wavelength corresponding to the bromine element absorption edge (0.9200 eV). Under these experimental conditions, an additional signal can be observed for each bromine atom. Hence, for about half of the compounds in each mixture, the presence of the bromine signal facilitates interpretation of the electron density difference map for bound fragment identification. Control experiments were conducted to optimize detection of the anomalous dispersion signal, without significant bromine atom displacement during the time required for X-ray diffraction data collection.

A typical shape-diverse mixture is depicted in Fig. 11.1A with calculated (theoretical) electron density features corresponding to each mixture component. Examination of the mixture reveals the utility of incorporating bromine atoms into the library as a source of diversity. Specifically, comparison of compounds 4, 8, and 9 shows similar three dimensional shapes, which may be differentiated by variation in both the substitution pattern of the bromine atoms and in the number of bromine substituents.

Figure 11.1B illustrates the outcome of crystallographic fragment screening, wherein the mixture depicted in Fig. 11.1A was soaked into pre-formed crystals of the kinase domain of a well validated oncology target, BCR-Abl [28]. Crystal screens were conducted with crystals of both wild-type and imatinib-resistant T315I mutant [29] kinase domains. In all, we discovered 11 fragments that interact with both wild-type and drug-resistant BCR-Abl, four of which were taken forward into fragment optimization. One lead series is currently in late pre-clinical development.

Examination of the $(2|F_{observed}|-|F_{calculated}|)$ and anomalous $(|F_+|-|F_-|)$ difference Fourier syntheses, or electron density maps, illustrated in Fig. 11.1B indicates that a compound from the mixture is bound at the active site (i.e., a fragment "hit" was detected) and that this compound contains one bromine atom. Further inspection of the electron density maps indicates that the bromine atom is *para*- to a bulky substituent. Compound 10 (Fig. 11.1A) represents the best match of the electron density features shown in Fig. 11.1B. Hence, the identity of the compound was established both by the presence of a bromine signal and by the shape of the observed electron density feature in the BCR-Abl active site.

Crystallographic screening of shape-diverse mixtures has proven to be a powerful technique, in part because it largely eliminates the need for time-consuming mixture deconvolution experiments. Rarely, however, have we encountered cases in which more weakly binding ligands were "masked" by competition with a

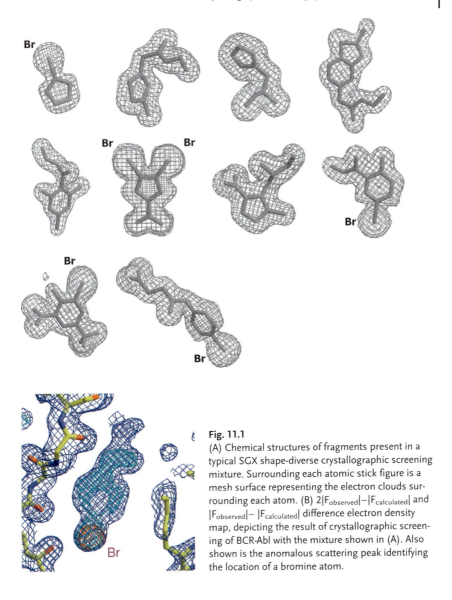

Fig. 11.1
(A) Chemical structures of fragments present in a typical SGX shape-diverse crystallographic screening mixture. Surrounding each atomic stick figure is a mesh surface representing the electron clouds surrounding each atom. (B) $2|F_{observed}|-|F_{calculated}|$ and $|F_{observed}|-|F_{calculated}|$ difference electron density map, depicting the result of crystallographic screening of BCR-Abl with the mixture shown in (A). Also shown is the anomalous scattering peak identifying the location of a bromine atom.

more potent fragment hit within the shape-diverse mixture. In practice, such masked fragments can be detected by re-screening each mixture, giving a screening hit after exclusion of the crystallographically detected bound fragment (i.e., the most potent component). Alternatively, this problem can be overcome by conducting individual crystal-soaking experiments with the other compounds comprising the shape-diverse mixture.

The process of crystallographic screening can be broken down into multiple steps, including: (1) obtaining the initial target protein structure, (2) enabling the target for crystallographic screening, (3) screening of the fragment library, and (4) analysis of screening results. Each of these steps is discussed in detail below, followed by an example of our experience with crystallographic screening of the coagulation enzyme Factor VIIa, which has been targeted by SGX for treatment of cardiovascular disease.

11.4.2
Obtaining the Initial Target Protein Structure

At SGX, *de novo* protein crystal structures are determined using a gene-to-structure platform that was developed to process multiple proteins in parallel. This platform consists of modular robotics and a comprehensive laboratory information management system (LIMS) that facilitates data entry and electronic data capture at all stages of the process. The LIMS system also permits comprehensive data mining for troubleshooting and project management. The SGX gene-to-structure platform has facilitated high-resolution (typically better than ~2 Å) structure determinations for a large number of drug discovery targets, including more than 50 unique human protein kinases and a large number of nuclear hormone receptor ligand-binding domains. Successes include many targets not represented in the public domain Protein Data Bank [30], some of which are regarded as being extremely difficult if not "impossible" to express, purify, and crystallize.

SGX platform robotics encompasses gene cloning, protein expression and purification, crystallization, and structure determination. Most of this work is conducted using 96-well format liquid-handling devices to process multiple expression constructs for many protein targets in parallel. Multiple constructs for a given target typically express various truncations of the N- and C-terminus and/or loop deletions. Precise truncations are defined with the results of bioinformatics analyses of target protein sequences and/or by experimental domain mapping via limited-proteolysis combined with mass spectrometry [31]. For a typical target of unknown structure, a minimum of 20–30 constructs are prepared in multiple expression vectors (encoding N- and C-terminal hexa-histidine tags and a removable N-terminal hexa-histidine Smt3 tag). Well expressed, soluble forms of the target protein are purified in parallel and then tested for crystallization, using a predetermined set of ~1000 crystallization conditions.

This modular, high-capacity platform has produced high resolution X-ray structures for many challenging drug discovery targets. Our many successes with technically challenging human drug discovery targets are a direct consequence of this impressive bandwidth. Rapid, fine sampling during the early stages of the process allows us to express the right truncated form(s) of a difficult target that crystallize, thereby enabling structure determination. At the other extreme, we exploit the bandwidth of the platform to approach a very large number of targets in parallel. The latter strategy proved particularly successful for our studies of the human kinome, described below.

The SGX gene-to-structure platform has been applied to the human kinome with the goal of enabling *de novo* crystal structure determinations of drug targets in this large family. To date, we have determined more than 50 unique human protein kinase structures, approximately half of which are not found in the public domain. Most of these structures are of important drug discovery targets, such as the chronic myeloid leukemia target BCR-Abl (see above), the asthma/inflammation target SYK (discussed below), Met, Aurora, and Ras. The remainder represent important kinase off-targets that are used during lead optimization to guard against incorporation of undesirable cross-kinome effects.

11.4.3
Enabling Targets for Crystallographic Screening

Once an initial crystal structure is obtained, additional experiments are conducted to enable the target for *FAST* fragment screening. This process encompasses the transition from small-scale crystal growth and data collection, required to determine a new crystal structure, to a robust large-scale process, required for target screening using crystallography. Typical requirements for crystallographic screening include the ability to routinely produce and soak crystals on a large-scale (~200 diffraction quality crystals/screen) and to obtain reproducible diffraction data to better than 2.5 Å resolution. In most cases, the crystal form used for *de novo* structure determination suffices for crystallographic screening of the fragment library.

In extreme cases, the process may require using information from the newly determined structure to engineer a new crystal form. Such protein re-engineering may be necessary to improve the crystal stability and/or the packing of target molecules in the crystals, such that the screen may be conducted by soaking compounds through solvent channels in the crystal to the active site. Re-engineering may also be used to improve crystal diffraction quality.

After obtaining a suitable crystal form, the system is validated by soaking "control" compounds known to bind and/or inhibit the target of interest. In the absence of reference inhibitors, substrate analogs, cofactors, or other known ligands (i.e., ATP analogs and staurosporine for protein kinases) serve as controls. If the reference compound(s) is visible in difference electron density maps (e.g., as illustrated in Fig. 11.1B), the soaking system is considered validated. After validation of the crystal form, the ability to soak mixtures into the system is tested. In some cases, soaking conditions must be further optimized to permit efficient soaking of mixtures. Once the soaking process and crystal form are fully validated, the fragment library (~1000 compounds) is soaked into the crystals using 100 mixtures of ten shape-diverse compounds.

11.4.4
Fragment Library Screening at SGX-CAT

Once a target is enabled for crystallographic screening, crystals are prepared for data collection. Crystals are soaked with compound mixtures (typically with each fragment present at ~10 mM), flash-frozen, and stored in liquid nitrogen. All experiments are tracked in the SGX LIMS system, which is accessible from SGX-CAT at the APS. Direct T1 line connectivity permits rapid data transfer between the two sites. Once the frozen crystals are transported to SGX-CAT by courier, pertinent sample information is accessed from the LIMS system and the samples are loaded into data collection carousels. Multiple data collection carousels may be stored in liquid nitrogen and queued for automated data collection. When a carousel is ready for analysis, it is transferred from the storage dewar to the Mar crystal mounting robot [32].

Figure 11.2A shows the SGX-CAT X-ray diffraction facility on the 31-ID beamline at the APS, which includes X-ray optical elements (for focusing and wavelength selection), beam carriage tubes, a crystal mounting robot, cryogenic nitrogen gas stream for crystal cooling, and a MarCCD detector. To facilitate unattended data collection, crystal centering software was developed by SGX in conjunction with Mar Research. Figure 11.2B illustrates a screen shot of a crystal undergoing automated crystal-centering. The X-ray beam is coincident with the convergence of the large cross-sight. The crystal is mounted in the loop at the top of the panel, where it has been identified by software and marked with the small cross-sight. At the end of the automated centering process, the small and large cross-sights are brought into coincidence, resulting in placement of the crystal within the X-ray beam (~0.01–0.05 mm in diameter).

Data collection/processing parameters are retrieved from the SGX LIMS system to control both the progress of the diffraction experiment and data processing in real time. Reduced diffraction data are automatically transferred back to SGX headquarters in San Diego via our dedicated T1 line; and experimental parameters are captured by the SGX LIMS database. This system permits routine, unattended data collection from approximately 50 crystalline samples per day, enabling data acquisition for the entire SGX fragment library in a matter of 1–2 days. Fragment screening results are analyzed automatically using a 400-CPU linux cluster located at SGX–San Diego.

11.4.5
Analysis of Fragment Screening Results

Automated processing of diffraction data is performed using a system that combines proprietary SGX software and the CCP4 [33] program package. For each screening attempt with a ten-compound, shape-diverse mixture, the structure of the target protein is automatically re-determined by molecular replacement using a reference target structure pre-defined in the SGX LIMS. The reference structure is generally the best representative for that target, as defined by resolution limit, R-factor, and overall data/structure quality. Once this step is complete, difference Fourier syntheses are calculated to reveal any superficial electron density features that cannot be explained by either the protein or water molecules (see Fig. 11.1B for an example). For each unexplained electron density feature, an attempt is made to automatically identify the fragment within the mixture that best corresponds to the shape of the electron density. The output of this analysis may then

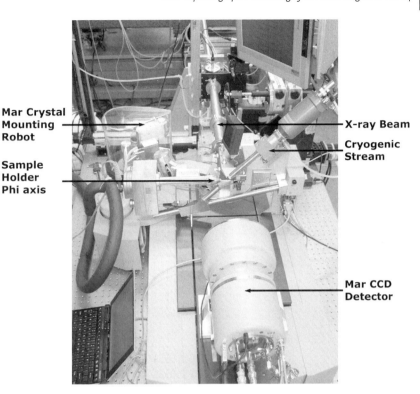

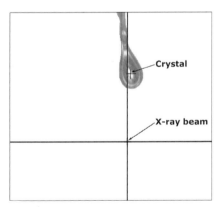

Fig. 11.2
(A) SGX-CAT beamline data collection apparatus at the Advanced Photon Source (APS). Shown are the X-ray beam carriage tubes, the cryogenic gaseous nitrogen stream, the sample stage, the Mar sample stage/automated sample changer, and the Mar CCD X-ray area detector. (B) Screen-shot depicting a frozen crystal mounted on the sample stage. The crystal is located at the small cross-sight and the location of the X-ray beam at the large cross-sight. Crystal-centering software automatically determines the location of the crystal and directs the Mar sample stage to make it coincident with the X-ray beam.

be viewed for each screening attempt using the Xtalview/Xfit crystallographic visualization package [34].

Once the automated processing/fragment identification is complete, "snapshots" of each difference electron density feature, with accompanying ligand atomic stick-figure structures are accessible via the SGX LIMS system. Visual inspection of these images represents the first point at which manual intervention is required.

Figure 11.3 depicts a sub-set of results for a crystallographic screen of the serine protease factor VIIa to illustrate the outcome of our approach. Difference electron density maps are displayed in a pre-determined target-specific orientation. For factor VIIa, initial electron density maps are shown for the primary specificity (or S_1 pocket). Examination of Fig. 11.3 shows significant electron density in panels 2, 4, 6, 8, and 9, suggesting the presence of bound fragment hits. Examination of panels 1, 3, 5, and 7 shows features consistent with only water molecules bound in the S_1 site (i.e., empty active sites). Visual scanning of the electron density images facilitates prioritization of the three-dimensional map viewing process and assessment of the results of the automated fragment fitting routine. This combination of proprietary and public domain software tools provides an efficient process for analyzing the results of SGX fragment library crystallographic screening.

11.4.6
Factor VIIa Case Study of SGX Fragment Library Screening

Factor VIIa (FVIIa) is a trypsin-like serine protease responsible for activation of Factors IX and X, both of which are essential for efficient blood clotting. Small molecule inhibitors of FVIIa and FXa have been sought as potential antithrombotic agents for treatment of deep venous thrombosis (a.k.a., economy class syndrome) [35, 36]. Development of orally active inhibitors of these targets has proved extremely challenging, because of the apparent requirement of a positively charged group to interact with the S1 pocket of the enzyme, leading to poor oral bioavailability. The SGX fragment screen was conducted for FVIIa to discover novel, non-basic fragments that would bind at the S_1 site.

We screened the SGX fragment library by soaking fragment mixtures against crystals of apo des-Gla FVIIa [37]; 15 fragment hits (~1.5–2.3 Å resolution) were found at four sites (Fig. 11.4): 11 hits at S1, three at a previously identified exosite [38], one at a separate previously identified exosite [39], and two at a novel site adjacent to S1, which includes the oxyanion hole. Fragment IC_{50} values ranged from 190 µM to ~50 mM. The two previously identified exosites were initially detected during screens of phage-display peptide libraries that identified certain large, cyclic peptides as potent FVIIa inhibitors [38, 39]. To the best of our knowledge, the SGX compounds depicted in Fig. 11.4 represent the only non-peptide inhibitors that bind to either of these two exosites.

S1 fragment hits include basic aromatic compounds that are typical of serine protease inhibitors plus several neutral haloaromatic fragments. Inhibitors con-

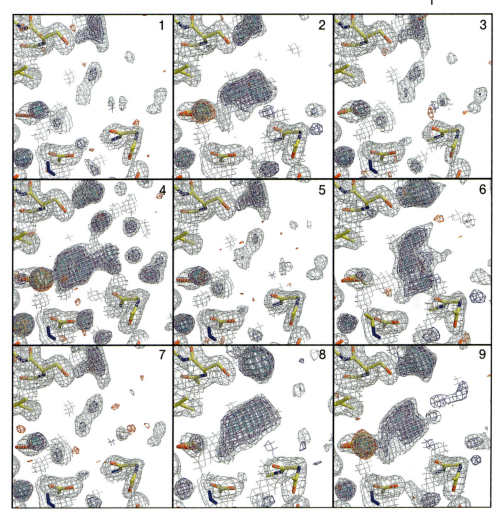

Fig. 11.3
Automated graphical output analysis for part of the crystallographic fragment screen of Factor VIIa. Each panel shows a common, predetermined view of the S_1 pocket of the enzyme active site. $2|F_{observed}|-|F_{calculated}|$ and $|F_{observed}|-|F_{calculated}|$ difference electron density maps are shown for nine different crystals, showing the specificity pocket and features corresponding to either bound fragment hits (panels 2, 4, 6, 8 and 9) or water molecules (i.e., empty active sites).

taining similar haloaromatic S1 groups have been reported for FXa [40–42] and thrombin [43–45], but not for FVIIa. In our work, we found that the haloaromatic S1 fragment hits exploit FVIIa binding modes that resemble those observed for the haloaromatic portions of FXa and thrombin inhibitors, all of which displace a conserved water molecule at the bottom of the S1 pocket with a halogen. The sidechain of Ser-190 (FXa has an alanine at residue 190) rotates about the CA–CB

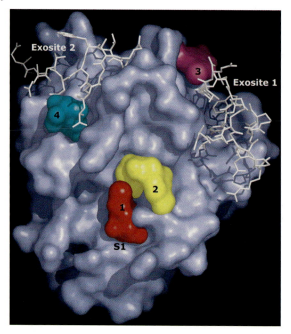

Fig. 11.4
Pymol molecular surface [68] of Factor VIIa, displaying the locations of the 15 bound fragments in four separate sites revealed by crystallographic screening. Each site is depicted by the combined molecular surface of the fragments bound at that site: 11 fragments at S1 (Site 1), two fragments at a site which includes the oxyanion hole (Site 2), one fragment at an exosite (Site 3; Exosite 1), and three fragments at the other exosite (Site 4; Exosite 2). Phage-display cyclic peptides binding to the exosites [38, 39] are shown as atomic stick figures (PDB codes: 1DVA and 1JBU).

bond to accommodate the halogen. Figure 11.5 shows overlaid views of FVIIa with a SGX fragment, 2-hydroxy-3-bromo-5-chloropyridine, and benzamidine binding in the S1 pocket.

Fragment screening of FVIIa documented that small organic molecules can indeed bind at the previously identified peptide-binding exosites, identified a new site adjacent to S1, and showed for the first time that haloaromatics bind to the FVIIa S1 pocket, analogous to FXa and thrombin. These fragment hits provided new starting points for synthesis of novel FVIIa inhibitors.

11.5
Complementary Biochemical Screening of the SGX Fragment Library

In addition to crystallographic screening, SGX conducts biochemical screening of the ~1000-compound fragment library, using a Beckman BioMek FX liquid-handling system equipped with a Sagian Rail. The entire SGX fragment library can be

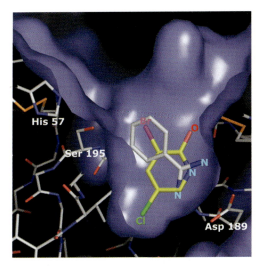

Fig. 11.5
Molecular surface of the S1 pocket of Factor VIIa, comparing binding of a brominated SGX fragment (2-hydroxy-3-bromo-5-chloropyridine; yellow bonds) with that of benzamidine (overlayed gray bonds).

screened one compound at a time via the appropriate biochemical assay in a matter of a few hours to complement results from crystallographic screening.

It is often challenging to use biochemical assays to characterize weakly binding ligands accurately. Specifically, many organic ligands cause spectroscopic interference at detection wavelengths typically used for biochemical assays. At low ligand concentrations in conventional HTS, spectral interference is usually not a significant problem. However, for detection of weakly bound fragment hits (typical IC_{50} values range from ~10 μm to > 500 μM), compounds must be assayed at substantially higher concentrations. We routinely screen the SGX fragment library at 500 μM ligand concentration, using biochemical assays formatted to minimize spectral interference, while maximizing throughput. IC_{50} values are determined for all biochemical hits ($\geq 50\%$ inhibition). Ligands with IC_{50} values in excess of 500 μM cannot be detected in our biochemical assays. In practice, we use crystallographic screening to detect such weakly bound fragments, which can be routinely optimized to single-digit nanomolar IC_{50} values for the target of interest.

Biochemical screening at conventional HTS concentrations is plagued by false positives [46]. Screening at elevated ligand concentrations further increases the probability of false positives. To address these limitations of biochemical screening, fragments that show target inhibition in our biochemical assays that went undetected in the crystallographic screen are examined individually by soaking them into crystals as single compounds. In most cases, the fragment is not detected in the crystal screen, indicating that the biochemical result is either a false-positive or that the target protein crystal lattice cannot accommodate that particular ligand.

(The latter explanation is unlikely given the size of the crystal solvent channels and our strict requirements for crystallographic screening enablement.) In either case, the fragment is abandoned in favor of fragments that do show target binding in the crystal. In rare cases, however, a biochemically active compound will exhibit target binding in the crystal. This infrequent failure of the initial fragment mixture screen can usually be explained by the presence of a more potent compound in the shape-diverse fragment screening mixture, which precluded crystallographic binding of the lower-affinity true biochemical positive. Thus, combining our crystallographic screening strategy with a high ligand concentration biochemical screen represents a powerful system for detecting and validating weakly binding ligands.

11.6
Importance of Combining Crystallographic and Biochemical Fragment Screening

In practice, crystallographic fragment screening often detects ligands that bind so weakly that they cannot be detected by our complementary biochemical assays. Furthermore, for some targets all crystallographic screening hits have proven undetectable in our biochemical screen. The sensitivity of our crystallographic screening approach derives in large part from the high local concentration of protein (~1 M) within the crystal. Typically, such high protein concentrations are not stable in solution for most targets. Therefore, stabilization of ultra-high protein density by the crystal lattice makes detection of weakly binding ($IC_{50} > 5$ mM) ligands possible. Conversely, the biochemical screen directly addresses two intrinsic limitations of the crystallographic screen. First, the crystallographic screen provides no information about target binding affinity. In the absence of biochemical information, selection of fragments for optimization would rest solely on the available co-crystal structures and accessibility of chemical handles for synthesis. Potent fragments could be overlooked. Second, biochemical screens can be conducted very quickly and do not require "enabled" crystallographic systems or even a crystal structure of the target.

At times, we have used the biochemical screen as a pre-screen during the crystallographic enablement phase of a project. We do not, however, proceed with fragment hit optimization in the absence of structural information. Once the crystallographic screen is enabled, the screen is conducted and fragments are selected for optimization from among the crystallographic hits and any structurally validated biochemically active fragments. The following section discusses criteria used to select fragment hits for structure-guided optimization.

11.7
Selecting Fragments Hits for Chemical Elaboration

As discussed above, a typical crystallographic screen yields approximately 10–50 hits per target with biochemical activities (IC_{50}) ranging from low μM to low mM. A fragment hit is only useful if it can be elaborated through efficient synthesis in directions that rapidly lead to dramatic improvements in activity. Computational prediction of which fragments represent the best candidates for optimization is not feasible, because of the huge number of possible analogs that can be generated from each fragment and the computational time required for predicting binding free energies. Instead, we select four to five of the most promising fragments to optimize in parallel. Our experience has shown that careful selection and prioritization of fragment hits typically provides two to three orders of magnitude enhancement in activity in the first round of library design and synthesis.

Because the library was designed to provide starting points suitable for elaboration of fragments into early lead compounds, the quality of the fragment hit is judged primarily by the co-crystal structure of the target-fragment complex. Hence, a primary determinant for choosing a fragment for chemical elaboration is a high-quality, unambiguous crystal structure of the target-fragment complex at better than 2.5 Å resolution, which clearly reveals the orientation and conformation of the bound ligand. Fragment hits are prioritized for synthetic elaboration based on the following criteria: location of the fragment binding site, fragment binding mode, structural accessibility of handles for synthesis, ligand efficiency, preliminary evaluation of synthetically accessible virtual libraries, novelty, and biochemical activity.

Ideal fragment hits are located at the active site or a known allosteric site. Fragments that bind at a previously unknown site remote from a lattice packing interface represent important opportunities for discovery of novel/selective lead compounds, but such sites do require validation through fragment elaboration into more potent compounds. The fragment binding mode must orient synthetic handles towards pockets or subsites. If the built-in synthetic handles of the fragment are oriented only towards solvent or are sterically blocked, alternative handles may be found by searching for available fragment analogs or introduced via synthesis of a new fragment analog. Synthetic feasibility is assessed by considering the diversity of available reagents that are compatible with the fragment hit and related synthons.

Ligand efficiency is assessed by examining the ratio of biochemical activity to the size of the fragment [47]. More efficient ligands involve more of their atoms in productive interactions with the binding site, thereby providing better starting points for elaboration and subsequent optimization. Novelty is evaluated by the combination of the parent fragment with its binding mode, which directs the assessment of possible virtual libraries. A familiar fragment can be observed to bind in an unusual way, which can provide novel elaboration opportunities. Observing a common binding mode for similar fragments sometimes provides an initial SAR and gives support, albeit indirect, for fragment hit selection. Previous experience with the same or a related target and the same or a similar scaffold represented by the fragment hit can also help support the choice of a fragment hit. Fragment biochemical activity is usually less important than the previously discussed criteria, because an active fragment that is poorly oriented in the site or inefficient will be difficult if not impossible to optimize.

A future challenge will be more thorough computational evaluation of the optimization potential for fragment hits. Our current binding free energy calculation (Section 11.8.3) is applied to compare different virtual libraries from the selected fragment hits, select the best of these libraries, and to select the top-scoring virtual library members for synthesis. This scoring approach is currently too expensive from the computational standpoint to apply to all of the possible virtual libraries for all fragment hits.

11.8
Fragment Optimization

Our goals for the first stage of fragment optimization are to improve upon parent fragment activity by $\geq$ 100-fold (IC$_{50}$ = ~1–10 mM → ~10–100 µM), to validate the selected fragment by establishing an initial SAR with small linear libraries at each available synthetic handle, and to correlate this SAR with observed co-crystal structures and computational predictions of potency. The following summary of our experience with spleen tyrosine kinase (Section 11.8.1) serves as an instructive example, before describing our approach at each stage of lead discovery and optimization in detail (Sections 11.8.2, 11.8.3, 11.8.4).

11.8.1
Spleen Tyrosine Kinase Case Study

Spleen tyrosine kinase (Syk) is a non-receptor tyrosine kinase required for signaling from immunoreceptors in hematopoietic cells. Syk is an inflammatory disease target that controls degranulation of mast cells in asthma [48, 49].

The crystal structure of Syk was obtained by SGX at 2.5 Å resolution [50] in the context of our human protein kinase pipeline. While initial crystal forms of Syk were undergoing optimization for crystallographic fragment screening, it was noted that the active site of Syk closely resembled that of Pak4 [51, 52]. Pak4 is a serine-threonine kinase oncology target, which also underwent structure determination at SGX (unpublished data). Polypeptide chain backbone atoms for 35 residues constituting the active sites of Syk and Pak4 superimpose with a root-mean-square deviation (rmsd) of ~0.6 Å (Fig. 11.6). For comparison, the calculated rmsd for all equivalent alpha carbon atomic pairs for the kinase domains of Syk and Pak4 is ~2.0 Å. This result came as something of a surprise to us, because Syk and Pak4 are members of different protein kinase families and occupy radically different locations in the evolutionary dendrogram of the human kinome [53]. Biochemical screening of several inhibitors in different series with Pak4 and Syk

Fig. 11.6
Polypeptide chain backbone atoms for 35 residues constituting the active sites of Syk (light blue, ~2.1 Å resolution) and Pak4 (red, ~2.8 Å resolution), each bound to staurosporine (shown nearly edge-on in the center of the panel), superimposed with a root-mean-square deviation (rmsd) of ~0.6 Å.

revealed similar inhibition patterns. The combination of highly similar active site structures and similar biochemical inhibition patterns suggested that Pak4 could act as a crystallographic screening surrogate for Syk, while Syk crystals underwent further optimization for crystallographic fragment screening.

Screening of a small, prototype fragment library for Syk inhibitors therefore began with crystallographic screening of Pak4 and parallel biochemical screening of Pak4 and Syk. In all, crystallographic screening yielded five hits in the hinge region (Fig. 11.7). SGX-12981 (2-amino-3-methyl-5-bromopyridine; MW = 187), possessed two synthetic handles (bromine and methyl), both of which were ideally positioned within the fragment binding site for further elaboration (Fig. 11.8). SGX-12981 exhibited low millimolar biochemical inhibition of Syk. Some of the other fragment hits were considerably more active (IC_{50} = ~50 µM), but had no or only one sterically accessible handle, and none positioned their handles as optimally for elaboration within the site as seen for SGX-12981.

Several possible linear libraries were evaluated for elaborating the 3 and 5 positions of SGX-12981, using the process described in detail below (Section 11.8.3). Among the most favorable were a carboxamide library at the 3 position and an aryl library at the 5 position. Both linear libraries produced activity increases of ~300- to 500-fold as compared to the initial fragment hit. The most active analog at the 3 position (N-cyclopropylcarboxamide, SGX-64564) gave IC_{50} = 33 µM, while the best analog at the 5 position (*meta*-chlorophenyl, SGX-64535) gave IC_{50} = 20 µM. Co-crystal structures of 3-substituted fragment analogs, first with

236 | 11 Fragment-based Lead Discovery and Optimization

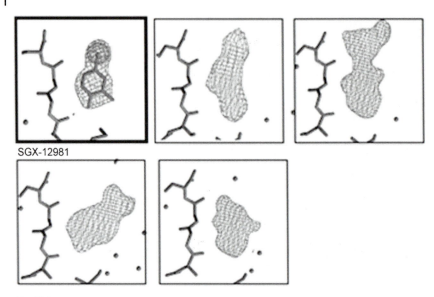

Fig. 11.7
Pilot study of crystallographic fragment screening with Pak4, showing five hits in the hinge region (hit rate = 5/85 = ~6%).

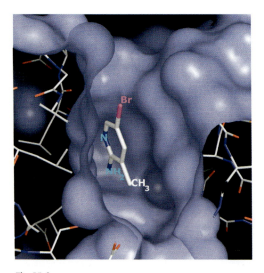

Fig. 11.8
Molecular surface of the hinge region of Pak4 showing a bound fragment hit (SGX-12981: 2-amino, 3-methyl-5-bromopyridine) detected in a crystallographic screening pilot study. The 3-methyl and 5-bromo substituents are ideally positioned for synthetic elaboration into the active site.

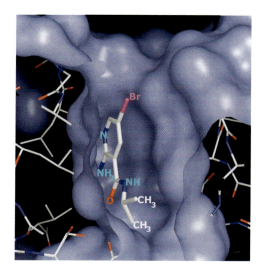

Fig. 11.9
Pak4 with SGX-14505 (2-amino-3-N-isopropylcarboxamide-5-bromopyridine), showing the results of linear library elaboration of the 3-methyl chemical handle of SGX-12981.

Pak4 and subsequently Syk, showed both the original mode of fragment binding (Fig. 11.9) and a flipped binding mode, where the 3 and 5 substituents exchange locations. Co-crystal structures of 5-substituted analogs consistently showed only the original fragment binding mode (Fig. 11.10). We focused our design efforts on the original binding mode, reasoning correctly that disubstituted analogs would not undergo flipping, because of the presence of the larger aryl ring at the 5 position, which is too large to fit into the pocket occupied by the 3 substituent in the original binding mode.

Using the process described in detail below (Section 11.8.4), a small, focused combinatorial library including these substituents and several other active substituents from the two linear libraries was designed and synthesized, producing another activity increase of ~50-fold. SGX-64926 (3-N-cyclopropylcarboxamide, 5-*meta*-methoxyphenyl) gave IC_{50} = 400 nM. Binding in the predicted conformation corresponding to that of the original fragment was confirmed by X-ray crystallography (Fig. 11.11).

Co-crystal structures documented that the 2-amino group hydrogen-bonded to a backbone carbonyl oxygen in the hinge region. Modifications to improve the strength of this hydrogen bond were considered, leading to a scaffold-swap from 2-aminopyridine to the corresponding 2-aminopyrazine (SGX-13573). Elaboration of this pyrazine demonstrated a similar SAR to that observed for the original pyridine series and converged quickly with synthesis of SGX-65372 [3-N-isopropylcarboxamide, 5-(3,4-dimethoxyphenyl), IC_{50} = 21 nM, MW = 316] and SGX-65373 (3-N-isopropylcarboxamide, 5-beta-naphthyl, IC_{50} = 24 nM, MW = 306), providing consistent potency increase of ~20-fold relative to the 2-aminopyridine series.

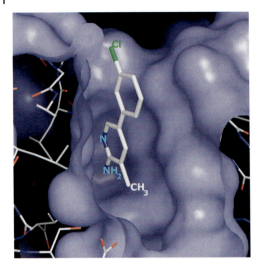

Fig. 11.10
Pak4 with SGX-64535 [2-amino-3-methyl-5-(meta-chlorophenyl)-pyridine], showing the results of linear library elaboration of the 5-bromo chemical handle of SGX-12981.

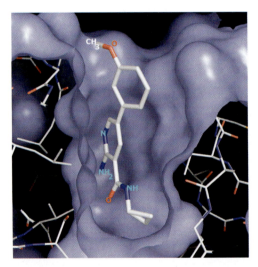

Fig. 11.11
Pak4 with SGX-64926 [2-amino-3-N-cyclopropylcarboxamide-5-(meta-methoxyphenyl)-pyridine], showing the results of combinatorial elaboration of the 3 and 5 positions of SGX-12981.

Fewer than 100 analogs were made over the course of ~10 weeks to produce SGX-65372 (IC$_{50}$ = 21 nM), which increased potency by >70 000-fold as compared to the original fragment hit. Co-crystal structures were obtained for SGX-65372 and SGX-65373 with both Pak4 and Syk, demonstrating preservation of the binding mode exhibited by the original fragment hit, SGX-12981 (Fig. 11.12). The similarity of the resulting co-crystal structures of SGX-65372 and SGX-65373 with Pak4 and Syk also validated earlier utilization of Pak4 as a crystallographic screening surrogate for Syk, while the Syk crystal system was undergoing optimization.

Cell-based assays demonstrated SGX-65372 inhibition of Syk target phosphorylation in human B-cells and inhibition of rat basophil degranulation. *In vivo* ADME studies in rats showed that SGX-65372 is bioavailable and has an acceptable half-life. Moreover, *in vitro* toxicity studies did not detect inhibition of either Cytochrome P450s or the hERG potassium channel. Target profiling studies with SGX-65372 documented strong selectivity for Syk compared to Zap70, the closest relative to Syk in the human kinome [53], and other tyrosine kinases. This project demonstrated the potential of the fragment-based approach to deliver potent and selective compounds with excellent "lead-like" properties.

This series also demonstrated additivity of substituent effects, as expected from the consistent binding modes exhibited by each of the series analogs. Elaboration of the 3-methyl of SGX-12981 to N-cyclopropylcarboxamide in SGX-64564 in-

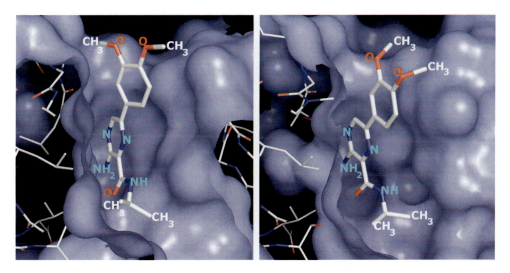

Fig. 11.12
SGX-65372 bound to Pak4 (left) and Syk (right), showing the results of combinatorial library elaboration of the 3-methyl and 5-bromo chemical handles of SGX-13573 (2-amino, 3-methyl-5-bromopyrazine). The optimized compound binds both Syk and Pak4 in the same fashion as observed for the parent fragment, SGX-12981 (Fig. 11.8).

creased activity ~45-fold. Elaboration of the 5-bromo of SGX-12981 to meta-methoxyphenyl in SGX-64536 increased activity ~75-fold. Simple additivity would predict that the corresponding disubstituted analog, SGX-64926, would improve activity relative to SGX-12981 by $45 \times 75 = 3375$-fold. The experimental increase in activity is about 3750-fold. Simple additivity such as this is not always observed. We have seen other series demonstrate super-additive effects that exceed estimates based on the potency enhancements derived at the linear library stage. In isolated cases, we have observed sub-additive effects going from linear to combinatorial libraries.

11.8.2
Fragment Optimization Overview

As discussed previously, optimization of fragment hits into early leads is expected to require an increase in binding energy of ~4–9 kcal mol^{-1} (3–6 orders of magnitude). Without appeal to structural information, this task would be daunting. However, with timely access to the right co-crystal structures, weakly binding fragments have been successfully optimized into potent early lead compounds. As summarized in a comprehensive review article by Rees et al. [54], two common methods have been employed to optimize fragment molecules: the fragment engineering method and the fragment linking method. Although fragment linking is very attractive in theory (Chapter 3) [55] and can yield dramatic increases in activity, the method can be quite challenging if a target screen does not yield adjacent fragment hits that provide appropriate geometry and substituents for chemical condensation. Fragment linking has been successfully applied and documented in fewer than ten reported cases [54], but may become more frequently applied through fragment-based approaches, due to increased availability of structurally validated hits.

To date, the most successful method at SGX has been the fragment engineering method, wherein optimization involves "growing" each fragment with "linear" libraries at each synthetic handle through automated parallel synthesis, followed by the synthesis of small, focused combinatorial libraries. This approach is stepwise, systematic, and lends itself to maximizing ligand efficiency.

Once fragments have been selected for synthesis, available reagents are assembled to generate a series of virtual linear libraries. Compounds are selected for synthesis from each fragment virtual library by predicting binding free energies based on parent fragment co-crystal structure(s). *In silico* docking of fragments is *not* part of the SGX fragment-based lead generation strategy. Experience has shown that much more reliable results can be obtained using experimentally determined structures of protein-fragment complexes as starting points for planning synthetic chemistry.

In principle, the best optimization approach from a statistical experimental design perspective [56–58] would be to construct combinatorial libraries based on the original fragment hit by simultaneously varying substituents at each of its synthetic "handles". This approach would reveal both additive and non-additive substituent effects. However, combinatorial libraries, even those based on ostensibly simple chemistry, are not always easy to make quickly and may require significant chemical methods development. Because the initial goal of our process is to rapidly validate the choice of fragment by establishing optimization potential through an initial SAR, it is more pragmatic to design and synthesize a series of small, one-dimensional ("linear") libraries from the original fragment hit, one chemical handle at a time. Following experimental validation of optimization potential, focused combinatorial libraries are designed and synthesized based on SAR established with linear libraries.

11.8.3
Linear Library Optimization

The first stage of fragment optimization proceeds through several steps:
1. Propose potential synthetic routes for linear libraries for each fragment hit based on chemical "room to move". Operational considerations include the number and type of available synthetic handle, fragment orientation in the binding site, and availability of related analogs or synthons with additional chemical handles. Multiple libraries at each handle are typically proposed (e.g., an amino chemical handle can yield carboxamides, sulfonamides, amines, etc.).
2. Select reagents compatible with each proposed synthesis from in-house and commercial sources. This step yields hundreds to up to ~40 000 possible reagents. Reagent lists are computationally filtered to remove groups that could interfere with synthesis and that are incompatible with "lead-like" properties. Several commercial software tools support these steps, including MDL Reagent Selector, the Available Chemicals Database [59], and proprietary tools developed using software toolkits from Daylight Chemical Information Systems [60] and Open Eye Scientific Software [61].
3. Generate virtual libraries of all synthetically accessible analogs from the original fragment or related synthons and corresponding reagent lists. Commercial software tools are also used for this step, including the Optive Research LibraryMaker [62], and the Daylight Reaction Toolkit [60].
4. Filter all analogs with a rapid, approximate method based on the co-crystal structure of the target-fragment complex to eliminate analogs of the fragment hit that are unlikely to bind. We use a computational approach based on the energy minimization and docking scoring functions from Northwestern Dock [63]. We constrain each analog to bind initially as observed for the fragment hit, sample multiple sterically allowed, low-energy conformers of its substituents using the Open Eye [61] OMEGA software [64], and perform a rapid energy minimization for each conformer against all available structures of the target protein (up to 50 in some cases). Multiple protein structures with differing conformations are included at this step, because our minimization/scoring approach maintains the protein structure as a rigid object and only allows ligand movement. Elaborated fragment conformations are only rejected if they prove sterically incompatible with ALL of the available target structures.

5. Predict the binding free energies of each surviving analog with a more accurate approach. Current docking scoring functions primarily provide only "yes" or "no" results, leading to a high false-positive rate and inaccurate rank ordering of relative binding affinities. With the utilization of additional CPU time, more accurate approaches are available, including free-energy perturbation [65, 66], and Molecular Mechanics/Poisson-Boltzmann Surface Area (MM/PBSA) [67]. Such approaches provide accuracy up to ~1.5–2.0 Kcal mol^{-1} (about 100-fold in activity) in relative binding free energy. We have focused on MM/PBSA because it is computationally more efficient than most other approaches, provides similar accuracy, and has been successfully applied to multiple classes of proteins and ligands. MM/PBSA uses a molecular dynamics simulation to generate an ensemble of varied protein–ligand conformations; and both the protein and the ligand are treated as mobile, flexible objects. We typically use one to two different representative target structures for MM/PBSA calculations for selection of top-scoring fragment analogs.
6. Select the majority of the library from the most diverse, top-scoring fragment analogs, plus some additional elaborations chosen to include more diversity or reflect medicinal chemistry experience. We include these additional analogs to allow for inaccuracies in predicted binding free energies and to allow for protein flexibility not reflected within the ensemble of target structures used in Step 4.
7. Each library is synthesized via automated parallel synthesis. All successfully synthesized analogs of the fragment hit are purified and characterized as discrete compounds with >80% purity by LC/MS, and, for a representative subset of each library, by proton NMR spectroscopy.
8. Each compound is biochemically evaluated for percent inhibition of the target (typical ligand concentrations = 50–500 µM). Compounds demonstrating >50% inhibition are subjected to IC$_{50}$ determination.
9. Co-crystal structures of the target bound to a representative subset of active compounds are determined to evaluate the consequences of fragment hit elaboration. In most cases, the mode of binding observed for the fragment hit is preserved. In the small minority (~5%) of cases in which the original mode of fragment binding is perturbed by the addition of a substituent to a chemical handle, the co-crystal structure provides alternative starting points for further chemistry.

11.8.4
Combinatorial Library Optimization

Additional fragment optimization continues with combinatorial libraries incorporating substitutions giving significant gains in ligand efficiency (not just activity) from the first round of linear library synthesis:
1. Focused combinatorial libraries of ~50 compounds are synthesized. We typically include the top few most efficient and/or active substituents at each position, plus a few additional substituents at each position to test for non-additiv-

ity. Additional substituents can be selected from less-active analogs, new diverse substituents, medicinal chemistry experience, or from more specialized computational predictions.
2. Each focused combinatorial library is synthesized, purified, characterized, assayed, and followed up with representative co-crystal structure determinations as outlined in Steps 6 through 9 above.
3. Cell-based assays are initiated when biochemical $IC_{50} \leq 1\ \mu M$.
4. Combinatorial library optimization continues for fragment series showing the best potential for cellular activity.
5. *In vitro* DMPK profiling (human plasma protein binding, rat and human liver microsome assays) is initiated at this stage to help select the best elaborated fragment series.

11.9
Discussion and Conclusions

Conventional lead discovery and optimization typically begins with relatively potent hits of ~10 µM IC_{50} or better, because lower potency hits are frequently difficult to optimize: they may be false-positives or have multiple binding modes, which complicates optimization due to inconsistent and confusing SAR. Traditional lead optimization strategies implicitly rely on a consistent binding mode and resulting SAR as analoging proceeds.

How can lead optimization possibly work starting with millimolar potency screening hits? Crystallographic detection of weakly bound fragment hits identifies only those ligands with well defined modes of target binding and eliminates most false-positives. Fragment optimization is focused on derivatives that are predicted to both retain the original binding mode and increase potency. Systematic crystallographic characterization of analogs in the fragment optimization series detects alternate binding modes. Not all fragment series optimize smoothly, but those that do not can be diagnosed quickly and set aside in favor of other series that are being pursued in parallel. Figure 11.13 shows the increase in activity from the parent fragment for SGX linear libraries for a variety of different target classes. There is no correlation between the measured potency of the original fragment hit and the outcome of linear library synthesis: potent biochemical activity of the starting compound is not required for successful optimization.

Conventional structure-based design usually starts with "inefficient" ligands from HTS campaigns or partially optimized ligands that require addition of groups to improve affinity and simultaneous removal of other portions of the molecule to minimize both MW and log P. Chemical starting points with poor "lead-like" properties typically start out with higher potency, but do not guarantee smooth optimization paths. Moreover, conventional approaches often employ structure-based design only in later stages of the process, when required affinity/activity gains are relatively modest (i.e., 10-fold transitions from IC_{50} values of

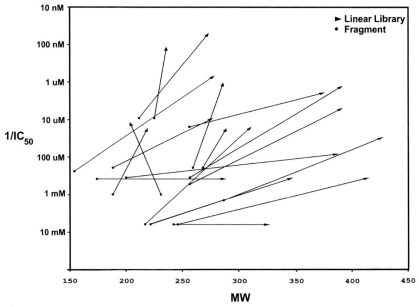

Fig. 11.13
The increase in activity from the crystallographic screening fragment hit (●) to the linearly elaborated fragment (▶) for various target classes is plotted versus the MW for each fragment hit/elaborated fragment pair. Each line connects a parent fragment to its corresponding linear library analog.

~100 nm to <10 nM). Given that uncertainties inherent in the best computational methods are typically ~100-fold, *in silico* tools are poorly suited to guide such modest affinity gains. The SGX fragment-based approach aims at large stepwise improvements in activity of 100-fold or greater, which helps to ensure that combined structural and computational approaches have the desired impact. Figure 11.13 documents that fragment optimization frequently achieves 100-fold or greater activity increases from the parent fragment with minimal increases in molecular weight, thereby maximizing ligand efficiency.

Finally, it should be noted that our objectives in fragment-based lead discovery go well beyond the discovery and optimization of novel hits. The overarching goal of our approach is the discovery of fragment hits with improved potential for successful optimization to *bona fide* clinical candidates. Fragment hits from crystallographic screening are inherently compatible with lead-like properties, wherein less truly is more.

Future research directions in our technology include evaluating the impact of more up-front effort into obtaining multiple crystal forms of the same target to increase the odds of successful soaks and to reveal new binding sites. We successfully identified multiple fragments from the same mixture that bind at adjacent sites on the surface of the target (e.g., Sites 1 and 2 in Fig. 11.4) and found cases where adjacent fragments bind independently or cooperatively (binding of a second fragment depends on binding of another). Systematic thermodynamic profiling of analogs during the optimization process may reveal whether or not more optimizable series have distinctive $\Delta G = \Delta H - T\Delta S$ profiles as compared to less successfully optimized series. More thorough computational evaluation of virtual libraries at the stage of fragment selection may also help identify fragments that are likely to optimize smoothly, rather than hitting an affinity plateau.

11.10
Postscript: SGX Oncology Lead Generation Program

A multi-target oncology lead discovery program was recently initiated at SGX. Expertise in high-throughput structure determination of historically challenging proteins, crystallographic fragment screening, and structure-guided fragment optimization are all critical to this effort. The *FAST* technology is being used to discover novel hits for a portfolio of more than 20 well validated oncology targets (including both kinases and non-kinase targets). Target proteins are processed simultaneously, with relative priorities formally re-evaluated on a quarterly basis. Because the strategy of this project is to prosecute multiple targets in parallel, they progress through the process at different rates with timelines for obtaining multiple lead compound series for each target ranging from 6 months to 24 months. Our strategy was designed to allow natural "evolution", whereby the "fittest" targets are completed rapidly, with more challenging (i.e., "less fit") targets following in turn. Thus far, multiple targets have yielded potent ($IC_{50} < 10$ nM) with cellular activity.

References

1 Hann, M. M., A. R. Leach, G. Harper. **2001**, Molecular Complexity and Its Impact on the Probability of Finding Leads for Drug Discovery. *J. Chem. Inf. Comput. Sci.* 41, 856–864.

2 Hann, M. M., T. I. Oprea. **2004**, Pursuing the leadlikeness concept in pharmaceutical research. *Curr. Opin. Chem. Biol.* 8, 255–263.

3 Oprea, T. I., A. M. Davis, S. J. Teague, P. D. Leeson. **2001**, Is There a Difference between Leads and Drugs? A Historical Perspective. *J. Chem. Inf. Comput. Sci.* 41, 1308–1315.

4 Teague, S. J., A. M. Davis, P. D. Leeson, T. Oprea. **1999**, The Design of Leadlike Combinatorial Libraries. *Angew. Chem. Int. Ed.* 38, 3743–3748.

5 Bohacek, R. S., C. McMartin, W. C. Guida. **1996**, The Art and Practice of Structure-Based Drug Design: A Molecular Modelling Perspective. *Med. Res. Rev.* 16, 3–50.

6 Kuntz, I. D. **1992**, Structure-Based Strategies for Drug Design and Discovery. *Science* 257, 1078–1082.

7 Kuntz, I. D., E. C. Meng, B. K. Shoichet. **1994**, Structure-Based Molecular Design. *Acc. Chem. Res.* 27, 117–123.

8 Caflisch, A., A. Miranker, M. Karplus. **1993**, Multiple copy simultaneous search and construction of ligands in binding sites: application to inhibitors of HIV-1 aspartic proteinase. *J. Med. Chem.* 36, 2142–2167.

9 Verlinde, C. L. M. J., G. Rudenko, W. G. J. Hol. **1992**, In search of new lead compounds for trypanosomiasis drug design: a protein structure-based linked-fragment approach. *J. Comput. Aided Mol. Des.*, 6, 131–147.

10 Allen, K. N., C. R. Bellamacina, X. Ding, C. J. Jeffery, C. Mattos, G. A. Petsko, D. Ringe. **1996**, An experimental approach to mapping the binding surfaces of crystalline proteins. *J. Phys. Chem.* 100, 2605–2611.

11 Shuker, S. B., P. J. Hajduk, R. P. Meadows, S. W. Fesik. **1996**, Discovering high affinity ligands for proteins: SAR by NMR. *Science* 274, 1531–1534.

12 Nienaber, V. L., P. L. Richardson, V. Klighofer, J. J. Bouska, V. L. Giranda, J. Greer. **2000**, Discovering novel ligands for macromolecules using X-ray crystallographic screening. *Nat. Biotechnol.* 18, 1105–1108.

13 Sanders, W. J., V. L. Nienaber, C. G. Lerner, J. O. McCall, S. M. Merrick, S. J. Swanson, J. E. Harlan, V. S. Stoll, et al. **2004**, Discovery of Potent Inhibitors of Dihydroneopterin Aldolase Using CrystaLEAD High-Throughput X-ray Crystallographic Screening and Structure-Directed Lead Optimization. *J. Med. Chem.* 47, 1709–1718.

14 Gill, A. L., M. Frederickson, A. Cleasby, S. J. Woodhead, M. G. Carr, A. J. Woodhead, M. T. Walker, M. S. Congreve, et al. **2005**, Identification of Novel p38α MAP Kinase Inhibitors Using Fragment-Based Lead Generation. *J. Med. Chem.* 48, 414–426.

15 Lesuisse, D., G. Lange, P. Deprez, D. Bénard, B. Schoot, G. Delettre, J.-P. Marquette, P. Broto, et al. **2002**, SAR and X-ray. A New Approach Combining Fragment-Based Screening and Rational Drug Design: Application to the Discovery of Nanomolar Inhibitors of Src SH2. *J. Med. Chem.* 45, 2379–2387.

16 Card, G. L., L. Blasdel, B. P. England, C. Zhang, Y. Suzuki, S. Gillette, D. Fong, P. N. Ibrahim, et al. **2005**, A family of phosphodiesterase inhibitors discovered by cocrystallography and scaffold-based drug design. *Nat. Biotechnol.* 23, 201–207.

17 Rockway, T. W., V. Nienaber, V. L. Giranda. **2002**, Inhibitors of the protease domain of urokinase-type plasminogen activator. *Curr. Pharm. Des.* 8, 2541–2558.

18 Lipinski, C. A., F. Lombardo, B. W. Dominy, P. J. Feeney. **1997**, Experimental and Computational Approaches to Estimate Solubility and Permeability in Drug Discovery and Development Settings. *Drug Deliv. Res.* 23, 3–25.

19 Congreve, M., R. Carr, C. Murray, H. Jhoti. **2003**, A "Rule of Three" for fragment-based lead discovery? *Drug Discov. Today* 8, 876–877.

20 Wenlock, M. C., R. P. Austin, P. Barton, A. M. Davis, P. D. Leeson. **2003**, A Comparison of Physiochemical Property Profiles of Development and Marketed Oral Drugs. *J. Med. Chem.* 46, 1250–1256.

21 Vieth, M., M. G. Siegel, R. E. Higgs, I. A. Watson, D. H. Robertson, K. A. Savin, G. L. Durst, P. A. Hipskind. **2004**, Characteristic Physical Properties and Structural Fragments of Marketed Oral Drugs. *J. Med. Chem.* 47, 224–232.

22 Hann, M., B. Hudson, X. Lewell, R. Lifely, L. Miller, N. Ramsden. **1999**, Strategic Pooling of Compounds for High-Throughput Screening. *J. Chem. Inf. Comput. Sci.* 39, 897–902.

23 Mason, J. S., I. Morize, P. R. Menard, D. L. Cheney, C. Hulme, R.F., Labaudiniere. **1999**, A new 4-point pharmacophore method for molecular similarity and diversity applications: Overview of the method and applications, including a novel approach to the design of combinatorial libraries containing privileged substructures. *J. Med. Chem.* 42, 3251–3264.

24 Elsevier MDL **2004**, Elsevier MDL, San Leandro, Calif.

25 Advanced Chemistry Development, **2004**, Advanced Chemistry Development, Toronto, available at: http://www.acdlabs.com/products/phys_chem_lab/logd/.

26 Fink, T., H. Bruggesser, J.-L. Reymond. **2005**, Virtual Exploration of the Small Molecule Chemical Universe Below 160 Daltons. *Angew. Chem. Int. Ed.* 44, 1504–1508.

27 Swayze, E. E., E. A. Jefferson, K. A. Sannes-Lowery, L. B. Blyn, L. M. Risen, S. Arakawa, S. A. Osgood,

S. A. Hofstadler, et al. **2002**, SAR by MS: A Ligand Based Technique for Drug Lead Discovery Against Structured RNA Targets. *J. Med. Chem.* 45, 3816–3819.

28 Sawyers, C. L. **1999**, Chronic myeloid leukemia. *N. Engl. J. Med.* 340, 1330–1340.

29 Gorre, M. E., M. Mohammed, K. Ellwood, N. Hsu, R. Paquette, P. N. Rao, C. L. Sawyers. **2001**, Clinical resistance to STI-571 cancer therapy caused by BCR-ABL gene mutation or amplification. *Science* 293, 876–880.

30 H.M.Berman, J.Westbrook, Z.Feng, G.Gilliland, T.N.Bhat, H.Weissig, I.N.Shindyalov, P.E.Bourne. **2000**, The Protein Data Bank. *Nucleic Acids Res.* 28, 235–242.

31 Xie, X., T. Kokubo, S. L. Cohen, U. A. Mirza, A. Hoffmann, B. T. Chait, R. G. Roeder, Y. Nakatani, et al. **1996**, Structural similarity between TAFs and the heterotetrameric core of the histone octamer. *Nature* 380, 287–288.

32 Marresearch **2005**, Marresearch, Norderstedt, Germany.

33 Collaborative Computational Project. **1994**, The CCP4 Suite: Programs for Protein Crystallography. *Acta. Cryst.* D50, 760–763.

34 McRee, D. E. **1999**, Xtalview/Xfit – A Versatile Program for Manipulating Atomic Coordinates and Electron Density. *J. Struct. Biol.* 125, 156–165.

35 Robinson, L. A., E. M. K. Saiah. **2002**, Anticoagulants: inhibitors of the factor VIIa/tissue factor pathway. *Annu. Rep. Med. Chem.* 37, 85–94.

36 Pauls, H. W., W. R. Ewing. **2001**, The Design of Competitive, Small-Molecule Inhibitors of Coagulation Factor Xa. *Curr. Top. Med. Chem.* 1, 83–100.

37 Kemball-Cook, G., D. J. Johnson, E. G. Tuddenham, K. Harlos. **1999**, Crystal Structure of Active Site-Inhibited Human Coagulation Factor VIIa (des-Gla). *J. Struct. Biol.* 127, 213–223.

38 Dennis, M. S., M. Roberge, C. Quan, R. A. Lazarus. **2001**, Selection and Characterization of a New Class of Peptide Exosite Inhibitors of Coagulation Factor VIIa. *Biochemistry* 40, 9513–9521.

39 Dennis, M. S., C. Eigenbrot, N. J. Skelton, M. H. Ultsch, L. Santell, M. A. Dwyer, M. P. O'Connell, R. A. Lazarus. **2000**, Peptide exosite inhibitors of factor VIIa as anticoagulants. *Nature* 404, 465–470.

40 Adler, M., M. J. Kochanny, B. Ye, G. Rurnennik, D. R. Light, S. Biancalana, M. Whitlow. **2002**, Crystal Structures of Two Potent Nonamidine Inhibitors Bound to Factor Xa. *Biochemistry* 41, 15514–15523.

41 Maignan, S., J.-P. Guilloteau, Y. M. Choi-Sledeski, M. R. Becker, W. R. Ewing, H. W. Pauls, A. P. Spada, V. Mikol. **2003**, Molecular Structures of Human Factor Xa Complexed with Ketopiperazine Inhibitors: Preference for a Neutral Group in the S1 Pocket. *J. Med. Chem.* 46, 685–690.

42 Matter, H., D. W. Mill, M. Nazare, H. Schreuder, V. Laux, V. Wehner. **2005**, Structural Requirements for Factor Xa Inhibition by 3-Oxybenzamides with Neutral P1 Substituents: Combining X-ray Crystallography, 3D-QSAR, and Tailored Scoring Functions. *J. Med. Chem.* 48, 3290–3312.

43 Tucker, T. J., S. F. Brady, W. C. Lumma, S. D. Lewis, Stephen J. Gardell, A. M. Naylor-Olsen, Y. Yan, J. T. Sisko, et al. **1998**, Design and Synthesis of a Series of Potent and Orally Bioavailable Noncovalent Thrombin Inhibitors That Utilize Nonbasic Groups in the P1 Position. *J. Med. Chem.* 41, 3210–3219.

44 Young, M. B., J. C. Barrow, K. L. Glass, G. F. Lundell, C. L. Newton, J. M. Pellicore, K. E. Rittle, H. G. Selnick, et al. **2004**, Discovery and Evaluation of Potent P1 Aryl Heterocycle-Based Thrombin Inhibitors. *J. Med. Chem.* 47, 2995–3008.

45 Hartshorn, M. J., C. W. Murray, A. Cleasby, M. Frederickson, I. J. Tickle, H. Jhoti. **2005**, Fragment-Based Lead Discovery Using X-ray Crystallography. *J. Med. Chem.* 48, 403–413.

46 McGovern, S. L., E. Caselli, N. Grigorieff, B. K. Shoichet. **2002**, A Common Mechanism Underlying Promiscuous Inhibitors from Virtual and High-Throughput Screening. *J. Med. Chem.* 45, 1712–1722.

47 Hopkins, A. L., C. R. A. Groom. **2004**, Ligand efficiency: a useful metric for lead selection. *Drug Discov. Today* 9, 430–431.

48 Zhang, J., E. H. Berenstein, R. L. Evans, R. P. Siraganian. **1996**, Transfection of

Syk protein tyrosine kinase reconstitutes high affinity IgE receptor-mediated degranulation in a Syk-negative variant of rat basophilic leukemia RBL-2H3 cells. *J. Exp. Med.* 184, 71–79.

49 Costello, P. S., M. Turner, A. E. Walters, C. N. Cunningham, P. H. Bauer, J. Downward, V. L. J. Tybulewicz. **1996**, Critical role for the tyrosine kinase Syk in signalling through the high affinity IgE receptor of mast cells. *Ocogene* 13, 2595–2605.

50 Atwell, S., J. M. Adams, J. Badger, M. D. Buchanan, I. K. Feil, K. J. Froning, X. Gao, J. Hendle, et al. **2004**, A novel mode of Gleevec Binding is Revealed by the Structure of Spleen Tyrosine Kinase. *J. Biol. Chem.* 279, 55827–55832.

51 Gnesutta, A., J. Qu, A. Minden. **2001**, The Serine/Threonine Kinase PAK4 prevents caspase activation and protects cells from apoptosis. *J. Biol. Chem.* 276, 14414–14419.

52 Abo, A., J. Qu, M. S. Cammarano, C. Dan, A. Fritsch, V. Baud, B. Belisle, A. Minden. **1998**, PAK4, a novel effector for Cdc42Hs, is implicated in the reorganization of the actin cytoskeleton and in the formation of filopodia. *EMBO J.* 17, 6527–6540.

53 Manning, G., D. B. Whyte, R. Martinez, T. Hunter, S. Sudarsanam. **2002**, The Protein Kinase Complement of the Human Genome. *Science* 298, 1912–1934.

54 Rees, D. C., M. Congreve, C. W. Murray, R. Carr. **2004**, Fragment-Based Lead Discovery. *Nat. Rev. Drug Discov.* 3, 660–672.

55 Murray, C. W., M. L. Verdonk. **2002**, The consequences of translational and rotational entropy lost by small molecules on binding to proteins. *J. Comput. Aided Mol. Des.* 741, 741–753.

56 Linusson, A., J. Gottfries, T. Olsson, E. Ornskov, S. Folestad, B. Norden, S. Wold. **2001**, Statistical Molecular Design, Parallel Synthesis, and Biological Evaluation of a Library of Thrombin Inhibitors. *J. Med. Chem.* 44.

57 Blaney, J., M. A. Siani, D. Spellmeyer, A. Wong, W. Moos. **1995**, Measuring diversity: Experimental Design of Combinatorial Libraries for Drug Discovery. *J. Med. Chem.* 38, 1431–1436.

58 Box, G. E. P., W. G. Hunter, J. S. Hunter. **1978**, *Statistics For Experimenters. An Introduction To Design, Data Analysis, And Model Building*, John Wiley & Sons, New York.

59 MDL, Elsevier MDL, San Leandro, CA.

60 Daylight Chemical Information Systems **2005**, Daylight Chemical Information Systems, Mission Viejo, Calif.

61 OpenEye **2005**, OpenEye Scientific Software, Santa Fe, N.M.

62 Optive Research **2005**, Optive Research, Austin, Tex.

63 Lorber, D. M., B. K. Shoichet. **1998**, Flexible ligand docking using conformational ensembles. *Protein Sci.* 7, 938–950.

64 Bostrom, J., J. R. Greenwood, J. Gottfries. **2003**, Assessing the performance of OMEGA with respect to retrieving bioactive conformations. *J. Mol. Graphics Model.* 21, 449–462.

65 Kollman, P. **1993**, Free Energy Calculations: Applications to Chemical and Biochemical Phenomena. *Chem. Rev.* 93, 2395–2417.

66 Pearlman, D. A., P. S. Charifson. **2001**, Are Free Energy Calculations Useful in Practice? A Comparison with Rapid Scoring Functions for the p38 MAP Kinase Protein System. *J. Med. Chem.* 44, 3417–3423.

67 Kollman, P. A., I. Massova, C. Reyes, B. Kuhn, S. Huo, L. Chong, M. Lee, T. Lee, et al. **2000**, Calculating Structures and Free Energies of Complex Molecules: Combining Molecular Mechanics and Continuum Models. *Acc. Chem. Res.* 33, 889–897.

68 DeLano, W. L. **2002**, South San Francisco.

12
Synergistic Use of Protein Crystallography and Solution-phase NMR Spectroscopy in Structure-based Drug Design: Strategies and Tactics

Cele Abad-Zapatero, Geoffrey F. Stamper, and Vincent S. Stoll

12.1
Introduction

Structural biology is an integral part of the drug discovery process in the pharmaceutical industry. Two complementary methods dominate among the experimental techniques capable of providing detailed three-dimensional (3D) structural information of therapeutically relevant targets, most often with bound ligands: protein crystallography and solution-phase NMR spectroscopy. The structural information these techniques provide continues to grow in depth and breadth, impacting every step of drug discovery from lead identification to preclinical testing (Fig. 12.1). Used together, protein crystallography and solution-phase NMR can significantly increase both the pace of discovery and the quality of the resulting compounds that are moved forward into development.

The benefits to drug discovery of rapidly available, high-resolution 3D structural information provided by protein crystallography are well established [1, 2]. While the technique's success depends on diffraction-quality protein crystals, significant advances in protein biochemistry and protein crystallization technologies have dramatically increased the number of targets amenable to structure determination by X-ray diffraction. Driven in large part by a number of worldwide structural genomics efforts underway in both academia and industry [3], these advances include robotic systems for cloning, expression, protein purification and crystallization [4]. In addition, continued hardware and software developments, both at synchrotron and in-house X-ray sources, have decreased the time of *de novo* structure solution from months to days, while co-crystal structures are routinely obtained within hours. Thus, once diffraction-quality crystals are obtained, structure determination is not a bottleneck in the discovery process. In fact, a number of drugs on the market were discovered in large part by routine use of structural information provided by protein crystallography [2, 5], a clear indication that this technique has become a driver of the design process and not a retrospective analysis tool.

In academia, solution-phase NMR spectroscopy continues to develop as a technique for obtaining 3D structural information of macromolecules including target–ligand complexes. Recent advances have made NMR structural work possible

Fragment-based Approaches in Drug Discovery. Edited by W. Jahnke and D. A. Erlanson
Copyright © 2006 WILEY-VCH Verlag GmbH & Co. KGaA, Weinheim
ISBN: 3-527-31291-9

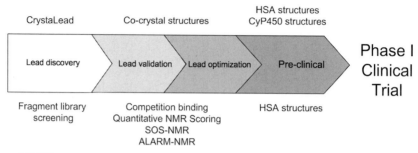

Fig. 12.1
A schematic diagram of how protein crystallography and solution-phase NMR spectroscopy interact in the drug discovery process. The segmented arrow represents the different parts of drug discovery from lead discovery to pre-clinical testing. Above the arrow, techniques where protein crystallography impacts each segment are highlighted. Beneath the arrow, the NMR methods used in each phase of the process are shown. Though the primary role of protein crystallography is in lead validation and optimization, the technique is also used for lead identification, via CrystaLead [38], and is beginning to be utilized commercially to design away pharmacokinetic problems via structure determinations of compounds with both human serum albumin [39] and cytochrome P450s [40]. Likewise, solution-phase NMR impacts all phases of drug discovery. Details of these methods have been presented elsewhere [41, 42]. More recently, ALARM-NMR, developed at Abbott Laboratories, has proven to be a valuable technique for lead validation [43].

on larger proteins [6] and have improved the accuracy of the resulting models [7]. However, the technique's most effective role in the pharmaceutical industry is in lead discovery and validation. Indeed, solution-phase NMR is now the established method for small-fragment library screening for lead identification [8, 9]. Further, integration of NMR-based screening with traditional high-throughput screening (HTS) and affinity-based mass spectrometry (AS/MS) methods [10] provides critical corroborative evidence of ligand binding to the target molecule[11], an essential component of lead validation.

Figure 12.1 illustrates the impact that both protein crystallography and solution-phase NMR have on the drug discovery process, revealing the myriad of ways in which the techniques could be used individually or synergistically to drive structure-based drug design. However, there is also an explicit redundancy. In a resource-limited environment, choices must be made as to where each technique will be employed in the most efficient way. Abbott Laboratories has solved this problem by integrating the two techniques, exploiting the strengths of each technique for maximum impact. This approach to integration of the two techniques is illustrated in Fig. 12.2.

Typically, lead compounds from either HTS or AS/MS are first evaluated by solution-phase NMR. As indicated by the dashed line in Fig. 12.2, hits from HTS

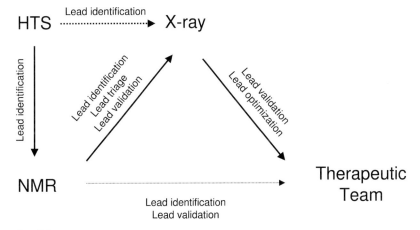

Fig. 12.2
A schematic representation of the interactions among the key participants in a structure-based drug discovery effort. Compounds identified by HTS or other methods are fed into protein crystallography or solution-phase NMR spectroscopy for further validation. While both methods can in principle validate these leads, it is more efficient to triage the leads first by NMR (solid arrow) and then further examine the most promising candidates by protein crystallography. Protein crystallography then begins the iterative lead optimization process with the project team, providing the bulk of the 3D structural information of target–ligand complexes. In this phase of discovery, solution-phase NMR spectroscopy plays a more complementary role (dashed arrow), generally providing structural information in specific cases [20].

methods are rarely sent directly for co-crystal structure analysis. Rather, solution-phase NMR provides a level of triage to these leads to further validate the binding observed by the more high-throughput methods. Solution-phase NMR may independently identify additional leads from fragment libraries. Information gathered at this stage is given to both the therapeutic team for activity analysis as well as the protein crystallography effort to establish the experimental structural bases for subsequent structure-based lead optimization.

The purpose of this chapter is to illustrate how our laboratory has integrated protein crystallography and solution-phase NMR to drive structure-based drug design. To do this, we present two case studies, each describing how the use of both techniques synergistically was paramount in successfully identifying, validating and optimizing potential drug candidates. The first case study highlights our discovery of compounds that target the human protein tyrosine phosphatase 1B, an important target for type II diabetes. Although the unique characteristics of the target did not permit the discovery of a clinical candidate, the combination of protein crystallography and solution phase NMR-based screening methods was critical to the discovery of compounds with cellular activity. The second case study highlights the discovery of compounds targeting the UDPMurNac-D-alanyl-D-alanine adding enzyme (MurF), a potential target for antibiotics. By combining the strengths of solution-phase NMR spectroscopy and protein crystallography, this discovery effort rapidly yielded compounds with 40-fold improved potency over the lead compounds in less than 12 months, allowing for efficient target validation [37].

12.2
Case 1: Human Protein Tyrosine Phosphatase

12.2.1
Designing and Synthesizing Dual-site Inhibitors

12.2.1.1 The Target

During the past decade, protein tyrosine phosphatases (PTPs) have been considered attractive targets for therapeutic intervention because of their involvement in regulating cell function as counterbalance to the phosphate donating and more abundant protein kinases. Among them, human protein tyrosine phosphatase 1B (PTP1B) has received considerable attention as a validated target for non-insulin dependent diabetes mellitus (NIDDM) or type II diabetes. Approximately 130 million people are currently afflicted by NIDDM and the patient population is still expected to more than double by 2025 [12]. Existing therapies alone or in combination are not fully satisfactory and more efficacious agents are needed as well as a better understanding of the underlying causes of the disease [12].

Several lines of evidence have provided strong support for the direct role of PTP1B in the de-phosphorylation of the insulin receptor and the corresponding downregulation of insulin signaling [13–17]. Moreover, knockout mouse studies of the PTP1B gene from two different laboratories [18, 19] showed that the resulting mice were healthy and lean. They displayed enhanced insulin sensitivity and resistance to diet-induced obesity. These studies combined suggest that inhibition of PTP1B would be an effective diabetes therapy.

The catalytic activity of PTP1B permitted a fast and reliable assay that provided robust numbers for the inhibition constants (K_i) of the compounds by measuring the rate of hydrolysis of the surrogate substrate *p*-nitrophenyl phosphate. The relative inhibition constant of the successive compounds against other phosphatases was monitored using a panel of phosphatases that included the closest homolog (T-Cell PTP, TCPTP). Crystallographic protocols have been discussed in detail previously [20].

12.2.1.2 Initial Leads

Extensive HTS screening of PTP1B using conventional methods did not provide any viable leads. The screening results and subsequent enzymatic analysis revealed the presence of several small isoquinoline diol compounds (~160 Da) that appeared to be inhibitors of PTP1B at the low micromolar level. Repeated experiments using soaking and co-crystallization protocols failed to show the presence of any of this class of compounds in the active site. In contrast, the active site residues appeared to be in a different conformation. More detailed work demonstrated that these compounds inhibited PTP1B in an irreversible manner by oxidizing the active site Cys215. Controlled experiments in terms of soaking time and variable amounts of DTT permitted the characterization of the three different states of oxidation (i.e. sulfenic, sulfinic and sulfonic) of the thiolate ion as was described by [21]. The sulfenic acid intermediate was shown to be rapidly converted into a unique sulphenyl-amide species, which might play a role in the regulation of PTP1B *in vivo* [22].

In order to identify viable and robust reversible inhibitors as leads for subsequent optimization, NMR-based screening was undertaken using a library of approximately 10 000 compounds. The screening protein sample was a truncated version of PTP1B containing 288 [20] residues as opposed to the 1–321 construct used in the crystallography work [23]. The NMR protein was either uniformly labeled with ^{15}N or selectively labeled with δ-^{13}CH$_3$ Ile. Ligand binding was monitored by observing chemical shift changes in ^{1}H/^{15}N or ^{1}H/^{13}C-HSQC spectra in the presence of test compounds [24, 25]. The experimental conditions permitted the characterization of a high-resolution HSQC spectrum of the protein where ligand binding could be recognized readily (Fig. 1 in [20]). This NMR screen identified a diaryloxamic acid (compound **1** in Fig. 12.3) as a ligand for PTP1B. The observed shifts were similar to the ones observed for pTyr. Significant perturbations were observed in the ^{15}N amide resonances for Val49, Gly220 and Gly218, all defining the active site, and a dramatic shift for the δ-^{13}CH$_3$ of Ile219. An approximate K_d for the compound was estimated to be around 100 µM, using an NMR titration curve that was of the same of order of magnitude as the one obtained by the *p*-nitrophenylphosphate hydrolysis assay (K_i ~300 µM). These data suggested that the diaryloxamate (**1**) was a *bona fide* reversible inhibitor of PTP1B binding to the active site and could be used as a pTyr mimetic for further optimization as suggested by similar findings with monoaryloxamates [20]. Initial soaking and co-crystallization experiments with this initial lead did not provide conclusive evidence of the mode of binding in PTP1B. Only the synthesis of a ten-fold more potent napthyloxamic acid analog validated the series (**2**). This more potent analog had a NMR-measured K_d of approximately 30 µM, which correlated well with the results obtained in the pNPP hydrolysis assay (K_i ~40 µM). The kinetic analysis provided strong evidence for a competitive reversible inhibition and the X-ray structure validated the series unambiguously (Fig. 12.4a.).

The X-ray structure revealed the essential features of the binding mode for this class of inhibitors, as shown in Fig. 12.4a. Compound **2** (napththyloxamic acid) binds to an open form of PTP1B, distinctively different from the binding mode observed in pTyr and analog compounds where the WPD loop (residues Trp179-Pro180-Asp181-Phe182-Gly183-Val184; WPD loop for short) closes down and snugly encloses the substrate. The second carboxylate in the benzoic acid of the compound appears to hold the "jaws" of the WPD loop open (Fig. 12.4a) by interacting with Arg221 and, in doing so, opens a larger space for future exploration. Significant Van der Waals interactions are the benzene making contact with the hydrophobic side-chain of Gln262 and the napthyl ring with the side-chain of

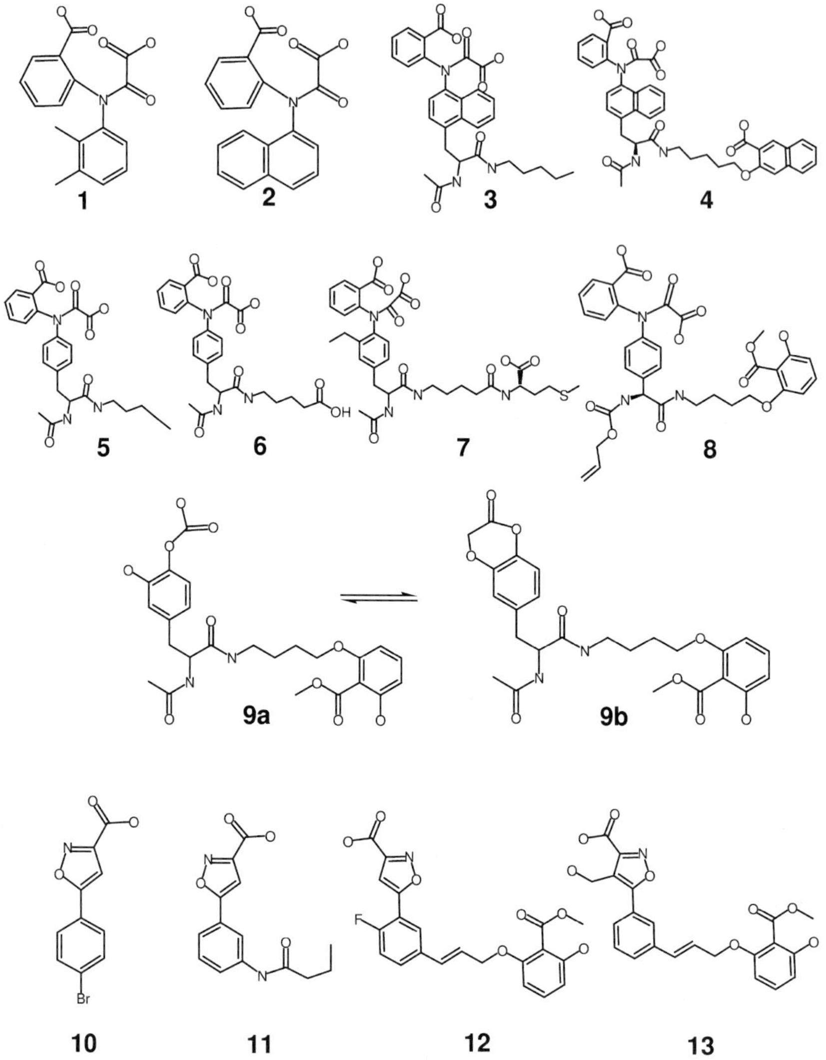

Fig. 12.3
Schematic diagrams of the structures of the inhibitors of PTP1B discussed in the text (compounds **1–13**).

Ile219. The napthyl ring could provide some binding energy by its proximity to Tyr46 but its orientation is not optimal.

12.2.1.3 Extension of the Initial Fragment

Once the initial lead (fragment **1**) has been validated, any fragment-based approach to inhibitor design needs to consider: (a) the point of attachment to the in-

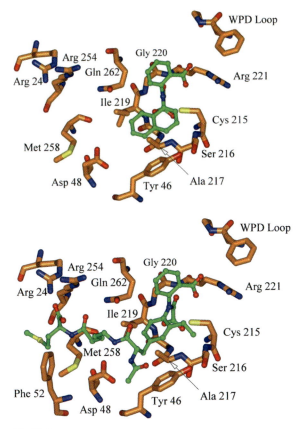

Fig. 12.4
Crystal structures of PTP1B complexed with inhibitors of the napthyl oxamate series. (a) Structure of PTP1B with inhibitor **2**, showing the environment of the binding pocket for this class of inhibitors. Notice the WPD loop (discussed in the text) in the up position and interaction of the carboxylate of the benzoate moiety interacting with Arg221. Reprinted in part from [19], with permission from the American Chemical Society, copyright 2003. (b) Structure of PTP1B with inhibitor **7**, showing the similarity of binding of the oxamate head on the first site (discussed in text), the linker fragment and the environment and binding of the second site group, specially the carboxylic acid of the Met-like amino acid with Arg24 and Arg254 on site 2 of PTP1B. Color scheme: orange = carbon, red = oxygen, blue = nitrogen, yellow = sulfur; inhibitor carbon atoms are shown in green for clarity. Reprinted from [25], with permission from Elsevier, copyright 2003.

itial lead (fragment 1) of any other additional chemical groups (fragment 2) and (b) the anchoring and extending groups that will provide the scaffold for the subsequent attachment. The superposition of the naphthyloxamic acid (**2**) with the canonical substrate (pTyr) provided initial suggestions for those three unknowns (Fig. 4 of ref. [20]). Although binding in a different mode, compound **2** and phosphotyrosine presented very important similarities. First, the phosphate and the oxamate groups of the two entities were similarly positioned. Second, the naphthalene ring shares the same hydrophobic space as the benzene ring of pTyr. These coincidences suggested that a diamide extension could provide an anchor for the napthyloxamate at the active site and provide additional potency by interacting in a mode analogous to the phosphotyrosine-containing peptide. Initial attempts using 5- or 6-napthyl substitutions did not provide any additional activity. However, incorporation of the diamido chain at the 4-position yielded compound **3** (Fig. 12.3) with an approximate 40-fold gain in potency ($K_i \sim 1$ µM). The active site contacts characteristic of the phosphate-mimic oxamate group were preserved in compounds **2** and **3**, but the X-ray structure documented a critical fact: the napthyl ring system had flipped 180 degrees (Fig. 3b of ref. [20]). Now the diamide chain extended out of the active site and hydrogen-bonded to Asp48 in a bidentate interaction similar to the one observed in the canonical pTyr substrate. Moreover, the pentyl chain extension curled gently along a narrow passage and pointed the way towards the second, noncatalytic phosphotyrosine site. The X-ray structures of **2** (Fig. 12.4a) and **3** (Fig. 3b of ref. [20]) complexed with PTP1B provided critical pieces of information by validating the initial fragment, showing the binding mode and providing an initial extension group for attaching any second-site ligands.

12.2.1.4 Discovery and Incorporation of the Second Fragment

The available amino acid sequence data suggested that the second site was much less conserved among PTPs and thus could provide a handle on selectivity, especially against TCPTP [26]). Some initial crystallographic experiments with some extended di-fluorophosphonates had suggested that reaching the second, noncatalytic phosphotyrosine site was possible (unpublished data). However, whether this was possible to achieve using the naphthyloxamate head was still an open question.

The ensuing strategy was to use the diamide anchor and the alkyl extension, characterized earlier, to position groups that could interact with the critical residues in the second site. NMR screening monitoring the labeled ^{13}C-methionine at Met258 was used to identify compounds capable of specific binding near site 2. Met258 was chosen because of its strategic position in site 2 (Fig. 12.4a). The screening was performed on a collection of approximately 10 000 compounds. Screening hits for this site consistently exhibited much weaker potency ($K_d > 1$ mM) than the ones for site 1 and soaking experiments in PTP1B crystals resulted in weak electron density peaks near site 2, but with very low occupancy. Typical compounds found in this round of screening were small fused ring aromatic acids, such as 2-benzofuran, 2-benzothiophene and 2-quinolinecarboxylic acids (Fig. 5 of ref. [20]). The X-ray soaking experiments suggested that some appear to lie "flat" in the shallow pocket that existed in the second ligand site near Met258 (Fig. 12.4b; data not shown).

The inference was made that, given the compounds found to be active in the NMR screen, it was reasonable to assume that a napthoic acid could also be a ligand for site 2. Given the chemical feasibility and ease of synthesis, compound **4** was synthesized with full stereocontrol, starting from N-Boc-(S)-3-iodoalanine methyl ester [20]. The K_i for inhibitor 4 was approximately 22 nM (N=3 measurements). Thus, the second-site ligand provided a 25-fold improvement in inhibitory activity. The conformation and position of the active-site fragment as well as the diamide linker region were identical to the ones found in the independent fragment (compound 3) and the additional napthoic acid made interactions with Arg254 (2.6 Å) and Arg24 (2.8 Å) in site 2 (Fig. 3c of ref. [20]). Although modest, the double-site compound showed a two-fold selectivity *in vitro* over TCPTP (Table 12.1), providing an initial proof of concept and validity of the fragment-based approach to inhibitor design for PTP1B.

Table 12.1 Summary of the relative inhibition constants (K_i) between PTP1B and TCPTP for selected inhibitors of PTP1B. Data extracted and adapted from refs. [20, 27–31].

Phosphatase (µM)	Compound								
	1	2	3	4	6	7	9	12	13
PTP1B	290.0	40.0	1.1	0.02	0.08	0.02	9.0	7.0	0.9
TCPTP	160.0	40.0	1.1	0.05	0.40	0.07	180.0	160.0	20.0

12.2.1.5 The Search for Potency and Selectivity

Compound 4 demonstrated that the concept of double-site ligand with an initial selectivity was possible, additional potency and a more robust selectivity factor were the next goals. Further structure-based efforts led to the discovery of compound 6, which provided an extended alkyl carboxylic acid quite suitable for high-throughput amide bond formation. The presence of two arginine residues (Arg24 and Arg254) combined with the results of the second site screening suggested that a negatively charged (acid) functionality might be an important part of any site 2 ligand. This preference was confirmed by some initial chemical couplings to compound 3, where it was found that the amino acid phenylalanine was 24-fold times more potent than the phenyl group alone (Table 2 of ref. [27]). This initial observation prompted the preparation of a library of different L-amino acids to be coupled to the core template (6). Among them, L-methionine (7) was the most potent of all (K_i = 80 nM) and exhibited a superior *in vitro* (5-fold) selectivity over the proof-of-concept compound (4). The result also confirmed the importance of the acid functionality since the corresponding ester was 7-fold less potent [27]. The X-ray crystal structure of PTP1B complexed with 7 demonstrated unambiguously these inferences (Fig. 12.4b) and showed the mode of binding of the free carboxyl of L-Met to the second site pocket that included Arg24. In addition, the structure showed that the lipophilic side-chain of L-Met extended along the hydrophobic surface provided by Phe52 and Met258.

12.2.2
Finding More "Drug-like" Molecules

12.2.2.1 Decreasing Polar Surface Area on Site 2

The milestone compounds previously discussed (**4**, **7**) exhibited significant potency (20–80 nM) and the latter one had an acceptable initial selectivity *in vitro*. However, both of them were marked by very large polar surface area (PSA) dominated by the exposed carboxylic groups as well as significant molecular weight issues (MW > 500 Da). The second-site NMR screening discussed earlier that unveiled a variety of small organic compounds, mostly acid-containing, also revealed two salicylic acids and a quinaldic acid (Fig. 2 and compounds **3–5** of ref. [28]). The dissociation constants for these compounds (K_d) were in the sub-millimolar to low-millimolar range. The salicylic-based ligands identified by the earlier NMR screen were considered particularly attractive because of their superior affinity and wider structural diversity.

The X-ray structure of PTP1B with the amino-acid L-Met in the second site (**7**) discussed above (Fig. 12.4b) provided critical information to guide the strategy to link this salicylic acid analogs to the existing scaffold. The amide of the methionine could be replaced by an ether linkage to provide the same hydrogen bonding interaction with Gln262. In order to preserve the salicylic acid feature of the site 2 ligands, a C2 symmetrical methyl 2,6-dihydrobenzoate was selected for the initial ether formation. The resulting inhibitor (**8**) exhibited good inhibitory potency and acceptable selectivity (Table 12.1) and, more importantly, the PSA had been reduced by 15%, by replacing the carboxylic acid in the second site by an ether-linked salicylate.

12.2.2.2 Monoacid Replacements on Site 1

After the efforts described above, the design and synthesis of new pTyr mimetics with a mininum number of ionizable groups at physiological pH turned out to be a tortuous path. A few analogs with no ionizable group were explored (Table 12.1; compounds **9–12** of ref. [29]). However, although they offered the potential to hydrogen bond with the catalytic pocket, none of them showed any inhibitory activity. Thus, it appeared inevitable that any design should include at least one charged interaction that could provide an anchor to the phosphate-binding pocket of PTP1B. After several attempts, a 2-hydroxy phenoxy acetic acid analog on site 1 linked to the established salicylate moiety on site 2 (Fig. 12.3, **9a**) showed low micromolar potency (K_i = 9 µM) against PTP1B and approximately 20-fold selectivity over TCPTP in *in vitro* assays (Table 12.1). More importantly, this compound exhibited a high level of membrane permeability (>10 × 10^{-6} cm s^{-1}) in Caco-2 cell permeability assays. It should be pointed out that the presence of the 2-hydroxy in compound **9a** facilitates the formation of the lactone form, making possible the equilibrium between the lactone and acid forms (**9a**, **9b**; Fig. 3 of ref. [29]). The existence of this equilibrium could be of some utility in bypassing poor cell permeability in certain pharmacophores directed towards PTP1B. Nonetheless, the ratios of the two forms under various physiological conditions is uncertain and in this particular case the limited potency precluded further development.

The X-ray structure of compound **9a** established that the compound existed in the hydroxy acid form. The binding is rather tight, with the WPD flap fully closed in a manner analogous to other complexes reported where the flap is in the down position. The salicylate component at the second site binds in the way that was also found for other compounds (Fig. 3 of ref. [29]).

12.2.2.3 Core Replacement

In spite of the significant gains made in the "drug-like" properties of compounds **8–9a**, the latter compound still exhibited a rather large PSA and its MW was too large for further development. A replacement of the highly charged site 1 head group was desirable, as well as a more compact linker design. To achieve these goals focused libraries were designed, based on various heterocycle-containing monocarboxylic acids with low pK_a values. The compounds in the libraries were tested in the standard colorimetric pNPP hydrolysis assay and were found to be inactive when tested at 300 µM. The inhibition constants were estimated to be around 900 µM. Screening of these compound libraries by NMR using PTP1B (1–288) selectively labeled with ^{13}C at the methyl groups of isoleucine residues (δ1 only) unveiled an isoxazole carboxylic acid (**10**; Fig. 2 of ref. [30]) as a weak binder with an estimated K_d of 800 µM. The X-ray structures of **10** and a related analog (**11**, Fig. 12.5a) containing a four-atom extension off the *meta* position of the phenyl ring (K_i = 150 µM) established the mode of binding of this novel core to the PTP1B active site and a simple solution to link the isoxazole core to the methyl salicylate piece at the site 2 ligand. Further conformational modification of the original peptide bond linker led to the identification of **12**. This class of compounds represents a novel class of inhibitors of PTP1B consisting of smaller, monocarboxylic acid cores. The critical physico-chemical parameters of MW and PSA had been reduced from 682 Da and 220 Å^2 (for compound **7**) to 410 Da and 145 Å^2 (for compound **12**). The X-ray structure of **12** showed that the isoxazole carboxylic acid binds to the PTP1B active site with the WPD loop in the closed conformation, with a bidentate interaction between the carboxylic acid and the side-chain of Arg221. The entire phenyl-isoxazole ring system fits snugly in the hydrophobic pocket occupied by the phenyl ring of the pTyr substrate. The remaining interactions observed in earlier compounds of the salicylate-containing classes are maintained. The X-ray structure of the most potent compound of the series, compound **13**, is presented in Fig. 12.5b using a surface representation. Consistent with the more favorable physico-chemical parameters, compound **12** showed acceptable cell permeability in Caco-2 cell membrane assays. More importantly, using COS-7 cells transiently transfected with exogenous PTP1B, the cells exhibited a dose-dependent effect in the reversal of dephosphorylation of STAT3 (signal transducer and activator of transcription 3) induced by the overexpression of PTP1B. Further extension of the SAR of this novel class of isoxazole carboxylic acid inhibitors of PTP1B using structure-based methods lowered the potency to a K_i of 0.9 µM, while retaining 20-fold selectivity over TCPTP and cellular activity in the COS-7 cell assay (**13**) [31].

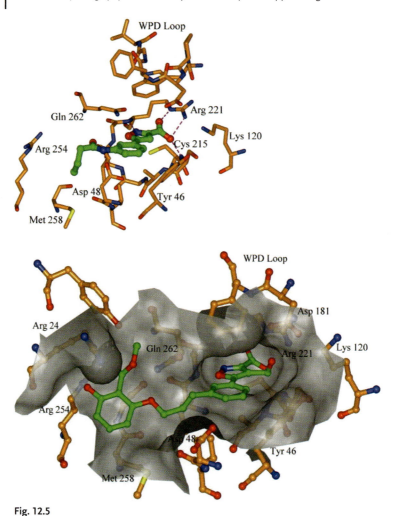

Fig. 12.5
Crystal structures of PTP1B complexed with inhibitors of the isoxazole series. (a) Structure of PTP1B with inhibitor **11** showing the environment of the binding pocket for this class of inhibitors. Notice the WPD loop in the down position, covering the inhibitor and the bidentate, coplanar, interactions of the isoxazole head with Arg221. (b) Structure of PTP1B with inhibitor **13** in a transparent surface representation showing the close interactions on site 1 (isoxazole head), the short linker to site 2 and the different interactions of the salicylate moiety with Arg24 and Arg254 on the second site (compare with Fig. 12.4b). Color scheme as in Fig. 12.4. Fig. 12.5b reprinted from [29], with permission from Elsevier, copyright 2004.

12.3
Case 2: MurF

12.3.1
Pre-filtering by Solution-phase NMR for Rapid Co-crystal Structure Determinations

12.3.1.1 The Target
MurF catalyzes the ATP-dependent formation of the UDP-MurNAc-pentapeptide. This reaction is the last step in the first stage of the peptidoglycan biosynthetic pathway [32]. The Mur enzymes in this pathway, which include all the murein enzymes, are potential targets for antibacterial design; further, the success of a commercial product targeting the MurA enzyme provides validation for this pathway as a source for the discovery of new antibiotics [33].

12.3.1.2 Triage of Initial Leads
As illustrated in Fig. 12.2 by the dashed line, normally initial hits identified by HTS and or AS/MS screens are validated by solution-phase NMR prior to co-crystallization attempts. Done largely for practical reasons, attempts to co-crystallize every hit from HTS would be resource-heavy. Therefore, identified hits from high-throughput methods are further validated by solution-phase NMR and priorities for crystallization are determined. In the case of MurF, initial screening for compounds that bind to MurF was done by AS/MS. Eleven unique compounds were identified by this method as potential inhibitors. Fortunately, MurF was very amenable to NMR techniques, providing readily interpretable $^1H/^{13}C$-HSQC spectra [34]. Thus, solution-phase NMR was used to triage the list of AS/MS hits and corroborated the binding of two (Fig. 12.6) of the 11 hits. This information led to the co-crystal structure determination of both of these lead compounds, individually, by protein crystallography [34].

The crystal structure of MurF from *E. coli* is known [35]; however, this unliganded form of the enzyme presents a very different conformational picture compared to the *Streptococcus pneumoniae* enzyme with either lead compound bound. MurF is a large three-domain enzyme that, in the unliganded form, exhibits an extended, open conformation. However, the co-crystal structures obtained with the two lead compounds (see Fig. 12.6) reveal a "closed" conformation. Further, Yan and coworkers noted difficulty in soaking ATP into the formed crystals of the unliganded enzyme [35]. Attempts to obtain unliganded crystals of the *S. pneumoniae* enzyme were unsuccessful: crystals of the enzyme could only be obtained in the presence of reasonably potent compounds (IC_{50} better than 150 µM). As a consequence, all subsequent crystal structure determinations of MurF in complex with inhibitors were done using crystals obtained by co-crystallization experiments.

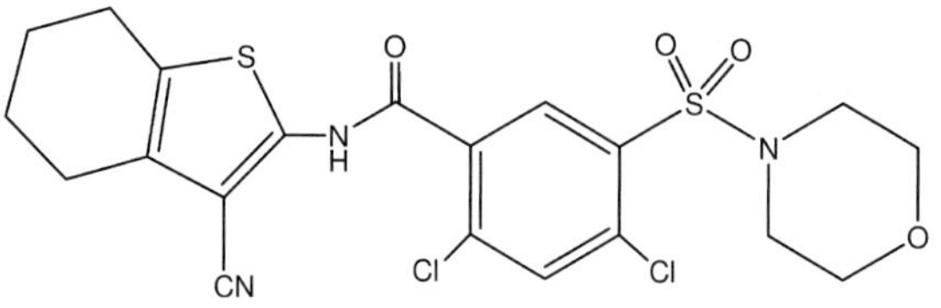

Fig. 12.6
Schematic chemical structures of the two lead compounds for the MurF program verified by solution-phase NMR to be valid hits.

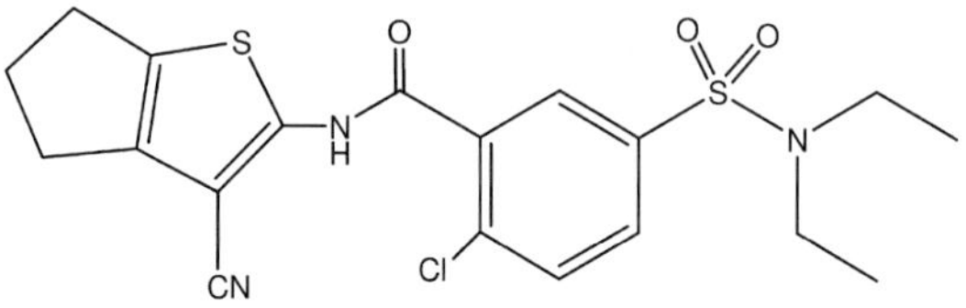

12.3.1.3 Solution-phase NMR as a Pre-filter for Co-crystallization Trials

Protein crystallography may act as a guide for medicinal chemistry efforts focused on lead optimization, provided rapid access to target-ligand structural information as well as a robust activity assay are attainable. As noted, MurF crystals could only be obtained by co-crystallization with reasonably potent compounds. For routine target–ligand complex structure determinations, this method is more resource- and time-consuming than traditional soaking techniques. In addition to this rate-limiting step, the activity assay for the enzyme was difficult and throughput was limited. The assay used is a modified radioactive assay [36] that requires three substrates, of which one had to be obtained from crude bacterial extracts. Initially, single-point assays were used to provide an initial indication of the potency, but this imprecise method led to several unsuccessful crystallization attempts with compounds that were thought to be reasonably potent.

In order to maintain throughput consistent with providing guidance to the ongoing medicinal chemistry efforts, solution-phase NMR spectroscopy was used as a parallel filter for co-crystallization. Given the interpretable ^{1}H/^{13}C-HSQC spectra that could be obtained for MurF in solution, as many as 50 compounds day^{-1} could be screened for confirmation of binding. The results of the solution-phase NMR experiments were used successfully as a predictor for co-crystallization. Further, these same data helped corroborate the IC$_{50}$ data. In general, good correlation between the binding was observed, as determined by NMR and IC$_{50}$ values determined from the activity assay (see Fig. 12.7). This combination of solution-phase NMR pre-screening and subsequent target–ligand co-crystal structure determination allowed for rapid structure determinations, which could in turn be used

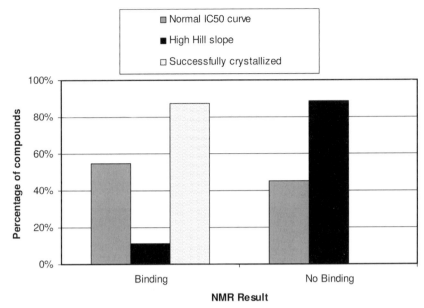

Fig. 12.7
Bar graph comparing the percentage of compounds that were shown to bind by NMR, had a normal IC_{50} curve and co-crystallized with MurF. Good agreement was observed for compounds that were shown to bind by NMR, exhibited normal IC_{50} curves and successfully produced co-crystals. A small percentage of compounds that exhibited high Hill slopes also showed binding by NMR. Of the compounds that did not bind in the solution-phase NMR experiment, a significant percentage had normal IC_{50} curves, underlying the importance of having the corroborating information from solution-phase NMR.

to guide medicinal chemistry efforts, creating an efficient optimization process resulting in significant (>40-fold) potency improvement over the original lead compounds [37]. The compounds' lack of antibacterial activity in bacterial growth assays allowed for prompt termination of the project.

12.4
Conclusion

A brief review of the strategies and tactics used in our laboratory that combine the use of solution-phase NMR and protein crystallography in structure-based drug design is presented. The interwoven roles of these two critical structural methods are discussed, with special emphasis on their use in lead identification and validation, lead optimization and the discovery of compounds with favorable PK properties. Examples from PTP1B and MurF are used to illustrate the strategies taken.

Low-affinity fragments discovered by NMR screening were progressively optimized using linking strategies and structure-based methods to design and synthesize nanomolar inhibitors of PTP1B. Leads with superior PK properties were found by more sensitive NMR-based screening methods, using selective ^{13}C labeling of methyl groups of isoleucine residues. Structure-based optimization of these more drug-like molecules using fragment assembly resulted in cell-active PTP1B inhibitors with limited *in vitro* selectivity against TCPTP.

Structure-based drug design programs often encounter targets that are difficult to crystallize, have a complicated assay, or both. Integrating the structural biology effort in the manner shown in Fig. 12.2 greatly increases the chances for successful structure-based design, even in these more difficult situations. As demonstrated in Case 2, using solution-phase NMR spectroscopy to verify the binding of MurF inhibitors prior to co-crystallization kept the protein crystallography and medicinal chemistry efforts tightly coupled during the optimization process. This allowed the 3D structural information provided by crystallography to be used as a guide for chemical modifications of the parent compound, resulting in the successful optimization of the lead series.

Acknowledgments

Philip Hajduk is acknowledged for his helpful comments and critical reading of the manuscript. Bruce Szczepankiewicz and Gang Liu are acknowledged for helpful comments and discussions, and Charlie Hutchins for assistance with the PTP1B figures.

References

1 Blundell, T.L., H. Jhoti, C. Abell **2002**, High-throughput crystallography for lead discovery in drug design. *Nat. Rev. Drug Discov.* 1, 45–54.

2 Congreve, M., C.W. Murray, T.L. Blundell **2005**, Structural biology and drug discovery. *Drug Discov. Today* 10, 895–907.

3 Nature **2000**, Supplement. *Nat. Struct. Biol.* 7, 927–994.

4 Stewart, L., R. Clark, C. Behnke **2002**, High-throughput crystallization and structure determination in drug discovery. *Drug Discov. Today* 7, 187–196.

5 Hardy, L.W., A. Malikayil **2003**, The impact of structure-guided drug design on clinical agents. *Curr. Drug Discov.* 2003, 15–20.

6 Riek, R., et al. **2002**, Solution NMR techniques for large molecular and supramolecular structures. *J. Am. Chem. Soc.* 124, 12144–12153.

7 Tolman, J.R. **2001**, Dipolar couplings as a probe of molecular dynamics and structure in solution. *Curr. Opin. Struct. Biol.* 11, 532–539.

8 Stockman, B.J., C. Dalvit **2002**, NMR screening techniques in drug discovery and drug design. *Prog. NMR Spectrosc.* 41, 187–231.

9 Meyer, B., T. Peters **2003**, NMR spectroscopy techniques for screening and identifying ligand binding to protein receptors. *Angew. Chem. Int. Ed. Engl.* 42, 864–890.

10 Comess, K.M., M.E. Schurdak **2004**, Affinity-based screening techniques for enhancing lead discovery. *Curr. Opin. Drug Discov. Dev.* 7, 411–416.

11 Hajduk, P.J., D.J. Burns **2002**, Integration of NMR and high-throughput screening. *Comb. Chem. High Throughput Screen* 5, 613–621.

12 Kiberstis, P. **2005**, A survey of suspects. *Science* 307, 369–384.
13 Kenner, K.A., et al. **1996**, Protein-tyrosine phosphatase 1B is a negative regulator of insulin- and insulin-like growth factor-I-stimulated signaling. *J. Biol. Chem.* 271, 19810–19816.
14 Chen, H., et al. **1997**, Protein-tyrosine phosphatases PTP1B and syp are modulators of insulin-stimulated translocation of GLUT4 in transfected rat adipose cells. *J. Biol. Chem.* 272, 8026–8031.
15 Ahmad, F., et al. **1995**, Osmotic loading of neutralizing antibodies demonstrates a role for protein-tyrosine phosphatase 1B in negative regulation of the insulin action pathway. *J. Biol. Chem.* 270, 20503–20508.
16 Chen, H., et al. **1999**, A phosphotyrosyl mimetic peptide reverses impairment of insulin-stimulated translocation of GLUT4 caused by overexpression of PTP1B in rat adipose cells. *Biochemistry* 38, 384–389.
17 Walchli, S., et al. **2000**, Identification of tyrosine phosphatases that dephosphorylate the insulin receptor. A brute force approach based on "substrate-trapping" mutants. *J. Biol. Chem.* 275, 9792–9796.
18 Elchebly, M., et al. **1999**, Increased insulin sensitivity and obesity resistance in mice lacking the protein tyrosine phosphatase-1B gene. *Science* 283, 1544–1548.
19 Klaman, L.D., et al. **2000**, Increased energy expenditure, decreased adiposity, and tissue-specific insulin sensitivity in protein-tyrosine phosphatase 1B-deficient mice. *Mol. Cell Biol.* 20, 5479–5489.
20 Szczepankiewicz, B.G., et al. **2003**, Discovery of a potent, selective protein tyrosine phosphatase 1B inhibitor using a linked-fragment strategy. *J. Am. Chem. Soc.* 125, 4087–4096.
21 van Montfort, R.L., et al. **2003**, Oxidation state of the active-site cysteine in protein tyrosine phosphatase 1B. *Nature* 423, 773–777.
22 Salmeen, A., et al. **2003**, Redox regulation of protein tyrosine phosphatase 1B involves a sulphenyl-amide intermediate. *Nature* 423, 769–773.
23 Puius, Y.A., et al. **1997**, Identification of a second aryl phosphate-binding site in protein-tyrosine phosphatase 1B: a paradigm for inhibitor design. *Proc. Natl. Acad. Sci. USA* 94, 13420–13425.
24 Shuker, S.B., et al. **1996**, Discovering high-affinity ligands for proteins: SAR by NMR. *Science* 274, 1531–1534.
25 Hajduk, P.J., et al. **2000**, NMR-based screening of proteins containing 13C-labeled methyl groups. *J. Am. Chem. Soc.* 122, 7898–7904.
26 Iversen, L.F., et al. **2002**, Structure determination of T cell protein-tyrosine phosphatase. *J. Biol. Chem.* 277, 19982–19990.
27 Xin, Z., et al. **2003**, Potent, selective inhibitors of protein tyrosine phosphatase 1B. *Bioorg. Med. Chem. Lett.* 13, 1887–1890.
28 Liu, G., et al. **2003**, Selective protein tyrosine phosphatase 1B inhibitors: targeting the second phosphotyrosine binding site with non-carboxylic acid-containing ligands. *J. Med. Chem.* 46, 3437–3440.
29 Xin, Z., et al. **2003**, Identification of a monoacid-based, cell permeable, selective inhibitor of protein tyrosine phosphatase 1B. *Bioorg. Med. Chem. Lett.* 13, 3947–3950.
30 Liu, G., et al. **2003**, Fragment screening and assembly: a highly efficient approach to a selective and cell active protein tyrosine phosphatase 1B inhibitor. *J. Med. Chem.* 46, 4232–4235.
31 Zhao, H., et al. **2004**, Isoxazole carboxylic acids as protein tyrosine phosphatase 1B (PTP1B) inhibitors. *Bioorg. Med. Chem. Lett.* 14, 5543–5546.
32 Anderson, M.S., et al. **1996**, Kinetic mechanism of the *Escherichia coli* UDP-MurNAc-tripeptide D-alanyl-D-alanine-adding enzyme: use of a glutathione S-transferase fusion. *Biochemistry* 35, 16264–16269.
33 El Zoeiby, A., F. Sanschagrin, R.C. Levesque **2003**, Structure and function of the Mur enzymes: development of novel inhibitors. *Mol. Microbiol.* 47, 1–12.
34 Longenecker, K. L., Stamper, G. F., Hajduk, P. J., Fry, E. H., Jacob, C. G., Harlan, J. E., Edalji, R., Bartley, D. M., Walter, K. A., Solomon, L. R., Holzman, T. F., Gu, Y. G., Lerner, C. G., Beutel, B. A., Stoll, V. S. **2005**, Structure of MurF from Streptococcus pneumoniae co-crystal-

lized with a small molecule inhibitor exhibits interdomain closure. *Protein Sci.* 14, 3039–3047.
35 Yan, Y., et al. **2000**, Crystal structure of *Escherichia coli* UDPMurNAc-tripeptide D-alanyl-D-alanine-adding enzyme (MurF) at 2.3 A resolution. *J. Mol. Biol.* 304, 435–445.
36 Seals, J.R., et al. **1978**, A sensitive and precise isotopic assay of ATPase activity. *Anal. Biochem.* 90, 785–795.
37 Stamper, G.F., Longenecker, K. L., Fry, E. H., Jakob, C. G., Florjancic, A. S., Gu, Y. G., Anderson, D. D., Cooper, C. S., Zhang, T., Clark, R. F., Cia, Y., Black-Schaefer, C. L., Owen MacCall, J., Lerner, C. G., Hajduk, P. J., Beutel, B. A., Stoll, V. S. **2005**, Structure-based optimization of MurF Inhibitors. *Chem. Biol. Drug. Res.* 67, 58–65.
38 Nienaber, V.L., et al. **2000**, Discovering novel ligands for macromolecules using X-ray crystallographic screening. *Nat. Biotechnol.* 18, 1105–1108.
39 New Century Pharmaceuticals, Inc., available at: www.newcenturypharma.com.
40 Cytochrome P450 programme, available at: www.astex-technology.com.
41 Huth, J.R., et al. **2005**, Utilization of NMR-derived fragment leads in drug design. *Methods Enzymol.* 394, 549–571.
42 Hajduk, P.J., et al. **2004**, SOS-NMR: a saturation transfer NMR-based method for determining the structures of protein–ligand complexes. *J. Am. Chem. Soc.* 126, 2390–2398.
43 Huth, J.R., et al. **2005**, ALARM NMR: a rapid and robust experimental method to detect reactive false positives in biochemical screens. *J. Am. Chem. Soc.* 127, 217–224.

13
Ligand SAR Using Electrospray Ionization Mass Spectrometry

Richard H. Griffey and Eric E. Swayze

13.1
Introduction

High-throughput screening for the identification of hits appropriate for further medicinal chemistry is now well established in the drug discovery process. For many classes of targets, screening for active site inhibitors yields a sufficient number of hits such that additional selection criteria can be applied, such as drug-likeness and pharmacokinetic properties. However, classes of targets with open interaction surfaces, such as protein–protein complexes and RNA, produce low initial hit rates. Also, effective inhibition of these targets may require multiple interactions with the ligand; and the resulting leads may have complex structures. Such leads are unlikely to be discovered directly, and various strategies to identify and link hits for generation of leads have been devised (as discussed elsewhere in this book).

We describe a generic and efficient tool for the identification and elaboration of hits against both classes of target. Electrospray ionization mass spectrometry (ESI-MS) provides extremely gentle ionization conditions that allow low-affinity (~100–250 µM) protein–ligand and RNA–ligand complexes to transit from solution into the gas phase. These complexes can be detected using many types of mass analyzers, including time-of-flight, quadrupole, ion trap, or Fourier transform spectrometers.

ESI-MS methods have many advantages for the characterization of hits and leads. The utility of ESI-MS for the ionization, detection, and characterization of non-covalent complexes of nearly every type of biomolecule has now been described in >400 publications and more than 20 review articles [1–4]. Mass spectrometry has a major advantage over other screening methods: the identities and abundances of different complexes can be determined from direct observation, since the mass of every molecule serves as the intrinsic detection "label". ESI-MS has been used to characterize various features of protein–protein, protein–DNA, protein–RNA, and DNA–DNA complexes, including solution binding affinities, macromolecular and ligand binding stoichiometry, and competitive versus concurrent binding. Molecular interactions with dissociation constants ranging from

---

Fragment-based Approaches in Drug Discovery. Edited by W. Jahnke and D. A. Erlanson
Copyright © 2006 WILEY-VCH Verlag GmbH & Co. KGaA, Weinheim
ISBN: 3-527-31291-9

nanomolar to millimolar can be characterized using ESI-MS. Once isolated in the gas phase, non-covalent complexes can be interrogated via dissociation (MS/MS) to determine binding sites or can be probed for structural features using ion–molecule reactions (such as hydrogen–deuterium exchange). It is important to remember that target–ligand interactions are not at equilibria in the gas phase and do not reform if they dissociate in the MS instrument. Hence, care must be taken to chose and optimize the proper combination of instrument operation and solution conditions to insure success.

ESI-MS screening can be performed with a variety of ligand classes. The binding of hydrophobic and charged ligands can be detected. The size of the protein or RNA target is limited only for direct observation, where the increased energy required for desolvation may disrupt non-covalent complexes. ESI-MS has advantages over other methods for speed, specificity, sensitivity, and the ability to observe directly the stoichiometry of ligand binding. Here, we note additional advantages of ESI-MS/MS to determine the structure of a ligand and to map the binding site on the target. We do not cover the detection of non-covalent complexes using matrix-assisted laser-desorption ionization mass spectrometry (MALDI).

13.2
ESI-MS of Protein and RNA Targets

13.2.1
ESI-MS Data

Signals from peptides, proteins, and nucleic acids are distributed in an ESI-MS spectrum as a function of the macromolecule's mass-to-charge ratio (m/z), rather than the mass. Even large proteins and their complexes (up to 1 MDa) generate signals that can be detected with high sensitivity by a variety of mass analyzers, such as quadrupole, ion trap, Fourier transform ion cyclotron resonance (FT-ICR), and time-of-flight (TOF) spectrometers. Typically, complexes of folded proteins or nucleic acids in solution generate signals at a limited number of charge states (e.g., m/z values) once moved into the gas phase. These multiple signals improve the accurate measurement of the molecular mass, especially when a signal from the free protein or nucleic acid is present as an internal calibrant. Several review articles have been written describing ESI-MS parameters important for the observation of non-covalent complexes using ESI-MS [5]. We touch on these instrumental parameters and solution conditions that affect detection of non-covalent complexes below.

13.2.2
Signal Abundances

Signal abundances for ligand complexes generally are dominated by the properties of the macromolecule. Hence, a comparison of the abundance of signal from free protein/nucleic acid and a ligand complex reflects solution equilibrium. A variety of methods can be used to determine dissociation constants, including Scatchard analysis or fitting to a binding polynomial [6].

Solution Conditions are Critical

Most ESI-MS instruments can produce signals from low-nanomolar concentrations of proteins, and ~50–150 nM concentrations of RNA. However, ESI-MS experiments must be performed using volatile buffers that prevent non-specific adduction of counterions (i.e., phosphate, sodium) to the molecule of interest. Acceptable buffers include ammonium or alkylammonium acetate or formate; and a volatile zwitterionic buffer such as glycine may also be used. Typical buffer concentrations range from 10 mM to 150 mM, as required to insure proper folding in solution. Metals required for activity, such as zinc or magnesium, may be employed at concentrations (10–100 µM) below their non-specific K_D for the target. Low concentrations of reducing agents such as dithiothreitol may also be added to the solutions. The addition of polyethylene glycol should be avoided. For RNA and DNA, an organic co-solvent such as methanol or isopropanol may be added to enhance the rate of macromolecular desolvation. Care must be taken to insure that the mixture of solvent, salt, and buffer does not alter the solution conformation of the macromolecule of interest. However, on most types of mass spectrometers, the ESI-MS conditions can be adjusted to take a "snapshot" of a solution binding equilibrium. Experiments should be performed with either the target or the ligand concentrations below the anticipated K_D, to insure that a true equilibrium binding is being measured. This limits the utility of ESI-MS for the characterization of high-affinity interactions, but is ideal for detecting weak complexes.

Description of Basic ESI-MS Parameters

Owing to the large number of instrument configurations and electrospray source designs presently in use, it is difficult to describe a universal set of parameters to be used for the structure–activity relationship (SAR) studies of complexes. We describe, in general terms, the electrospray source parameters that must be properly adjusted to achieve the requisite "gentle" source conditions to avoid unintentional dissociation of non-covalent complexes in the electrospray source. These parameters are divided into three general categories; and each has a significant influence on the extent to which non-covalent complexes survive the desolvation process. Because these parameters are interdependent, careful optimization is often an iterative process. Below, we discuss aspects of three key source parameters: capillary–skimmer potential difference, pressure, and temperature in the interface region.

Potential Difference

In most ESI source designs, there is a region of intermediate pressure in the first-stage vacuum region, operating at a few torr, where collisional activation can be induced when a potential difference is applied between adjacent ion-focusing elements. In ESI sources employing a heated metal capillary, the ESI emitter is normally run at a high potential (±1–5 kV) relative to the desolvation capillary operating at a modest potential (0–150 V). Immediately adjacent to the low-pressure end of the desolvation capillary is the first (and often only) skimmer cone. It is the potential difference between the desolvation capillary and this skimmer cone that controls the extent of collisional activation the ions experience as they traverse the first vacuum stage. Similarly, while ESI source designs utilizing a glass capillary often operate with the ESI emitter grounded and the front of the capillary biased at a high potential to establish the electrospray plume, the voltage on the low-pressure side of the capillary is independently adjusted to control the extent of collisional activation induced in the capillary–skimmer region. Careful control of the capillary–skimmer potential difference allows one, for example, to detect intact transition metal ion complexes, while an excessive potential difference can result in complete dissociation.

Other factors, such as the amount of buffering agent present in solution, the temperature of the interface, and the pressure in the capillary–skimmer region also can have a significant influence on the behavior of the source, *vide infra*. The reader is encouraged to become proficient with an appropriate model system before attempting studies of unknown or putative complexes.

Pressure in the Interface Region
Another important parameter that typically is *not* adjustable on commercial ESI sources is the effective gas pressure in the interface region between the capillary inlet and the mass detector. The large pressure drop in this region leads to a supersonic expansion of the incoming ions that can result in significant ion heating (or cooling), depending on the effective pressure. Manipulating the pressure in the first vacuum region directly affects the number of collisions the ions experience as they traverse the capillary–skimmer junction. Control of the pressure in this region, either by throttling the rough pump that pumps the first vacuum stage or by leaking-in additional buffer gas, can profoundly influence the extent to which non-covalent complexes survive the desolvation process [7].

Capillary Heating
All ESI-MS instruments provide a mechanism to effect desolvation through heating the charged droplets. While the hardware designs and underlying approaches to heating the ESI source vary substantially from vendor to vendor, the common goal of these configurations is to allow the operator to control a key parameter that directly influences the rate and extent of desolvation of the electrospray droplets. With the heated metal capillary interface, there is generally a minimum temperature that must be employed to effect efficient desolvation of macromolecules. Excessive heat disrupts non-covalent complexes and, in extreme cases, disrupts covalent bonds via thermally induced dissociation. It should be noted that the "right" temperature depends on the size of the non-covalent complex, buffer system, and instrument. In many cases, conditions can be identified where a very weak interaction such as binding of water molecules or ammonium/acetate ions can be used as a measure of the "harshness" of the ESI ion desolvation process.

Additional Considerations

In the gas phase, it is possible to detect ligand–target interactions stabilized by many types of molecular interactions. Coulombic interactions among charged species readily move from solution to the gas phase. However, dipole–dipole and H-bond–mediated interactions also are detected easily, as they are strengthened in the absence of waters of hydration. The stringency of the mass spectrometry assay can be adjusted as described above to eliminate or facilitate detection of low-affinity ligands.

Proper design of control experiments is also integral to interpretation of the data. Low-affinity ligands can bind to the surfaces of charged macromolecules in a non-specific fashion. Such interactions are detected readily by counterscreening against a control target with a different geometry or structure. As discussed below, other methods may be employed to assess binding specificity at the correct site, such as competitive screening in solution with a known inhibitor or determination of activation energies for gas-phase dissociation.

13.3
Ligands Selected Using Affinity Chromatography

ESI-MS has established utility as a multichannel, parallel detector for the identification of individual constituents in a mixture, based on the accurate measurement of molecular mass. ESI-MS has been combined with a variety of chromatographic techniques to identify ligands that bind protein targets, including frontal affinity chromatography, capillary zone electrophoresis, pulsed ultrafiltration chromatography, and immunoaffinity chromatography. The choice of an appropriate solution–phase separation technique is a function of many considerations, but any technique must yield fractions of compounds from a mixture in a form suitable for ESI-MS. Hence, methods such as ion exchange chromatography that require non-volatile salts are not easily interfaced to ESI-MS instruments, while reversed-phase and affinity-based selection techniques with tagged proteins are adapted readily.

Many of the concepts important for MS-based methods were highlighted by Wieboldt et al. in studies of compounds mixtures binding to diazepam antibodies [8]. Pools of known and unknown benzodiazepines were prepared (each at 1 μM concentration) and screened for binding to five antibodies (100 nM solution) raised against specific compounds. The low-affinity ligands were removed using an ultrafiltration membrane; and bound ligands were eluted from the protein using a trifluoroacetic acid wash step. This wash was fractionated using reversed-phase chromatography and individual compounds eluting from the column were identified via ESI-MS using a tandem mass spectrometer. The capture efficiency was evaluated as a function of protein concentration, competitive inhibition of binding by a preferred ligand, and changes in the relative affinity of the ligands for the target. ESI-MS/MS also was used to identify an unknown component that bound with high affinity from a mixture.

13.3.1
Antibiotics Binding Bacterial Cell Wall Peptides

Complexation between vancomycin, ristocetin A, teicoplanin and two bacterial cell-wall analogues, Ac_2-L-Lys-D-Ala-D-Ala and Ac-D-Ala-D-Ala in aqueous solutions was examined by positive-ion electrospray mass spectrometry (ESI-MS) and capillary zone electrophoresis (CZE/ESI-MS) [9]. The ESI-MS data demonstrated that simple complexes between monomeric antibiotics and either peptide could be observed, along with complexes ranging from the simple homodimer of the antibiotic to more complex associations of the type $[(antibiotic)^2 + (tripeptide)^3]$. The same data also demonstrated that the homodimers of the investigated antibiotics are significantly suppressed in the presence of the tripeptides. CZE/ESI-MS was used to confirm that the complexes between the antibiotic and the tripeptide were present in the solution prior to their introduction into the mass spectrometer.

Jorgensen et al. studied the effects of peptide length, stereochemistry, and pH on the solution affinities of vancomycin and ristocetin using ESI-MS [10]. They observed a strong correlation between solution affinities measured using CD and ESI-MS as a function of peptide composition, length, and solution pH. The effect of capillary desolvation conditions was determined; and a narrow operating range was observed where the complex was desolvated, versus conditions where the abundances of the peptide–antibiotic complex (and measured association constants) were reduced due to gas-phase, collisionally activated dissociation.

Van de Kerk-van Hoof and Heck measured solution phase affinity constants for α- and β-avoparcin for a series of bacterial cell wall receptor-mimic peptides using ESI-MS [11]. The affinity constants were similar to those observed for vancomycin, though β-avoparcin displayed higher affinity than α-avoparcin. Given the similarities in structure with vancomycin, these results support the hypothesis that the appearance of vancomycin-resistant enterococci might be linked to the widespread use of avoparcin in agricultural settings.

13.3.2
Kinases and GPCRs

Annis et al. used affinity chromatography and ESI-MS to identify ligands that bind competitively to the Akt-1 and Zap-70 kinases and a G protein-coupled receptor [12]. Their method uses two chromatography steps. An initial affinity selection is performed using a rapid (15–20 s) size-exclusion chromatography (SEC), where compounds binding the target of interest are separated from free ligands. The complex is identified from the UV absorption and loaded automatically onto a C18 column. A high-temperature, reversed-phase chromatography step separates ligands from the complex and each other prior to mass spectrometry detection. The SEC is performed at several concentrations of a known inhibitor; and the ESI-MS response for the ligand of interest is plotted as a function of the inhibitor's concentration. Competitive inhibitors can be distinguished from allosteric inhibitors or compounds that bind concurrently from the shape of the binding curve. The experiment can be performed with several compounds in parallel; and their relative affinities for the target can be ranked. The authors demonstrated mixed mode inhibition (direct competition and non-competitive binding) for a Merck compound with the ATP binding site of Akt-1. They also identified high affinity ligands for the acetylcholine receptor M2, a G protein-coupled receptor. The members of the ligand pool were competed off the receptor with increasing concentrations of atropine. Among a series of related compounds, the highest affinity ligand detected with MS also showed activity in tissue, and a SAR relationship could be established.

13.3.3
Src Homology 2 Domain Screening

Kelly et al. [13] were among the first to use affinity chromatography to separate ligand complexes for the Src homology 2 domain of phosphatidylinositol 3-kinase from combinatorial pools of compounds, coupled with ESI-MS for identification of the ligands. They demonstrated the critical importance of placing a methionine residue at +3 in a 361-member hexapeptide library derived from a phosphonodifluoromethylphenylalanine residue. Initial fractions were collected from a high salt wash at pH 7.5 (unbound ligand), followed by three washes at pH 4.5, pH 3.5, and with TFA. The pH 3.5 wash contained the majority of high-affinity ligands and was selected for further analyses. Several mass combinations could be assigned to isobaric sets of potential ligands; and the authors used MS/MS to identify the high-affinity ligands, based on their fragmentation pattern.

Lyubarskaya et al. [14] demonstrated the utility of capillary isoelectric focusing coupled with ESI-MS detection for screening a mixture of phosphorylated peptides for binding to the Src homology 2 domain. The solution mixture of potential ligands and protein was separated and concentrated using CIF. Following ionization, the high-affinity complexes were further separated, based on mass in the gas phase, and ligands were identified via dissociation of the complexes in the gas phase. In another paper, Dunayevskiy et al. simultaneously measured dissociation constants for 19 FMOC tetrapeptides to vancomycin, using affinity capillary electrophoresis with ESI-MS detection, and observed a good correlation with results obtained using capillary electrophoresis with UV detection [15].

Bligh et al. [16] used ESI-MS to measure the dissociation constants between three classes of ligands and the Src SH2 domain. The ligands containing phosphonate groups bound with dissociation constants of 3–9 µM, comparable to values obtained using fluorescence depolarization. The compound with a sulfonate group bound more weakly (~100 µM). Two binding sites, with K_D values of 9.3 µM and 193 µM, were detected for one of the phosphonate ligands. The correct binding model could only be selected based on the ligand stoichiometry data obtained with ESI-MS.

13.3.4
Other Systems

A combined affinity purification-mass spectrometry approach has been used to analyze protein–peptide interactions in a mixture [17]. The affinities of mixtures of ligands were determined for Mena protein and a tubulin-specific antibody. Peptides that bound Mena were selectively purified from a tryptic digest of ActA, and could then be mapped back onto the ActA sequence. Similarly, peptides were identified from a digest of tubulin that bound with high affinity to a tubulin-specific antibody. Affinity mass spectrometry allowed the mapping of sequential binding motifs from two interacting proteins.

Frontal affinity chromatography coupled online to mass spectrometry (FAC/MS) has been used to estimate binding constants for individual protein ligands present in mixtures of compounds [18]. In this study, FAC/MS is used to determine enzyme substrate kinetic parameters and binding constants for enzyme inhibitors. Recombinant human N-acetylglucosaminyltransferase V was biotinylated and adsorbed onto immobilized streptavidin in a microcolumn (20 µl). The enzyme was shown to be catalytically competent, transferring GlcNAc from the donor UDP-GlcNAc to beta-D-GlcpNAc-($1 \rightarrow 2$)-alpha-D-Manp-($1 \rightarrow 6$)-beta-D-Glcp-OR acceptor giving beta-D-GlcpNAc-($1 \rightarrow 2$)-[beta-D-GlcpNAc-($1 \rightarrow 6$)]-alpha-D-Manp-($1 \rightarrow 6$)-beta-D-Glcp-OR as the reaction product. The kinetic parameters K_m and V_{max} for the immobilized enzyme could be determined by FAC/MS and were comparable to those measured in solution. Analysis of a mixture of eight trisaccharide analogs in a single run yielded K_D values for each of the eight compounds ranging from 0.3 µM to 36 µM. These K_D values were two to ten times lower than the inhibition constants, K_I values, determined in solution using a standard radiochemical assay. However, the ranking order of K_D was the same as the ranking of K_I values. FAC/MS assays can therefore be employed for the rapid estimation of inhibitor K_D values, making it a valuable tool for enzyme inhibitor evaluations.

Synthetic RS20 peptide and a set of its point-mutated peptide analogs have been used to analyze the interactions between calmodulin (CaM) and the CaM-binding sequence of smooth-muscle myosin light-chain kinase, both in the presence and the absence of Ca^{2+} [19]. Particular peptides, which were expected to have different binding strengths, were chosen to address the effects of electrostatic and bulky mutations on the binding affinity of the RS20 sequence. Relative affinity constants for protein/ligand interactions have been determined using electrospray ionization and Fourier transform ion cyclotron resonance mass spectrometry. The results evidence the importance of electrostatic forces in interactions between CaM and targets, particularly in the presence of Ca^{2+}, and the role of hydrophobic forces in contributing additional stability to the complexes both in the presence and the absence of Ca^{2+}.

Zechel et al. used time-resolved ESI-MS to monitor formation of a covalent complex between a xylanase and a UV-labeled substrate [20]. ESI-MS is very powerful for this type of study, since signals can be observed from the free protein, the complex between the protein and substrate, the protein and the intermediate product, and the protein and the final product. A time resolution of 10–50 msec can be realized with the ESI-MS instrument. It could be demonstrated that the steady-state concentration of the enzyme-intermediate complex reached 95% of the available protein at saturating concentrations of substrate. The kinetic parameters determined using ESI-MS were identical to those determined using stopped-flow UV spectroscopy.

13.4
Direct Observation of Ligand–Target Complexes

The earliest reports describing the potential of ESI-MS for screening of combinatorial libraries were a series of papers from the laboratories of Smith and Whitesides [21, 22]. They described the competitive binding of a series of aryl sulfonamides to bovine carbonic anhydrase II, using FT-ICR ESI-MS. They showed that the observed relative ion intensities from the complexes correlated with the known solution affinities (nanomoles to micromoles). Further, the complexes were not detected when the protein was denatured prior to the ESI-MS experiment. Signals were detected from individual compounds in the mixture, based on accurate measurement of exact mass, even for mass differences of <14 Da in a complex with a molecular mass >25 kDa. Further, they could observe dissociation of the ligands from the complex. They studied series of compounds derived from amino acids; and they were able to establish the identity of the best inhibitors by performing MS/MS on ligands dissociated from the ions of the complex with carbonic anhydrase initially isolated in the gas phase using selective ion accumulation methods.

In subsequent papers, Gao et al. studied the binding of much larger libraries (256 compounds) to carbonic anhydrase (CA) [23, 24]. The first paper describes the use of electrospray ionization–mass spectrometry (ESI-MS) to screen two libraries of soluble compounds to search for tight binding inhibitors for CA. The two libraries, $H_2NO_2SC_6H_4C(O)NH$-AA_1-AA_2-$C(O)NHCH_2CH_2CO_2H$ where AA_1 and AA_2 were L-amino acids (library size: 289 compounds) or D-amino acids (256 compounds), were constructed by attaching tripeptides to the carboxyl group of 4-carboxybenzenesulfonamide. Screening of both libraries yielded, as the tightest binding inhibitor, compound 1 (AA_1 = AA_2 = L-Leu; binding constant K_A = 1.4×10^8 M^{-1}). The second paper describes the use of ESI-FTICR-MS to study the relative stabilities of noncovalent complexes of carbonic anhydrase and benzenesulfonamide inhibitors in the gas phase. Sustained off-resonance irradiation collision-induced dissociation was used to determine the energetics of dissociation of these sulfonamide complexes in the gas phase. When two molecules of a benzenesulfonamide were bound simultaneously to one molecule of carbonic anhydrase, one of them was found to exhibit significantly weaker binding upon sustained off-resonance irradiation. In solution, the benzenesulfonamide group coordinated as an anion to a zinc ion bound at the active site of the enzyme. The gas phase stability of the complex with the weakly bound inhibitor was the same as that of the inhibitor complexed with apoCA (CA with the zinc ion removed from the binding site). These results indicate that specific interactions between the sulfonamide group on the inhibitor and the zinc ion on CA were preserved in the gas phase. Experiments also showed a higher gas phase stability for the complex of para-nitro-benzenesulfonamide-CA than that for ortho-NO_2-benzenesulfonamide-CA complexes. This result is consistent with steric interactions of the inhibitors with the binding pocket of CA paralleling those in solution.

Jorgensen et al. studied the correlation between ESI-MS determined K_A values and those obtained using solution circular dichroism (CD) [10]. They demonstrated that vancomycin and ristocetin bound a series of D and L di- and tripeptides with affinities that matched the solution values. They derived equations to allow the calculation of binding constants for each member of a mixture of ligands and showed that the affinities were similar for a range of peptide concentrations around the solution K_A values for vancomycin. Further, the pH-dependence of the binding was identical for the ESI-MS and CD methods. In addition, they studied the effect of varying the skimmer voltage on the apparent ESI-MS K_A. The apparent K_A values were a function of the dissociation energy; and operating conditions should be selected that optimize desolvation without inducing dissociation of the non-covalent complex.

13.4.1
Observation of Enzyme–Ligand Transition State Complexes

Borchers et al. [25] used FTICR-ESI-MS to detect water molecules bound with high stability to the dimer of *E. coli* cytidine deaminase complexed with a transition state analog. The native protein binds two zinc atoms in the active site and forms a tight complex with 5-fluoropyrimidine-2-one (MW 246.07 Da). Through accurate measurement of molecular mass, they determined that two waters of hydration remained bound to the active protein dimer of the 31.5 kDa protein in the gas phase. Complexes with one and two bound inhibitors could be observed. Modeling shows these waters occupy the site normally filled by the departing ammonia molecule. The data suggests that improved inhibitors of the enzyme could be designed through incorporation of a group to fill the space of the bound water molecule.

13.4.2
Ligands Bound to Structured RNA

The Tar RNA stem-loop of HIV is an attractive target for drug discovery. During the viral life cycle, the Tar RNA binds *tat* protein and facilitates assembly of a competent transcriptional complex. Mei and co-workers screened >100 000 compounds for inhibitors of the Tar-*tat* interaction. They identified four ligands that bound the RNA and studied them using ESI-MS [26]. The compounds were shown to make contacts with the three-base internal UCC bulge, through chemical replacement of the upper tetraloop [27].

Similar methods have been applied to drug discovery efforts directed toward the 5'-UTR region of the hepatitis C virus (HVC) genomic RNA. This region contains a highly conserved structured RNA shown to be crucial for viral replication and translation, presumably by serving as an internal ribosomal entry site (IRES). Screening of a library of compounds for binding to a particular subdomain of the HCV IRES led to the identification of a hit having relatively weak affinity and modest selectivity for the target RNA. Using an MS-guided chemical optimization approach, this weak hit was elaborated into a lead structure which had sub-micromolar affinity for the target RNA and activity in a cellular assay [28].

Binding of multiple 2-deoxystreptamine (2-DOS) ligands to a model of the bacterial 16S rRNA A site was described by Griffey et al. [29]. They detected two binding sites for the 2-DOS aminosugar to the RNA at high (100 µM) ligand concentrations. The binding affinity for the second site contained fewer hydrogen bonds to the RNA, as evidenced by differences in the energy required to effect collisionally activated dissociation (CAD) of the binary and tertiary complexes in the gas phase. These results were consistent with molecular models of 2-DOS binding to the RNA, where a second binding site was identified where the D-ring of paromomycin normally binds.

13.4.3
ESI-MS for Linking Low-affinity Ligands

The low hit rates for RNA targets in traditional HTS assay formats can be traced to difficulties in detecting and accurately measuring low-affinity interactions between small molecules and the RNA. We have developed a high-throughput MS-based assay that directly measures ligand affinities of 0.01–1000.0 µM for RNA targets. In contrast to traditional HTS assays, the MS-based assay accurately quantifies binding affinity, stoichiometry, and specificity over a wide range of ligand K_D values. This highly quantitative information allows a pattern of SARs to emerge, even among relatively weak binders, and guides elaboration to higher-affinity compounds.

These advances in screening methods allow extension of drug design approaches that exploit information derived from studying weak ligand–target interactions such as "SAR by NMR" to RNA targets employing the MS-based assay. This "SAR by MS" process (Fig. 13.1) begins by screening a set of compounds for prospective binding affinity for an RNA target of interest. The new "motifs" that bind to the RNA are identified and the specific nature of their binding is probed through chemical elaboration and/or additional MS experiments. Ultimately, the accumulated SAR for a particular target suggests a pharmacophore hypothesis comprising key structural features of two or more ligands. This, in turn suggests that these motifs be incorporated into a single chemical entity. The concept that appropriate linking of two or more weak binding motifs will provide a large gain in binding energy has been demonstrated in several applications of the SAR by NMR strategy, as well as a combinatorial chemistry based approach.

The SAR by MS method has been used to identify a new class of ligand for the 1061A region of the bacterial 23S rRNA [30]. This portion of the rRNA interacts

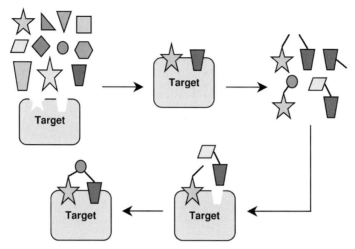

Fig. 13.1
The SAR by MS screening and lead optimization paradigm.

with the protein L11 and is the site of binding for the antibiotic thiostrepton. Our initial attempts at lead discovery for this antibacterial target via traditional HTS assays afforded extremely low hit rates. Although a crystal structure of the RNA–protein interaction is available, it is not amenable to traditional structure-based rational drug design approaches due to the large and complex nature of the interaction, making it an ideal target for the ligand-based approach offered by the SAR by MS strategy.

A screen of compound libraries revealed two classes of motifs (Fig. 13.2) that displayed an interesting SAR toward the U1061A RNA target (Fig. 13.3). The A-series ligands (**A1–A6**) are peptidic in nature, with the C-terminal and N-terminal ends of a D-amino acid moiety substituted with piperazine and carboxamide units, respectively. A positively charged side-chain is clearly beneficial for binding, as unsubstituted and neutral substituted derivatives display poor affinity. The N-substitution on the amino acid was less important, as **A1**, **A5**, and **A6** all showed similar affinity. The B-series motifs (**B1–B6**) contained piperazinyl-substituted aryl carboxylic acids. From this screen, the quinoxalin-2,3-dione unit of **B5** emerged as an important pharmacophore unit, as other aryl piperazines such as **B6** showed no affinity. We next prepared a series of related structures having various substitutions off the motif B quinoxalin-2,3-dione carboxyl group. The SAR indicates that diverse substitution off the quinoxalin-2,3-dione is tolerated and that there is substantial space for attaching pendant groups at the carboxylic acid position.

In order to further study the spatial relationships of motif binding to the U1061A construct, competition experiments were performed between several ligands that bind the RNA when examined singly. Because the ligand classes A (peptidic) and B (quinoxalin-2,3-dione) are distinctly different from a structural point of view, it was postulated that they occupy different binding locations on the

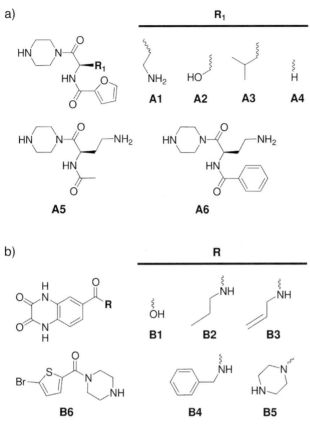

Fig. 13.2
Structures of selected motifs screened against the 23S rRNA L11 binding site subdomain. (a) Peptidic motif (series A). (b) Quinoxalin-2,3-diones (series B).

target RNA. That this is the case was made evident by a competition experiment, in which a ternary complex consisting of the U1061A RNA, **A1** and **B1** was observed (concurrent binding, Fig. 13.4). In contrast, the propyl-substituted ligand (**B2**) was completely displaced from the target RNA by the stronger-binding **A1** and no ternary complex was observed (competitive binding). Interestingly, the allyl-substituted ligand (**B3**) only binds the target RNA in the presence of **A1** (the ternary U1061A:**A1**:**B3** is observed), but no U1061A:**B3** complex is evident. This is suggestive of cooperative binding, as **B3** does not bind the target in a competitive setting *unless* **A1** is also bound. A possible explanation for the differences found between the allyl- and propyl-substituted ligands is that the alkene of the allyl group interacts favorably with the aromatic furan of **A1**, allowing for the formation of a ternary complex despite the close proximity of binding sites indicated by the competition of **A1** and **B2**.

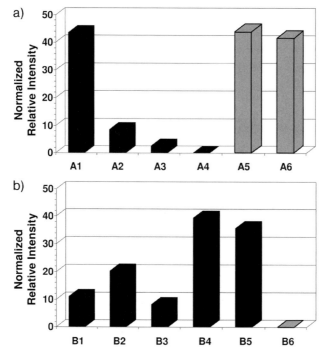

Fig. 13.3
Normalized binding intensities of the RNA:ligand complexes relative to the parent RNA. (a) Screens performed using 50 μM ligands + 2 μM RNA. (b) Screens performed using 150 μM ligands + 2 μM RNA.

The information derived from these simple competition experiments provides a "molecular ruler" with which to gauge the separation of the binding sites on the target RNA. Based on the binding data obtained, a linking hypothesis is suggested whereby motif **A1** is separated by three atoms from the quinoxaline-dione of motif **B1**. The similar affinity of **A1**, **A5**, and **A6** suggests that the aromatic portion of motif A is not critical, and may serve as an appropriate linking position. To test this hypothesis, several rigid and flexible linkers of the appropriate length were prepared via a solid phase synthesis, employing palladium-catalyzed cross-coupling reactions as the key linker assembly step. The linked compounds **10** were evaluated both for binding affinity to the U1061A target, as well as for their ability to inhibit bacterial transcription/translation in a cell-free functional assay (Fig. 13.5). The linked compounds **10a–10f** all bound markedly tighter to the target RNA than the parent motifs (estimated K_D values of 6–50 μM vs >100 μM for the motifs) and also displayed considerable sensitivity to linker size and orientation. Of particular note is the lower affinity for the flexible linker **10f**. This is consistent with the dynamic and flexible nature of RNA targets and highlights the importance of providing a rigid framework for binding elements. In contrast to the

13.4 Direct Observation of Ligand–Target Complexes

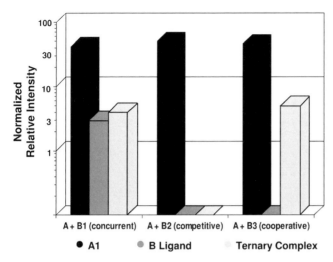

Fig. 13.4
ESI-MS competition experiments with ligands. Ligands **A1** and **B1** bind concurrently to the RNA target. In contrast, ligands **A1** and **B2** bind competitively, with only the complex due to binding of **A1** observed. Ligands **A1** and **B3** exhibit cooperative binding, as binding of **B3** to RNA is not observed unless **A1** is present. The propyl and allyl groups thus serve as "molecular rulers", with an optimal separation of ligands A and B consisting of three atoms.

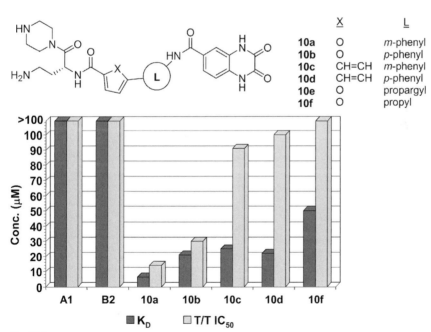

Fig. 13.5
Dissociation constant (K_D) and bacterial transcription/translation (T/T) IC_{50} for linked structures.

flexible linker, the rigidly linked **10a** displays a 20-fold enhancement in affinity for the RNA target relative to the motif ligands, with a K_D of 6.5 µM. Furthermore, this compound has similar activity (IC_{50} = 14 µM) in the related functional assay, indicating that **10a** is binding to the target RNA in a manner that interferes with bacterial translation.

13.5
Unique Features of ESI-MS Information for Designing Ligands

ESI-MS analysis provides unique information on the identity and binding stoichiometry of ligands for protein and RNA targets. ESI-MS can be used in an indirect mode to characterize ligands selected using a chromatography technique, or as a direct method to identify and characterize the binding locations, stoichiometry and affinity of ligands. ESI-MS can be used to detect water molecules bound in an active site of a protein or a protein–ligand complex, information of value for increasing the potency of a lead. The relative strength of interaction can be measured for multiple ligands, bound either competitively or non-competitively. ESI-MS screening methods have utility for drug discovery against both protein and RNA targets, providing quantitative information on ligand binding to direct lead optimization.

References

1 Smith, R. D., Light-Wahl, K. J., Winger, B. E., Loo, J. A. **1992**, *Org. Mass Spectrom.* 27, 811–821.
2 Smith, R. D., Bruce, J. E., Wu, Q. Y., Lei, Q. P. **1997**, *Chem. Soc. Rev.* 26, 191–202.
3 Loo, J. A. **1995**, *Bioconjug. Chem.* 6, 644–665.
4 Loo, J. A. **1997**, *Mass Spectrom. Rev.* 16, 1–23.
5 Hofstadler, S., Sannes-Lowery, K., Hannis, J. **2005**, *Mass Spectrom. Rev.* 24, 265–285.
6 Sannes-Lowery, K. A., Griffey, R. H., Hofstadler, S. A. **2000**, *Anal. Biochem.* 280, 264–271.
7 Sannes-Lowery, K. A., Hofstadler, S. A. **2000**, *J. Am. Soc. Mass Spectrom.* 11, 1–9.
8 Wieboldt, R., Zweigenbaum, J., Henion, J. **1997**, *Anal. Chem.* 69, 1683–1691.
9 Hamdan, M., Curcuruto, O., DiModugno, E. **1995**, *Rapid Commun. Mass Spectrom.* 9, 883–887.
10 Jorgensen, T., Roepstorff, P., Heck, A. **1998**, *Anal. Chem.* 70, 4427–4432.
11 van de Kerk-van Hoof, A., Heck, A. **1999**, *J. Antimicrob. Chemother.* 44, 593–599.
12 Annis, D. A., Nazef, N., Chuang, C. C., Scott, M. P., Nash, H. M. **2004**, *J. Am. Chem. Soc.* 126, 15495–15503.
13 Kelly, M., Liang, H., Sytwu, I., Vlattas, I., Lyons, N., Bowen, B., Wennogle, L. **1996**, *Biochemistry* 35, 11747–11755.
14 Lyubarskaya, Y., Carr, S., Dunnington, D., Prichett, W., Fisher, S., Appelbaum, E., Jones, C., Karger, B. **1998**, *Anal. Chem.* 70, 4761–4770.
15 Dunayevskiy, Y., Lyubarskaya, Y., Chu, Y., Vouros, P., Karger, B. **1998**, *J. Med. Chem.* 41, 1201–1204.
16 Bligh, S. W., Haley, T., Lowe, P. N. **2003**, *J. Mol. Recognit.* 16, 139–148.
17 Rudiger, A., Rudiger, M., Carl, U., Charkraborty, T., Roepstorff, P., Wehland, J. **1999**, *Anal. Biochem.* 275, 162–170.
18 Zhang, B., Palcic, M., Schriemer, D., Alvarez-Manilla, G., Pierce, M., Hindsgaul, O. **2001**, *Anal. Biochem.* 299, 173–182.

19 Nousiainen, M., Derrick, P., Lafitte, D., Vainiotalo, P. **2003**, *Biophys. J.* 85, 491–500.

20 Zechel, D. L., Konermann, L., Withers, S. G., Douglas, D. J. **1998**, *Biochemistry* 37, 7664–7669.

21 Bruce, J. E., Anderson, G. A., Chen, R., Cheng, X., Gale, D. C., Hofstadler, S. A., Schwartz, B. L., Smith, R. D. **1995**, *Rapid Commun. Mass Spectrom.* 9, 644–650.

22 Cheng, X., Chen, R., Bruce, J., Schwartz, B., Anderson, G., Hofstadler, S., Gale, D., Smith, R. **1995**, *J. Am. Chem. Soc.* 117, 8859–8860.

23 Gao, J., Wu, Q., Carbeck, J., Lei, Q. P., Smith, R. D., Whitesides, G. M. **1999**, *Biophys. J.* 76, 3253–3260.

24 Gao, J., Cheng, X., Chen, R., Sigal, G., Bruce, J., Schwartz, B., Hofstadler, S., Anderson, G., Smith, R., Whitesides, G. **1996**, *J. Med. Chem.* 39, 1949–1955.

25 Borchers, C. H., Marquez, V. E., Schroeder, G. K., Short, S. A., Snider, M. J., Speir, J. P., Wolfenden, R. **2004**, *Proc. Natl. Acad. Sci. USA* 101, 15341–15345.

26 Sannes-Lowery, K. A., Mei, H.-Y., Loo, J. A. **1999**, *Int. J. Mass Spectrom.* 193, 115–122.

27 Sannes-Lowery, K. A., Hu, P., Mack, D. P., Mei, H.-Y., Loo, J. A. **1997**, *Anal. Chem.* 69, 5130–5135.

28 Seth, P. P., Miyaji, A., Jefferson, E. A., Sannes-Lowery, K. A., Osgood, S. A., Propp, S. S., Ranken, R., Massire, C., Sampath, R., Ecker, D. J., Swayze, E. E., Griffey, R. H. **2005**, *J. Med. Chem.* 48, 7099–7102.

29 Griffey, R. H., Sannes-Lowery, K. A., Drader, J. J., Mohan, V., Swayze, E. E., Hofstadler, S. A. **2000**, *J. Am. Chem. Soc.* 122, 9933–9938.

30 Swayze, E. E., Jefferson, E. A., Scannes-Lowery, K. A., Blyn, L. B., Risen, L. M., Ara Kawa, S., Osgood, S. A., Hofstadler, S. A., Griffey, R. H. **2002**, *J. Med. Chem.* 45, 3816–3819.

14
Tethering

Daniel A. Erlanson, Marcus D. Ballinger, and James A. Wells

14.1
Introduction

One of the common challenges throughout this book is how to identify small chemical fragments that bind weakly to target biological molecules. Among fragment-based approaches, *Tethering* is unique in using a covalent, reversible bond to stabilize the interaction between a fragment and a target protein [1, 2].

The general process is outlined in Fig. 14.1. First, a cysteine residue is either co-opted or introduced in a target protein. Metaphorically, the cysteine residue serves as a fishing line to capture fragments (fish) that bind near the cysteine. The protein is incubated with pools of thiol-containing, small molecule fragments which are conjugated to a common, hydrophilic thiol (such as cysteamine) for improved water solubility. By controlling the redox conditions in the experiment with exogenous reducing agents, equilibria can be established so that the cysteine residue in the protein *reversibly* forms disulfide bonds with the individual fragments. In the absence of any affinity between a fragment and the protein, no fragment should bind more favorably than any other; a pool of fragments produces a statistical mixture of different protein–fragment complexes, as well as unmodified protein and cysteamine-modified protein. However, if a fragment has inherent affinity for the protein and binds near the cysteine residue, the fragment–protein conjugate is stabilized, and this complex predominates. A fragment thus selected can be easily identified by mass spectrometry of the equilibrium mixture, and if each fragment in a pool has a unique molecular weight, so do the resulting protein–fragment conjugates. While the identified fragments are often weak ligands, X-ray crystallography of the protein–fragment conjugates is often facilitated by the covalent bond. These captured fragments then serve as starting points for conversion to non-covalent ligands by chemical optimization and removal of the thiol functionality.

In the following pages, we present an overview of the theory and uses of Tethering. After first considering the basis of the technique in thermodynamic terms (section 14.2), we show how the technology can be used in the active sites of enzymes to identify fragments (section 14.4), which can then be elaborated to more potent inhibitors. Section 14.5 considers how Tethering can be used to not only

---

Fragment-based Approaches in Drug Discovery. Edited by W. Jahnke and D. A. Erlanson
Copyright © 2006 WILEY-VCH Verlag GmbH & Co. KGaA, Weinheim
ISBN: 3-527-31291-9

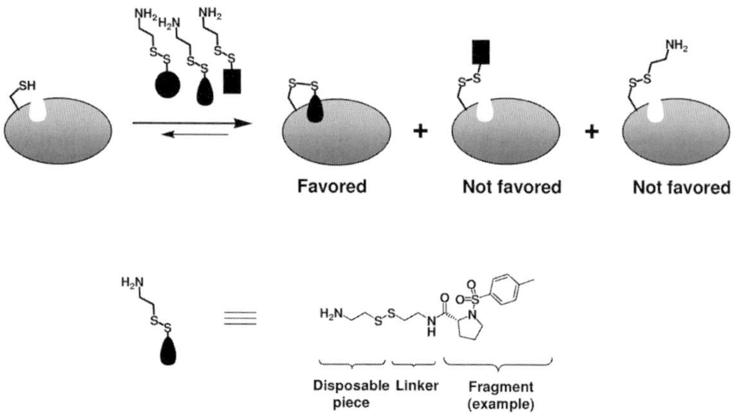

Fig. 14.1
Tethering schematic. A fragment is selected if it has inherent affinity for the protein and binds in the vicinity of the cysteine residue. An example disulfide-containing fragment is shown below, illustrating the variable "fragment" portion, the linker, and the cysteamine "disposable piece" lost when the fragment forms a disulfide bond with the protein.

identify fragments but also to link these fragments to more rapidly identify starting points for drug discovery. In section 14.6, we consider the uses of Tethering to identify allosteric sites as well as to probe targets typically considered unamenable to fragment-based techniques. Finally, we consider some variations of the technique and related technologies from other laboratories (section 14.7).

14.2
Energetics of Fragment Selection in Tethering

At the core of the technology lies the thermodynamics of the disulfide bond. This bond amplifies the binding energy between the fragment and the protein by raising the effective concentration of the fragment at the site of interest and so pays some of the entropic cost of fragment binding. As discussed in Chapter 3 of this book, the cost to non-covalent small-molecule ligand binding exacted by the loss of rigid-body entropy has been estimated at 4 kcal mol^{-1}, or three orders of magnitude at 298 K [3]. A powerful feature of Tethering is the ability to control the degree to which the disulfide aids fragment binding. By modulating the ratio of oxidized to reduced thiols in solution, the selection stringency in a screen can be adjusted for the strength of hits expected or desired.

During Tethering, a complex equilibrium is established in which the concentrations of reduced species and disulfide pairs are dictated by the reduction potentials of the thiols involved and by the ratio of oxidized to reduced species in solution. The reduction potential, or, in ΔG terms, the conjugation energy, of a fragment thiol to the target cysteine is determined by: (a) the inherent chemical energy (redox potential) of the fragment-cysteine disulfide bond, in thiol, (b) non-covalent interactions between the protein and fragment, and (c) the entropic and non-covalent interaction energies associated with the linkage between the protein and the bound fragment. The energies associated with protein–fragment conjugates have been presented in terms of these three factors in the context of describing the cooperativity of fragment binding to IL-2 [4]. For a single fragment–target protein pair, these can be written as in Eq. (1):

$$\Delta G_{conjugation} = \Delta G_{thiol} + \Delta G_{noncov} + \Delta G_{linker} \tag{1}$$

While relative $\Delta G_{conjugation}$ estimates might be obtained for selected fragment–protein pairs via titrations of either the reductant or fragment concentrations, estimates of the three components of this energy are not possible from these experiments alone.

ΔG_{thiol} depends on the inherent reduction potentials of the protein cysteine and fragment thiols and considers the chemical bond apart from any non-covalent interactions. To normalize out the contribution of ΔG_{thiol} to conjugation amongst the fragments in a screening experiment, the library of fragments can be constructed such that the thiol portion of the molecules is kept constant or very similar, e.g. $HS(CH_2)_nCONHR$, with $n = 1-3$ but constant in any given pool. Measurements of the reduction potentials of a variety of simple thiols indicate that the disulfide bonds formed by these groups have similar energies [5, 6]. They also have energies similar to the oxidized 2-mercaptoethanol or cystamine disulfide bonds; thus, apart from additional interactions between the fragments and the protein, the fragments are on equal footing with the simple buffering thiols in solution.

The ΔG_{noncov} component comprises the energies one expects to directly translate from fragment preference in a Tethering experiment to ligand affinity upon removal of the disulfide linker. This component may include contributions from hydrophobic interactions, hydrogen bonds, charge–charge interactions, etc. While predicting the binding affinities of non-covalently linked fragments to their target proteins based on Tethering data is desirable, this has been difficult in practice due to the contribution of the last component of the conjugation energy, ΔG_{linker}.

ΔG_{linker} includes any non-covalent interactions between the linker and the protein as well as the entropic effect of the tether. The non-covalent interaction is difficult to assign unequivocally to the "linker" or the "fragment", though this could in principle be assessed quantitatively by appropriate structure–activity relationship (SAR) experiments on non-covalent fragment ligands. The entropic effect of the linker refers to how the linker influences the probability that the fragment occupies its preferred site in its preferred orientation at a given time. This is dictated not only by the thiol linker on the fragment, but by the protein dynamics of the segment containing the target cysteine as well. The entropic component of ΔG_{linker} is likely to vary significantly from linker to linker and may have a strong impact on the overall conjugation strength.

Fig. 14.2
Three scenarios illustrating Tethering with the same fragment but with different linkers and different cysteine residues. In A, the linker is too short and the fragment does not bind effectively. In B, the linker is long and the fragment receives little entropic binding advantages. In C, the linker length provides a large entropic contribution to fragment conjugation energy.

This concept is illustrated by considering three possible Tethering scenarios (Fig. 14.2). In all three, the ΔG_{thiol} and ΔG_{noncov} are assumed to be the same and the linker is assumed to have no significant non-covalent interactions with the protein. In the first case (A), the fragment–protein linkage does not allow the fragment to nestle into its preferred binding mode; the ΔG_{linker} is expected to be highly unfavorable. In (B), the linker is long and floppy; while the fragment can bind in its preferred orientation, it is not well constrained to this position, and ΔG_{linker} is expected to be only modestly favorable. In (C), the target cysteine is in an ideal position, the fragment linker is small and more rigid, such that the fragment is constrained into its preferred binding mode, and ΔG_{linker} is expected to be highly favorable.

While the entropic effect of ΔG_{linker} is difficult to quantify, estimates may be possible in some cases, by using literature calculations of the entropic cost of freezing rotatable bonds [7]. These studies indicate that the magnitude of ΔG_{linker} may be dramatically affected by linker structure and length. Zhou has also reported a method for quantifying the entropic effect of linkers on binding interactions for linked protein ligands [8, 9].

Variable ΔG_{linker} values for different target cysteines and different fragment-binding modes severely hamper efforts to translate conjugation strengths to non-covalently ligand affinity in either an absolute or relative sense. However, for like-binding fragments, a relative translation is tenable. When such a series has the same linker and is conjugated to the same cysteine residue on a protein, the ΔG_{linker} values should be essentially constant. Thus chemical optimization of fragments for increased disulfide conjugation strength, as long as they do not affect the linker moiety, should yield similar improvement in non-covalently linked ligand affinity. In other words, the disulfide linkage does not preclude fragment optimization. Indeed, the linkage is necessary when a fragment's non-covalent affinity is still too weak to detect.

14.3
Practical Considerations

The reversible covalent bond formation in Tethering has both advantages and disadvantages. On the positive side, mass spectrometry detects hits quickly and, as a "positive" detection method, is less prone to false positives than typical inhibition assays [10]. In addition, covalent bond formation gives a rough indication of the site where the fragment binds, since it must be within several Ångstroms of the cysteine residue. The bond facilitates modeling and crystallography: if the fragment is not highly soluble by itself, crystallography of the non-covalent complex may be difficult, while the disulfide-bonded complex is more likely to yield a structure. A related advantage is that the stoichiometry of the fragment in the complex is exactly one-to-one.

Of course, the very ability to detect weak binders is a somewhat double-edged sword. While ΔG_{noncov} values for typical small fragments may initially be small, they can be improved through chemical optimization. Still, optimizing a very weakly binding fragment can be challenging, although a second-generation version of the technology, *Tethering with Extenders*, largely solves this problem (see below). One potential limitation of Tethering is the need to generate cysteine-containing mutant proteins, but with the modern tools of site-directed mutagenesis and protein expression, this is rarely a significant hurdle. Tethering does require sufficient knowledge about a protein's structure to inform where to place the cysteine; while a crystal structure is not required, a good model of the protein is essential. A more serious endeavor is the synthesis of the disulfide-containing fragment library: very few disulfide-containing fragments are commercially available, and introducing a disulfide onto a fragment requires at least one additional chemical step. And finally, although the mass spectrometry screen is rapid, it does require an expensive piece of equipment.

14.4
Finding Fragments

14.4.1
Thymidylate Synthase: Proof of Principle

We first applied Tethering to thymidylate synthase (TS). This enzyme converts deoxyuridine monophosphate (dUMP) to thymidine monophosphate (dTMP), an activity essential for DNA synthesis. The cancer drug 5-fluorouracil irreversibly inhibits TS, and a selective inhibitor of a non-human form of the enzyme could yield a new antibiotic or antifungal drug [11].

In addition to its biological interest, TS was ideally suited for developing Tethering [12]. It is well characterized both structurally and mechanistically, and the number of inhibitors developed for the enzyme demonstrates that it is a "druggable" target. Moreover, the active site contains a nucleophilic cysteine residue.

Although the E. coli version of the enzyme that we used contains four other cysteine residues, crystallography revealed these to be largely non-surface exposed, and they did not interfere with our experiments.

Initial experiments screened pools of ten compounds, each present in roughly ten-fold excess over TS, with a total disulfide concentration of about 2 mM and a reducing agent (2-mercaptoethanol) concentration of 1 mM. After screening about 1200 compounds, we saw a strong selection for N-phenyl-sulfonamide-substituted proline fragments, as represented by N-tosyl-D-proline. This fragment could be selected from a pool of 100 compounds, each present at roughly the same concentration as TS. However, larger pools have more compounds with similar molecular weights, which makes the data challenging to interpet. In practice, pools of 5–10 compounds strike a balance between throughput and unambiguous interpretation.

As stated previously, a critical feature of Tethering is that thermodynamics govern disulfide bond formation. This control was assured by adding a reducing agent (2-mercaptoethanol). Without reducing agent, the active-site cysteine reacts with whichever disulfide it encounters first, usually the solubilizing element common to all library members. Although even a small amount of reducing agent allows disulfide exchange, the N-tosyl-D-proline fragment could tolerate strongly reducing conditions. In fact, even in the presence of 20 mM of 2-mercaptoethanol, where the ratio of reductant to total thiol was 10:1, a prominent peak corresponding to N-tosyl-D-proline was still observable.

Screens with chemically similar fragments showed that, although substitutions around the aromatic moiety and in the stereochemistry of the proline residue preserved the fragment's conjugation strength, the proline residue itself was essential. Crystallography of N-tosyl-D-proline covalently linked to TS explained these SARs: the proline residue sits snugly within a hydrophobic pocket and one of the sulfonamide oxygen atoms makes a hydrogen bond to Asn 177 on the enzyme, but the phenyl ring is in a relatively open area (Fig. 14.3).

To learn whether the disulfide bond itself changed how the fragment binds, we determined the crystal structure of N-tosyl-D-proline bound non-covalently to TS.

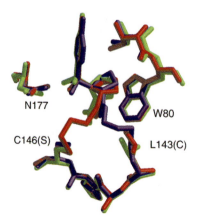

Fig. 14.3
Structures of TS with the N-tosyl-D-proline fragment bound through two different cysteine residues (red, blue) or non-covalently bound (green). Reprinted with permission from [2].

As shown in Fig. 14.3, the "free" fragment binds in a nearly identical manner to the disulfide-linked fragment, demonstrating that the covalent linkage does not affect how the fragment binds.

To test whether nearby cysteines would be suitable for Tethering, we mutated the active-site cysteine to a serine and introduced a new cysteine nearby (C146S, L143C). When we performed Tethering on this mutant enzyme, we also strongly selected N-tosyl-D-proline; and when we solved the X-ray crystal structure, we found that this fragment binds in a manner very similar to the other structures, despite the very different trajectories that the disulfide linkage takes (Fig. 14.3). The lack of influence of the disulfide attachment on the fragment's binding mode, along with the fact that the fragment could be strongly selected from more than one cysteine residue, suggested the inherent fragment affinity was more important energetically than the specifics of how it was linked to the protein.

Enzymatic assays determined the inhibitory potential of N-tosyl-D-proline: the fragment has a K_i of 1.1 mM, so weak that it likely would be missed in any conventional screen. However, the crystal structure shows that the phenyl group binds in a similar position to the *p*-amino-benzoic acid moiety of the natural co-factor, methylenetetrahydrofolate (mTHF); and by simply grafting the glutamate moiety from this co-factor onto N-tosyl-D-proline, we boosted the affinity 40-fold to 24 µM. A small library of compounds with substitutions off the proline yielded a compound with a K_i of 330 nM, three orders of magnitude more potent than the original fragment (Fig. 14.4). The crystal structures of both of these compounds revealed that the central phenyl-sulfonamide-proline moiety remains in place, while the added appendages reach out to make new contacts within the large active site of the enzyme.

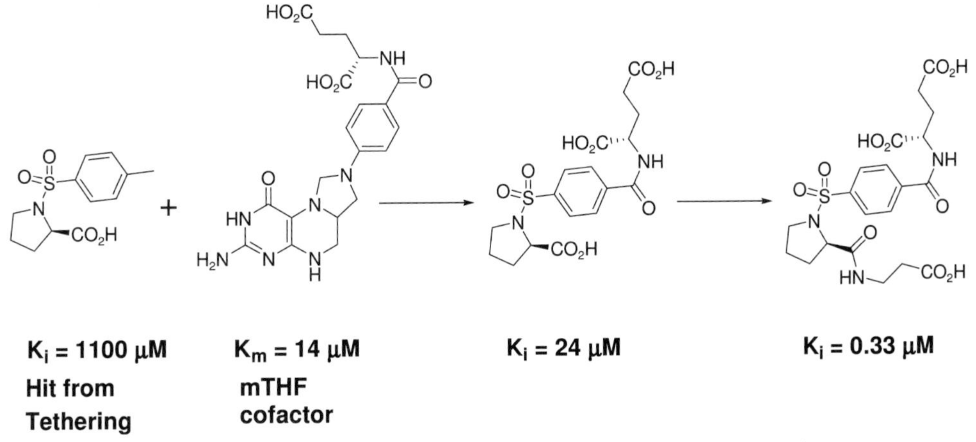

K_i = 1100 µM
Hit from Tethering

K_m = 14 µM
mTHF cofactor

K_i = 24 µM

K_i = 0.33 µM

Fig. 14.4
Improvements in potency of N-tosyl-D-proline. Structural analyses revealed that the glutamate moiety from the mTHF cofactor could be appended to the hit from Tethering, and further elaboration led to a sub-micromolar inhibitor.

14.4.2
Protein Tyrosine Phosphatase 1B: Finding Fragments in a Fragile, Narrow Site

TS offered an accessible, hardy active site. To see whether Tethering could work for more difficult sites, we turned to protein tyrosine phosphatase 1B (PTP1B). The enzyme PTP1B, a key regulator of metabolism, dephosphorylates the insulin receptor, in effect turning it off. Thus, the protein is a key drug target for both diabetes and obesity [13]. However, the enzyme has evolved to recognize phosphotyrosine, a highly negatively charged moiety for which there are very few drug-like isosteres. We wanted to investigate whether Tethering would allow us to find a replacement for phosphotyrosine that binds in the active site [14].

For TS, we used the catalytic cysteine within the active site of the enzyme to capture the fragments. PTP1B also contains an active-site cysteine residue, but it is located at the bottom of the very deep and narrow active site and is critical for interacting with the phosphotyrosine. Moreover, the residues immediately around the active site are highly conserved; and we found that mutating these to cysteine residues destroyed or severely degraded enzymatic activity. In continuing with the fishing metaphor, we could not get a good foothold on the "marshy shore" of the active site to cast a line.

This problem was solved by Tethering with Breakaway Extenders, illustrated in Fig. 14.5, in which a single molecule, the Breakaway Extender, both protects the active-site cysteine from modification and positions another thiol for Tethering. This molecule contains an internal thioester linkage as well as an alkylating functionality. The alkylating functionality modifies a cysteine introduced some distance from the active site. For PTP1B, this Breakaway Extender also contains a known phosphotyrosine mimetic that occupies the active site and sterically blocks alkylation of the active site thiol. After alkylation of the introduced cysteine residue, the thioester linkage is chemically cleaved with hydroxylamine and the phosphotyrosine mimic breaks away, revealing a new thiol near the active site. In effect, the scheme lengthens the introduced cysteine residue so the terminal thiol can access a fragile, conserved part of the target protein without disrupting it.

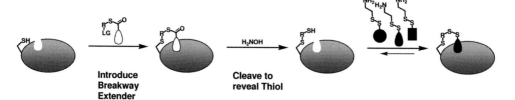

Fig. 14.5
Tethering with Breakaway Extenders. A Breakaway Extender is used to modify a cysteine residue some distance from the site of interest; this Extender can then be cleaved to reveal a new thiol for Tethering.

Several hits were identified when the modified PTP1B was screened against our library of disulfide-containing fragments. All selected fragments (not surprisingly) bore a negative charge. By far the most prominent hit was a pyrazine-carboxylic acid, which had a comparable potency to phosphotyrosine but had never been described as a phosphotyrosine mimetic. A crystal structure of this fragment covalently bound to PTP1B revealed that it binds in the active site by forcing open the so-called WPD loop to make room for its larger bulk. This form of PTP1B is less commonly observed than the catalytically active "closed" conformation. The fact that Tethering selected a molecule that binds in the open conformation is typical of the empirical nature of the approach: the protein adapts to suit the needs of the fragment. This means Tethering can identify binding modes that are unlikely to be predicted *a priori*. This ability has powerful implications for the discovery of novel inhibitors, as we see in the next section.

14.5
Linking Fragments

14.5.1
Interleukin-2: Use of Tethering to Discover Small Molecules that Bind to a Protein–Protein Interface

Protein–protein interactions regulate nearly every aspect of cell biology. Thus, small molecules that can bind at protein–protein interfaces have long been sought by the pharmaceutical industry. However, the interfaces in protein–protein interactions are often flat and lack deep pockets that can envelop a small molecule. Functional analysis of these interfaces shows that, despite the large contact interface, binding affinity is driven by a small subset of the residues, the so-called hot spots [15, 16]. Nonetheless, transferring hot-spot interactions from a protein to a small molecule is daunting. Unlike enzymes, which often offer small substrates to mimic, these targets have no such small molecule starting points. Extensive screening in the pharmaceutical industry has been remarkably unsuccessful in identifying validated hits.

Nevertheless, some potent inhibitors have been found for these difficult targets [17]. One of the best examples is the drug target interleukin-2 (IL-2), a powerful cytokine responsible for T-cell activation. Tilley and co-workers reported the first example of a low-micromolar compound (Ro26–4550) that bound to the cytokine with unit stoichiometry [18]. Peptides derived from the hot spot on IL-2 were altered by medicinal chemistry to find compounds that would bind to the IL-2 alpha receptor. Surprisingly, after several cycles of chemistry the compounds that most potently disrupted the interaction between the cytokine and the receptor bound not to the receptor but to IL-2 itself. Nevertheless, the work shows that low micromolar affinity compounds can bind stoichiometrically to such targets, even when they do not offer obvious binding sites.

To understand how this compound bound to IL-2, we solved the X-ray structure of the complex [19]. Surprisingly, the compound traps a conformation of the IL-2 receptor-binding site quite unlike the native structure of IL-2. Clearly this protein surface is dynamic, so rational design using a static structure of the native cytokine would not have proposed this molecule or exploited its binding site. Such molecules need to be found empirically.

Tethering was used to identify new fragments that target this site. Twelve mutants were prepared, each with a cysteine placed near the site that binds the IL-2 alpha receptor (Fig. 14.6 a); all the mutants were able to bind the IL-2 beta-receptor. These were individually screened using a disulfide fragment library containing about 7000 compounds. Each cysteine mutant selected a unique set of hits (Fig. 14.6 b), with some fragments shared by neighboring cysteines. The hits represented a mosaic of chemotypes that could bind the surface (Fig. 14.6 c). Several hits were chemically similar to fragments of Ro26–4550, including the guanido and bi-aryl functionalities. The more dynamic areas of IL-2 picked up more hits,

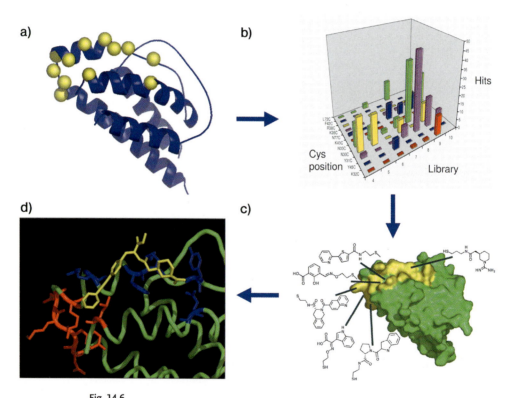

Fig. 14.6
(a) Location of the cysteine mutants introduced into IL-2. (b) A "chemoprint" illustrating the number of hits from different fragment libraries against different cysteine-containing mutants. This graphical representation illustrates that some cysteines select many fragments (for example, N33C) while others select very few (F42C). (c) Examples of some of the fragments selected from different cysteine residues, illustrating the range of functionalities selected. (d) Structure of Ro26–4550 bound to IL-2; residues colored blue selected relatively few fragments when mutated to cysteine, while residues colored red selected many.

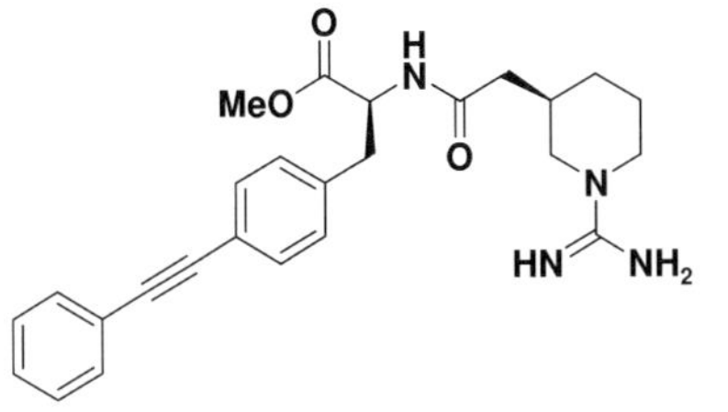

Fig. 14.7
Inhibitors of IL-2. See text for details.

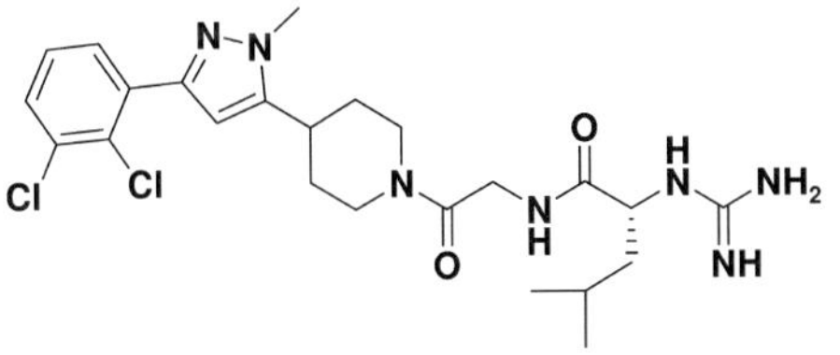

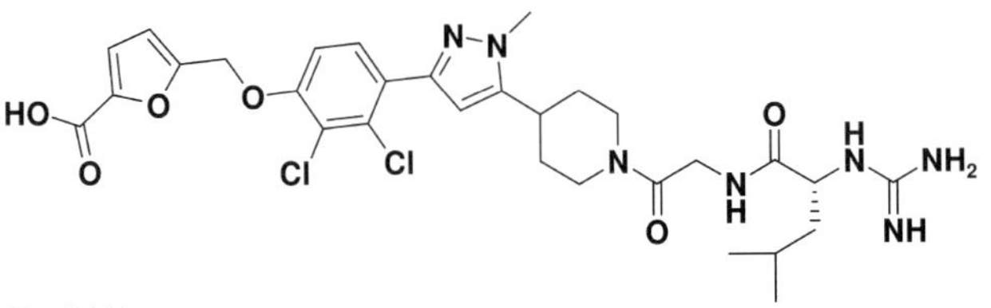

as expected since a more dynamic surface offers more opportunities to bind a small molecule (Fig. 14.6d).

Using information from Tethering, we assembled modules in a semi-combinatorial approach [20–22]. A bi-aryl acetylene piece was used to anchor a series of dipeptides to identify the optimal cyclohexyl guanido partner. The guanido fragment then served as an anchor to identify an optimal hydrophobic piece, a tri-cyclic pyrazole fragment. Further medicinal chemistry produced Compound 1 (Fig. 14.7), which has an IC_{50} of 3 µM for blocking the binding of IL-2 to the IL-2 alpha-receptor. After Tethering showed that IL-2 preferred aromatic acids next to the tricyclic pyrazole fragment, a 20-member library of aromatic acid fragments was appended to the terminal ring. This exercise identified eight compounds having sub-micromolar affinity. One compound containing a carboxyl-substituted furan, Compound 2, bound with unit stoichiometry to IL-2 and inhibited the interaction with IL-2 alpha receptor with an IC_{50} of 60 nM. This represents the highest-affinity small molecule compound yet reported for binding a cytokine. The compound blocked signaling in a cell line over-expressing the IL-2 receptor, with an IC_{50} of 3 µM; and it showed exquisite selectivity for IL-2 versus the most closely related cytokine, IL-15.

To better understand how this molecule bound to IL-2, we solved X-ray structures of the chemical intermediates leading up to Compound 2 (Fig. 14.8) [23]. The covalently linked guanido fragment bound with little structural rearrange-

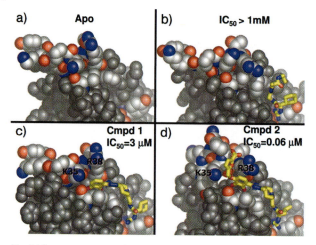

Fig. 14.8
(a) Structure of IL-2 only. (b) Structure of a covalently-linked guanidine-containing fragment bound to IL-2. (c) Structure of Compound **1** bound non-covalently to IL-2. (d) Structure of Compound **2** bound non-covalently to IL-2. Note the conformational shift in R38 and K35 to make room for the carboxylic acid.

ment to Glu-62 on IL-2 (Fig. 14.8b). Compound **1**, which fuses the guanido and tri-cyclic pyrazole, binds in a very similar manner to Ro26–4550 (Fig. 14.8c). These compounds require a rearrangement of side-chains to form a groove into which the tri-cyclic pyrazole binds. Compound **2** binds to a conformation that is even more dramatically different from free IL-2 (Fig. 14.8d). The loop region between helices 1 and 2 completely rearranges itself so that two positively charged residues (Lys-35 and Arg-38) sandwich the carboxy-furan piece of Compound **2**.

Several aspects of this example should apply to other protein–protein targets. The highly adaptive nature of the interface means that protein surfaces are not always as flat as they appear. Since one knows the site to engage but does not know the full range of chemotypes that can engage it, site-directed fragment approaches like Tethering are ideally suited to discover nucleating pieces that guide the design process. The Tethering approach was essential for designing Compound **2**, and its utility should extend to similar challenging targets.

14.5.2
Caspase-3: Finding and Combining Fragments in One Step

One of the biggest challenges in fragment-based drug discovery is not just finding fragments but figuring out what to do with them. In the case of IL-2 (above), we used fragments identified from Tethering to complement medicinal chemistry and structure-based drug design. However, other targets may not provide moderate affinity starting points nor the ability to generate additional structural informa-

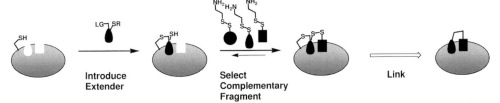

Fig. 14.9
Tethering with Extenders. An Extender is used to modify a residue in the protein; the Extender has some inherent affinity for the protein and also contains a thiol that can be used for Tethering. When a complementary fragment is identified, this can be linked with binding elements from the Extender to generate a potent inhibitor.

tion. To find a generalizable method to link fragments, we invented Tethering with Extenders [24].

Tethering with Extenders (Fig. 14.9) takes a fragment that binds a protein at a desired site and modifies it so that Tethering can be performed from the fragment. This fragment need only have modest affinity and could come from a previous Tethering experiment or other sources. The fragment is modified to contain an electrophile that reacts with a cysteine on the target protein, as well as a (possibly masked) thiol residue. The resulting modified fragment is called an *Extender*. After the Extender forms a covalent complex with the protein target, the thioester (if present) can be deprotected to reveal the thiol for Tethering. The two-dimensional connectivity between the Extender and any fragments identified from subsequent screens is known, even if the exact placement of both fragments is not. With this knowledge, binding elements from the Extender can be easily connected to newly discovered fragments. In theory, the resulting molecule, or diaphore, should have two separate binding elements and bind the target molecule more tightly than either fragment alone.

We tested this strategy on the enzyme caspase-3, a cysteine-aspartyl protease that is one of the central "executioners" of apoptosis. Excess apoptosis is attributed to a variety of diseases, from stroke to Alzheimer's disease to sepsis, making caspase-3 a popular drug target [25]. The enzyme also made an ideal starting point for constructing Extenders. It is well characterized both structurally and mechanistically and contains an active site cysteine residue that is irreversibly alkylated by small molecule inhibitors.

The first Extender constructed is shown in Fig. 14.10. Mass spectrometry showed we could modify caspase-3 cleanly and quantitatively with this molecule, even though the large subunit of the enzyme contains four other cysteine residues. We could also fully deprotect the thioester to reveal a free thiol. Screens against a library of about 7000 disulfide-containing fragments yielded one strong hit, a salicylic acid sulfonamide (Fig. 14.10). By simply replacing the disulfide bond with two methylenes and replacing the irreversible warhead with a reversible aldehyde, we created an inhibitor with $K_i = 2.8\ \mu M$. By rigidifying the linker,

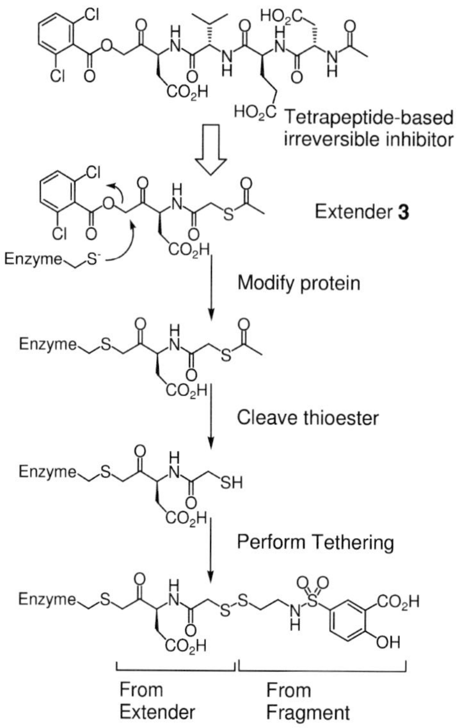

Fig. 14.10 Tethering with Extenders on caspase-3. Extender **3** covalently modifies the protein and can then be deprotected to reveal a thiol for Tethering. One of the strongest hits was the salicylic acid shown.

we boosted the affinity to 200 nM. Further medicinal chemistry allowed us to obtain 20 nM inhibitors (Fig. 14.11) [24, 26].

To ensure the technique could be generalized, we constructed a second Extender to explore a slightly different area of the protein. We modified caspase-3 with this Extender, deprotected the thioester, and screened the conjugate against our fragment library. We did not rediscover the salicylic acid hit from our first Extender screen, but we did identify several other hits, including a thiophene sulfone. When this fragment was linked to the Extender, the resulting inhibitor had K_i = 330 nM [24]. These examples illustrate the speed with which Tethering with Extenders can lead to potent inhibitors. Moreover, these inhibitors are non-peptidic and so are more useful as drug leads.

We used crystallography to understand the binding mode of these fragments. The structure of the salicylic acid fragment bound through the disulfide is shown in Fig. 14.12a, superimposed upon the structure of a tetrapeptide-based inhibitor. Significantly, the two inhibitors occupy roughly the same volume and make many of the same contacts, but do so using very different chemical moieties. Moreover, the S2 pocket in the salicylic acid structure is collapsed, while the S4 pocket expands to make room for the larger salicylic acid moiety. By introducing a substituent that binds in the S2 pocket, we boosted affinity by nearly two orders of magnitude (Fig. 14.11) [27].

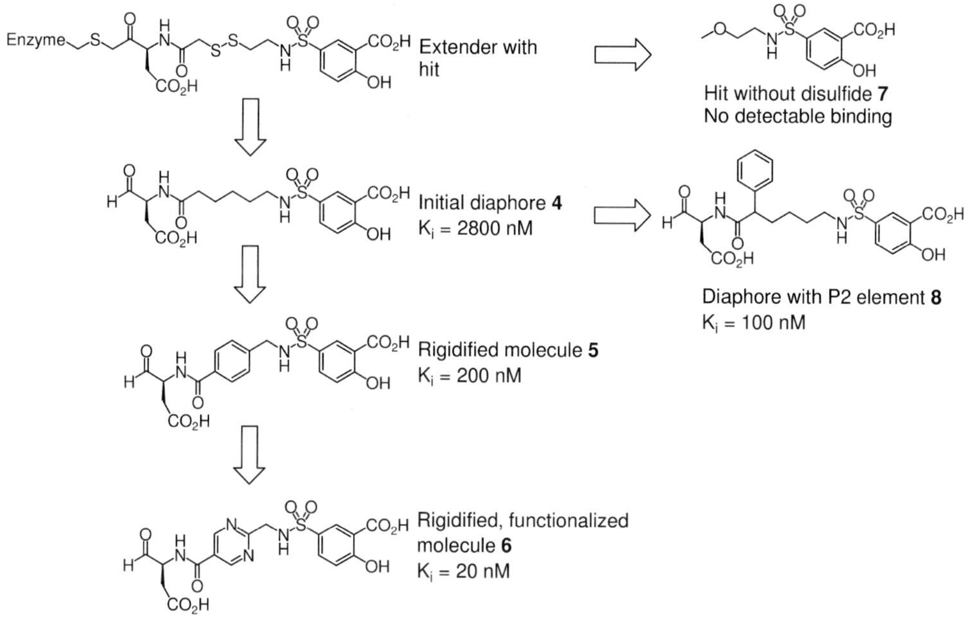

Fig. 14.11
Evolution of a fragment from Tethering with Extenders to a potent caspase-3 inhibitor. Simple replacement of the disulfide linker with an alkyl linker resulted in a low-micromolar inhibitor (**4**); and rigidification (**5**) and functionalization (**6** and **8**) of this linker led to increasingly potent inhibitors. The salicylic acid hit itself (**7**) had no detectable binding.

All of these features contrast with the structure of the second extender–fragment complex, shown in Fig. 14.12b. Here, the extender forces itself into the S2 pocket, but the disulfide linker then curves back to place the thiophene sulfone within the S4 pocket. The sulfone makes some of the same hydrogen bonds as the salicylic acid and the aspartyl residue in the tetrapeptide but with completely different chemistry. The flexibility of caspase-3 to accommodate different inhibitors is similar to that observed with PTP-1B and IL-2, and again emphasizes the ability of Tethering to identify fragments that would not have been easy to predict using structure-based design.

14.5.3
Caspase-1

The Tethering with Extenders approach was also used to identify inhibitors to the anti-inflammatory target caspase-1 [28]. In this case, one of the same Extenders previously designed for caspase-3 selected an entirely different set of fragments. This is consistent with the different substrate peptide sequence preferences: WEHD for caspase-1 versus DEVD for caspase-3 [29].

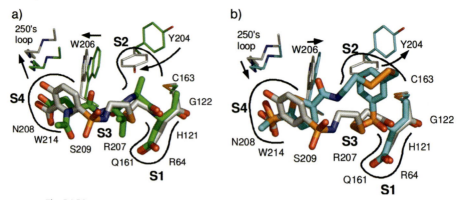

Fig. 14.12
(a) Structure of the salicylic acid fragment covalently bound to caspase-3 (gray), superimposed on a tetrapeptide-based inhibitor (green). Note the collapse of the S2 pocket and the widening of the S4 pocket to accommodate the salicylic acid moiety. (b) Structure of a second fragment covalently bound to caspase-3 (blue) superimposed on the salicylic acid fragment. Here the S2 pocket is intact and the linker takes an alternative path to arrive in the S4 pocket. Reprinted with permission from [24].

As with caspase-3, these hits were converted into potent, soluble inhibitors by replacing the disulfide linkage with a simple alkyl linkage (Fig. 14.13). Substituting a hydrophobic moiety at the S2 position gave a roughly 10-fold boost in potency. Several of these molecules demonstrated activity in cellular assays and selectivity for caspase-1 over the closely related caspase-5. Crystallography of several of these molecules in complex with caspase-1 revealed that they bind in an extended conformation as expected, but that the S2 pocket which collapses in caspase-3 remains open in caspase-1.

14.6
Beyond Traditional Fragment Discovery

14.6.1
Caspase-3: Use of Tethering to Identify and Probe an Allosteric Site

Active sites of enzymes have traditionally provided some of the best opportunities for drug discovery. Small substrates or cofactors serve as useful starting points for medicinal chemistry programs. Active sites also often provide cavities to shield the transition states from unwanted side reactions. These cavities are built to bind small molecules and so high-throughput screening can often identify hits that bind to active sites and inhibit the enzyme competitively.

Nonetheless, some active sites thwart drug discovery efforts. Many proteases, phosphatases, and lipases have highly charged or lipophilic substrates that lack drug-like properties. One example is the caspase class of aspartate cleaving pro-

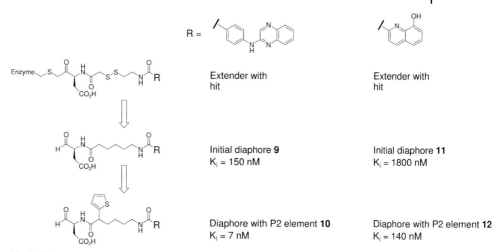

Fig. 14.13
Tethering with Extenders to identify caspase-1 inhibitors. Two of the hits from Tethering are shown, as are inhibitors derived from them. In these cases, addition of elements to fill the S2 pocket provided boosts in affinity.

teases. As discussed above, caspases have a strict requirement for an aspartyl P1 functionality and often require an electrophillic "warhead". This chemotype is undesirable from a pharmaceutical standpoint because of its instability and potential immunogenicity. Though we found non-peptidic inhibitors for caspase-3 and -1 using Tethering with Extenders, they contained a reversible aspartyl-aldehyde or ketone functionality. Extensive medicinal chemistry could not alter this without significantly reducing potency. Others in the pharmaceutical industry have reported similar experiences.

A Tethering screen of a 10 000-compound library on caspase-3 revealed two prominent hits that fully inhibited the enzyme in a stoichiometric fashion [30]. Peptide digestion and mass spectrometry showed that, surprisingly, the fragments were captured not by the active site cysteine on the large subunit, but rather by Cys290 on the small subunit located in a cavity at the dimer interface, some 15 Å from the active site (Fig. 14.14) [31]. Caspase-3 and caspase-7 share about 65% sequence identity, but are completely conserved in this cysteine-containing cavity. Indeed both compounds fully and stoichiometrically inhibit caspase-7 as well. The inhibition is fully reversed by adding high concentrations of reducing agent, releasing the fragments from their binding sites.

X-ray crystallography was used to understand the structural basis for inhibition for the allosteric compounds on caspase-7. Both compounds bind in a two-fold fashion to the cysteines, but with very different binding modes. The fluoro-indole compound (FICA) binds across the dimer interface and makes intermolecular hydrogen bonds to itself. The dichloro-phenoxy compound (DICA) binds to the same subunit to which it is covalently linked and does not make intermolecular

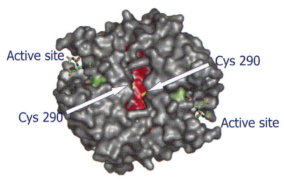

Fig. 14.14
Structure of caspase-3 showing the large central cavity with Cys290 located at the bottom. The enzyme is a dimer of dimers, accounting for the two active sites and two Cys290 residues.

contacts. The compounds seem to inhibit by causing a concerted structural rearrangement (Fig. 14.15). In particular, the compounds displace Arg187 so that the side-chain blocks substrate binding. This in turn causes a >5 Å shift in the active site Cys186 away from the catalytically competent position, while loops involved in binding substrate are no longer positioned to bind substrate.

This conformation may have a natural function. Caspase-3 and -7 are produced as inactive zymogens (pro-caspases) that are proteolytically cleaved to produce the active form. Indeed, the conformations of the allosterically inhibited and zymogen forms of the enzymes are very similar and very different from the active form of

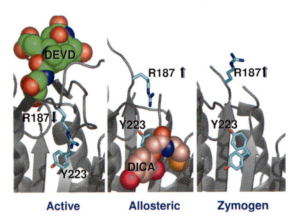

Fig. 14.15
(a) Structure of caspase-7 with a bound peptide-based active site inhibitor. (b) Structure of caspase-7 with a covalently bound allosteric site inhibitor (DICA). Note that Y223 and R187 have both shifted positions. (c) Structure of caspase-7 in the zymogen form, illustrating the different positions of Y223 and R187 compared to the active, inhibited form.

the enzyme (Fig. 14.15). Thus, we believe that the allosteric compounds trap the zymogen conformation – even in the absence of the pro-peptide. It seems plausible that the site has a natural regulator as yet unidentified. In fact, many of the other caspases contain a cavity at the dimer interface; in the case of caspase-1 this also changes when substrate binds [32].

About a dozen allosteric sites have been discovered from high-throughput screening and X-ray crystallography that can modulate enzyme activity [33]. Allosteric sites are targeted by many drugs, both approved and in clinical trials, notably the non-nucleoside reverse transcriptase drugs (NNRTs) to treat AIDS. Not only do allosteric sites provide additional chemical opportunities for drug discovery, these sites may not need to compete with high concentrations of substrates in vivo. Although allosteric regulation has been known for nearly 50 years, relatively few of these sites have been annotated in more recently discovered classes of signaling enzymes like kinases, phosphatases, and proteases with great pharmaceutical potential. Tethering could identify both these sites and nucleating fragments for drug discovery efforts.

14.6.2
GPCRs: Use of Tethering to Localize Hits and Confirm Proposed Binding Models

The examples of Tethering above relied on structural information to guide initial cysteine placement and fragment development. The site-directed aspect of Tethering can also probe binding sites in the absence of high-resolution structures.

We first used Tethering to support a model for how a ligand binds its receptor [34]. The G protein-coupled receptors (GPCRs) initiate cell-signaling cascades and are implicated in many diseases. Unfortunately, a high-resolution structure is available for only one member of this class, bovine rhodopsin [35]. This structure was used to construct a model of how the GPCR C5a receptor (C5aE) binds to its complement 5a-fragment (C5 a), a four-helix bundle protein. A model of the C5a receptor was built based on threading to the bovine rhodopsin structure. Considerable mutagenesis and ligand NMR studies led to a more refined model of how C5a may bind to the GPCR [36]. Mutagenesis and peptide analog studies show that a hexameric peptide derived from the C-terminal portion of the ligand is sufficient to bind and activate the receptor (EC_{50} ~100 nM). If the penultimate residue of the ligand is altered from cyclohexyl alanine to tryptophan, the effect of binding reverses: binding prevents signaling rather than activating it. Truncating the first three N-terminal residues of the hexamer reduces the binding 1000-fold, hampering efforts to home in further on the binding determinants.

Since Tethering excels at finding weak binding ligands, we reasoned that the trimer peptides from the C-terminus (EC_{50} ~ 150 µM) could bind more strongly to C5aR if covalently linked near their true binding site. We attached a cysteine onto the N-terminus of the trimer and also introduced cysteines into the receptor. The fact that the trimer sequence reverses the peptide from an agonist to an antagonist was used as a further functional test that the covalently linked peptides mimic the hexamer non-covalent peptides.

We introduced four cysteines at varying distances from the site where the model predicted the N-terminal cysteine would bind, choosing locations that mutational analyses suggested were close to the binding site but would not interfere with binding (Fig. 14.16). For example, F93C is expected to be >10 Å from the ideal disulfide distance (2.4 Å), L117C is somewhat intermediate at 2.8 Å, and P113C and G262C are ideal and at the same "depth" down their respective helices, three and six.

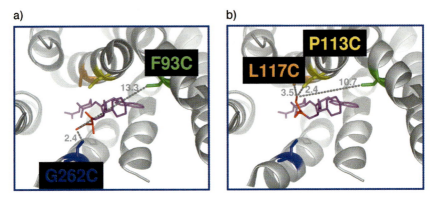

Fig. 14.16
(a, b) Models of C5aR showing likely distances between the cysteine mutations and the peptide. The views in both panels are the same, but are split for clarity. The molecular graphics in this and all other figures were done using the program PyMol [51].

The cysteine-containing mutants were expressed transiently in Cos-7 cells. None of the mutations had any significant effect on ligand binding or signaling with the full-length C5 a ligand or the hexameric peptides. However, the mutant proteins had dramatically different responses when covalently linked with the cysteine-containing trimeric peptides. For peptides blocking the interaction between the receptor and the ligand, the IC_{50} was the same for the most distal F293C variant and the wild-type receptor (IC_{50} ~150–200 µM). The potencies sharply increased with increasing proximity: from 45 µM for the L117C mutant to 7–8 µM for P113C and G262C. All reactions were performed under reducing conditions to achieve disulfide exchange and thermodynamic equilibrium. These reducing conditions had no impact on the wild-type receptor binding or signaling. The effects were fully reversible by further addition of 2-mercaptoethanol. As expected, the concentration of reductant needed to fully reduce the peptides was inversely related to their apparent affinities.

Tethering of peptides (Cys-dCha-Trp-dArg) generated from the non-covalent antagonist acted as antagonists in blocking inositol-3-phosphate accumulation by C5 a. In contrast, Tethering of agonist-derived peptides (Cys-dCha-Cha-dArg) functioned as agonists independent of C5 a [34]. Previous alanine-scanning mutational studies on the C5aR identified an I116A mutation that switches Trp-containing peptides from antagonists to agonists [36]. When this mutation was included with G262C, the covalently-linked Trp antagonist peptides acted as agonists [37]. All these data suggest the effects of the covalently linked peptides reflect the same binding mode and functional consequences as the larger non-covalent ligands. This is consistent with the examples (discussed above) where the structures of the covalently bound and free compounds bind in essentially the same mode.

To determine whether Tethering could be used to find small organic substitutes for the peptides [37], we screened a library of 10 000 disulfide-containing compounds in the presence of 10 mM 2-mercaptoethanol, using functional screening rather than mass spectrometry. Hits were defined as compounds that (a) could block the binding of ^{125}I-C5 a ligand by at least 15% when tested at 50 µM, (b) were specific to specific cysteine mutants, and (c) were fully reversible with the addition of high concentrations of 2-mercaptoethanol. As expected, we found the most hits at the same cysteines where the peptides bound most efficiently (~40–60 hits each at G262C and P113C, fewer hits at L117C, and no hits at F63C and wild-type C5aR). About 90% of the hits were unique for each site. This reflects similar experiences with Tethering at nearby sites in other targets (e.g. see the IL-2 example and the "chemoprint" in Fig. 14.6b).

The hits at G262C and P113C were triaged by further functional screening. About half the hits at each of these sites acted as agonists and the others as antagonists. The hits showed sharp SAR. For example, many of the hits contained a D-proline core. One containing a phenethyl–phenyl substituent stimulated the C5aR to the same extent as the natural C5 a ligand and was more potent than the covalently linked peptides. For example, in the presence of 1.5 mM 2-mercaptoethanol, the compound could block binding of ^{125}I-C5 a ligand with an IC$_{50}$ of 2 µM. Stereoisomers of this compound showed greatly reduced potency and functioned as antagonists. Moreover, the I116A mutant converted many of the antagonist hits to agonists. Larger substitutions at Ile116, such as I116F and I116W, converted the agonist hits and peptides to antagonists. Ile116 in the receptor appears to play a "gate keeper" function in regulating the conformational barrier between the receptor "on" and "off" states.

These results suggest that small organic mimics can be produced for these peptides. Moreover, Tethering provides a powerful tool for trapping conformational states in this GPCR. This is in principle similar to the trapping of "on" versus "off" states that rhodopsin, tethered via a Schiff's base, undergoes as light induces *cis–trans* isomerization [38]. It is also similar to the trapping of allosteric transitions that we observed with the caspases or the alternate conformations seen in IL-2 examples. The advantage of Tethering is that it provides distance constraints as well as mutational and redox controls to validate that the functional effects are from specific, site-defined interactions.

14.7
Related Approaches

This chapter has focused mainly on Tethering as invented and developed at Sunesis Pharmaceuticals. However, a number of related approaches have been reported recently, some of which use chemistry other than disulfides. In the following section we examine some of these, focusing on reversible bond formation under thermodynamic control. A broader review of the concept of site-directed ligand discovery, encompassing both reversible and irreversible bond formation, can be found elsewhere [39].

14.7.1
Disulfide Formation

The protein transthyretin (TTR) forms a homotetramer and transports two molecules of the small molecule hormone thyroxine. The small molecule binds and stabilizes the structure of TTR. If the protein unfolds, it can form amyloid fibrils, which have been linked to a variety of diseases. Although a number of small molecule inhibitors of TTR unfolding have been identified, these can bind to both thyroxine binding sites, and researchers have questioned whether occupancy of a single site would stabilize TTR sufficiently to prevent amyloidgenesis. Kelly and coworkers introduced a cysteine residue into TTR and tethered a known small molecule binder through a variety of different linkers [40]. Using a variety of biophysical techniques, they demonstrated that these covalent conjugates formed tetramers; and crystallographic structural determination revealed that the inhibitor bound as expected. In fact, covalently binding a single molecule could dramatically stabilize the protein against amyloidogenesis, although the magnitude of this effect was strongly dependent on the length of the linker. This finding has important implications for targeting the binding site therapeutically, since a single-site binder should prevent amyloidogenesis and could be administered at a lower dosage than an inhibitor that binds to both sites.

The mitochondrial protein Tom20 is a membrane-bound receptor responsible for binding to a short peptide sequence, the mitochondrial presequence, that targets proteins for mitochondrial import. There is little sequence homology among mitochondrial presequences, but a solution structure of the cytosolic domain of Tom20 in complex with a presequence peptide did reveal a hydrophobic binding groove on Tom20 that appears to recognize a pentapeptide [41]. Kohda and coworkers made a series of peptide libraries, with each peptide having a cysteine residue that could form a disulfide bond with a naturally occurring cysteine on Tom20 [42]. They then used matrix-assisted laser desorption ionization time-of-flight mass spectrometry (MALDI-TOF-MS) to determine which peptides bind most strongly to Tom20, and they determined that Tom20 actually recognizes a hexapeptide rather than a pentapeptide. They were also able to refine the consensus binding motif.

14.7.2
Imine Formation

The enzyme aspartate decarboxylase (ADC) contains an N-terminal pyruvoyl group (derived from a serine residue), which participates in the catalytic reaction by forming an imine with substrate L-aspartate. Abell and co-workers made use of this naturally occurring electrophile by screening the enzyme against a set of 55 separate amines and trapping the resulting imines with sodium cyanoborohydride [43]. Analysis by MALDI-TOF-MS then revealed which amines were able to form reversible covalent complexes. Two of the amines, including L-aspartate, were found to be decarboxylated by the enzyme, but several other amino acids were also found to bind without alteration, suggesting that these could be starting points for inhibitor design.

A series of elegant mechanistic and biophysical studies have focused on the bacterial transcription termination factor rho and its interaction with the natural product antibiotic bicyclomycin [44–46]. Bicyclomycin itself binds to rho with only modest affinity (Ki = 21 µM, K_d = 40 µM), but Kohn, Widger, Gaskell, and co-workers were able to introduce an aldehyde moiety into the molecule to form an imine with a lysine on the protein; this can be trapped with sodium borohydride. The aldehyde-containing molecule binds with more than ten-fold higher affinity (Kd = 3 µM, as judged by isothermal titration calorimetry) than unmodified bicyclomycin. Interestingly though, a crystal structure of the aldehyde-containing molecule bound to rho does not show the imine, but rather a hydrogen bond between the carbonyl of the aldehyde and the terminal amine of the lysine, suggesting that the origin of the increased affinity may not be entirely due to imine formation [47].

14.7.3
Metal-mediated

Carbonic anhydrases are active drug targets for a variety of diseases from glaucoma to cancer, and these zinc-containing metalloenzymes also contain a deep active site that can bind a variety of metal chelators. Simple molecules such as benzenesulfonamide bind to human carbonic anhydrase I (hCA-I) with low-micromolar affinity [48]. Researchers at North Dakota State University in Fargo found that they could enhance this affinity by two orders of magnitude by making "two-prong" inhibitors [49] containing both a benzenesulfonamide moiety and an iminodiacetate (IDA) moiety; this molecule only shows increased potency in the presence of copper (II) and the authors argue that this is due to the coordination of copper by surface histidine residues and the IDA functionality [48, 50].

14.8
Conclusions

As a fragment identification method, Tethering is one of many possible approaches. The technique is unique in requiring a covalent bond, which ensures that fragments bind in a stoichiometric fashion and also facilitates crystallography with low-affinity fragments. Tethering can target specific sites and wide-ranging conformations of a protein. Moreover, there is some evidence that Tethering can identify fragments that bind more weakly than those identified by other methods. For example, the guanido-fragment identified in IL-2, the salicylic acid sulfonamide fragment identified in caspase-3, the allosteric-binders identified in caspase-3, and the GPCR fragments showed no detectable inhibition by themselves. Although this increases the range of fragments accessible to the medicinal chemist, it does raise the question of whether some fragments may be so weak as to be essentially useless for further development. Nonetheless, given the success observed thus far, we believe there are many untapped opportunities for Tethering.

Acknowledgments

We thank all of our colleagues at Sunesis Pharmaceuticals for their contributions, without which Tethering would not have been possible, and Monya L. Baker for editorial assistance.

References

1 Tethering is a service mark of Sunesis Pharmaceuticals, Inc. for its fragment-based drug discovery.
2 Erlanson, D. A., Wells, J. A., Braisted, A. C. **2004**, Tethering: Fragment-based drug discovery. *Annu. Rev. Biophys. Biomol. Struct.* 33, 199–223.
3 Murray, C. W., Verdonk, M. L. **2002**, The consequences of translational and rotational entropy lost by small molecules on binding to proteins. *J. Comput. Aided Mol. Des.* 16, 741–753.
4 Hyde, J., Braisted, A. C., Randal, M., Arkin, M. R. **2003**, Discovery and characterization of cooperative ligand binding in the adaptive region of interleukin-2. *Biochemistry* 42, 6475–6483.
5 Keire, D. A., Strauss, E., Guo, W., Noszál, B., Rabenstein, D. L. **1992**, Kinetics and equilibria of thiol/disulfide interchange reactions of selected biological thiols and related molecules with oxidized glutathione. *J. Org. Chem.* 57, 123–127.
6 Millis, K. K., Weaver, K. H., Rabenstein, D. L. **1993**, Oxidation/reduction potential of glutathione. *J. Org. Chem.* 58, 4144–4146.
7 Mammen, M., Shakhnovich, E. I., Whitesides, G. M. **1998**, Using a convenient, quantitative model for torsional entropy to establish qualitative trends for molecular processes that restrict conformational freedom. *J. Org. Chem.* 63, 3168–3175.
8 Zhou, H. X. **2001**, The affinity-enhancing roles of flexible linkers in two-domain DNA-binding proteins. *Biochemistry* 40, 15069–15073.
9 Zhou, H. X. **2003**, Quantitative account of the enhanced affinity of two linked scFvs specific for different epitopes on the same antigen. *J. Mol. Biol.* 329, 1–8.
10 McGovern, S. L., Caselli, E., Grigorieff, N., Shoichet, B. K. **2002**, A common

mechanism underlying promiscuous inhibitors from virtual and high-throughput screening. *J. Med. Chem.* 45, 1712–1722.

11 Costi, M. P., Tondi, D., Rinaldi, M., Barlocco, D., Pecorari, P., Soragni, F., Venturelli, A., Stroud, R. M. **2002**, Structure-based studies on species-specific inhibition of thymidylate synthase. *Biochim. Biophys. Acta* 1587, 206–214.

12 Erlanson, D. A., Braisted, A. C., Raphael, D. R., Randal, M., Stroud, R. M., Gordon, E. M., Wells, J. A. **2000**, Site-directed ligand discovery. *Proc. Natl. Acad. Sci. USA* 97, 9367–9372.

13 Johnson, T. O., Ermolieff, J., Jirousek, M. R. **2002**, Protein tyrosine phosphatase 1B inhibitors for diabetes. *Nat. Rev. Drug Discov.* 1, 696–709.

14 Erlanson, D. A., McDowell, R. S., He, M. M., Randal, M., Simmons, R. L., Kung, J., Waight, A., Hansen, S. K. **2003**, Discovery of a new phosphotyrosine mimetic for PTP1B using breakaway tethering. *J. Am. Chem. Soc.* 125, 5602–5603.

15 Clackson, T., Wells, J. A. **1995**, A hot spot of binding energy in a hormone-receptor interface. *Science* 267, 383–386.

16 DeLano, W. L. **2002**, Unraveling hot spots in binding interfaces: progress and challenges. *Curr. Opin. Struct. Biol.* 12, 14–20.

17 Arkin, M. R., Wells, J. A. **2004**, Small-molecule inhibitors of protein-protein interactions: progressing towards the dream. *Nat. Rev. Drug Discov.* 3, 301–317.

18 Tilley, J. W., Chen, L., Fry, D. C., Emerson, S. D., Powers, G. D., Biondi, D., Varnell, T., Trilles, R., et al. **1997**, Identification of a small molecule inhibitor of the IL-2/IL-2Ra receptor interaction which binds to IL-2. *J. Am. Chem. Soc.* 119, 7589–7590.

19 Arkin, M. R., Randal, M., DeLano, W. L., Hyde, J., Luong, T. N., Oslob, J. D., Raphael, D. R., Taylor, L., et al. **2003**, Binding of small molecules to an adaptive protein-protein interface. *Proc. Natl. Acad. Sci. USA* 100, 1603–1608.

20 Braisted, A. C., Oslob, J. D., Delano, W. L., Hyde, J., McDowell, R. S., Waal, N., Yu, C., Arkin, M. R., et al. **2003**, Discovery of a potent small molecule IL-2 inhibitor through fragment assembly. *J. Am. Chem. Soc.* 125, 3714–3715.

21 Raimundo, B. C., Oslob, J. D., Braisted, A. C., Hyde, J., McDowell, R. S., Randal, M., Waal, N. D., Wilkinson, J., et al. **2004**, Integrating fragment assembly and biophysical methods in the chemical advancement of small-molecule antagonists of IL-2: an approach for inhibiting protein-protein interactions. *J. Med. Chem.* 47, 3111–3130.

22 Waal, N. D., Yang, W., Oslob, J. D., Arkin, M. R., Hyde, J., Lu, W., McDowell, R. S., Yu, C. H., et al. **2005**, Identification of nonpeptidic small-molecule inhibitors of interleukin-2. *Bioorg. Med. Chem. Lett.* 15, 983–987.

23 Thanos, C. D., Randal, M., Wells, J. A. **2003**, Potent small-molecule binding to a dynamic hot spot on IL-2. *J. Am. Chem. Soc.* 125, 15280–15281.

24 Erlanson, D. A., Lam, J. W., Wiesmann, C., Luong, T. N., Simmons, R. L., DeLano, W. L., Choong, I. C., Burdett, M. T., et al. **2003**, In situ assembly of enzyme inhibitors using extended tethering. *Nat. Biotechnol.* 21, 308–314.

25 O'Brien, T., Lee, D. **2004**, Prospects for caspase inhibitors. *Mini. Rev. Med. Chem.* 4, 153–165.

26 Choong, I. C., Lew, W., Lee, D., Pham, P., Burdett, M. T., Lam, J. W., Wiesmann, C., Luong, T. N., et al. **2002**, Identification of potent and selective small-molecule inhibitors of caspase-3 through the use of extended tethering and structure-based drug design. *J. Med. Chem.* 45, 5005–5022.

27 Allen, D. A., Pham, P., Choong, I. C., Fahr, B., Burdett, M. T., Lew, W., DeLano, W. L., Gordon, E. M., et al. **2003**, Identification of potent and novel small-molecule inhibitors of caspase-3. *Bioorg. Med. Chem. Lett.* 13, 3651–3655.

28 O'Brien, T., Fahr, B. T., Sopko, M., Lam, J. W., Waal, N. D., Raimundo, B. C., Purkey, H., Pham, P., et al. **2005**, Structural analysis of caspase-1 inhibitors derived from Tethering. *Acta Crystallogr. F* 61, 451–458.

29 Thornberry, N. A., Rano, T. A., Peterson, E. P., Rasper, D. M., Timkey, T., Garcia-Calvo, M., Houtzager, V., Nordstrom, P., et al. **1997**, A combinatorial approach defines specificities of members of the caspase family and Granzyme B. *J. Biol. Chem.* 272, 17907–17911.

30 Hardy, J. A., Lam, J., Nguyen, J. T., O'Brien, T., Wells, J. A. **2004**, Discovery of an allosteric site in the caspases. *Proc. Natl. Acad. Sci. USA* 101, 12461–12466.

31 Fuentes-Prior, P., Salvesen, G. S. **2004**, The protein structures that shape caspase activity, specificity, activation and inhibition. *Biochem. J.* 384, 201–232.

32 Romanowski, M. J., Scheer, J. M., O'Brien, T., McDowell, R. S. **2004**, Crystal structures of a ligand-free and malonate-bound human caspase-1: implications for the mechanism of substrate binding. *Structure* 12, 1361–1371.

33 Hardy, J. A., Wells, J. A. **2004**, Searching for new allosteric sites in enzymes. *Curr. Opin. Struct. Biol.* 14, 706–715.

34 Buck, E., Bourne, H., Wells, J. A. **2005**, Site-specific disulfide capture of agonist and antagonist peptides on the C5a receptor. *J. Biol. Chem.* 280, 4009–4012.

35 Palczewski, K., Kumasaka, T., Hori, T., Behnke, C. A., Motoshima, H., Fox, B. A., Le Trong, I., Teller, D. C., et al. **2000**, Crystal structure of rhodopsin: A G protein-coupled receptor. *Science* 289, 739–745.

36 Gerber, B. O., Meng, E. C., Dotsch, V., Baranski, T. J., Bourne, H. R. **2001**, An activation switch in the ligand binding pocket of the C5a receptor. *J. Biol. Chem.* 276, 3394–3400.

37 Buck, E., Wells, J. A. **2005**, Disulfide trapping to localize small-molecule agonists and antagonists for a G protein-coupled receptor. *Proc. Natl. Acad. Sci. USA* 102, 2719–24.

38 Cohen, G. B., Oprian, D. D., Robinson, P. R. **1992**, Mechanism of activation and inactivation of opsin: role of Glu113 and Lys296. *Biochemistry* 31, 12592–12601.

39 Erlanson, D. A., Hansen, S. K. **2004**, Making drugs on proteins: site-directed ligand discovery for fragment-based lead assembly. *Curr. Opin. Chem. Biol.* 8, 399–406.

40 Wiseman, R. L., Johnson, S. M., Kelker, M. S., Foss, T., Wilson, I. A., Kelly, J. W. **2005**, Kinetic stabilization of an oligomeric protein by a single ligand binding event. *J. Am. Chem. Soc.* 127, 5540–5551.

41 Abe, Y., Shodai, T., Muto, T., Mihara, K., Torii, H., Nishikawa, S., Endo, T., Kohda, D. **2000**, Structural basis of presequence recognition by the mitochondrial protein import receptor Tom20. *Cell* 100, 551–560.

42 Obita, T., Muto, T., Endo, T., Kohda, D. **2003**, Peptide library approach with a disulfide tether to refine the Tom20 recognition motif in mitochondrial presequences. *J. Mol. Biol.* 328, 495–504.

43 Webb, M. E., Stephens, E., Smith, A. G., Abell, C. **2003**, Rapid screening by MALDI-TOF mass spectrometry to probe binding specificity at enzyme active sites. *Chem. Commun.* 2003, 2416–2417.

44 Vincent, F., Openshaw, M., Trautwein, M., Gaskell, S. J., Kohn, H., Widger, W. R. **2000**, Rho transcription factor: symmetry and binding of bicyclomycin. *Biochemistry* 39, 9077–9083.

45 Vincent, F., Widger, W. R., Openshaw, M., Gaskell, S. J., Kohn, H. **2000**, 5a-formylbicyclomycin: studies on the bicyclomycin-Rho interaction. *Biochemistry* 39, 9067–9076.

46 Brogan, A. P., Widger, W. R., Bensadek, D., Riba-Garcia, I., Gaskell, S. J., Kohn, H. **2005**, Development of a technique to determine bicyclomycin-rho binding and stoichiometry by isothermal titration calorimetry and mass spectrometry. *J. Am. Chem. Soc.* 127, 2741–2751.

47 Skordalakes, E., Brognan, A. P., Park, B. S., Kohn, H., Berger, J. M. **2005**, Structural mechanism of inhibition of the Rho transcription termination factor by the antibiotic bicyclomycin. *Structure* 13, 99–109.

48 Banerjee, A. L., Eiler, D., Roy, B. C., Jia, X., Haldar, M. K., Mallik, S., Srivastava, D. K. **2005**, Spacer-based selectivity in the binding of "two-prong" ligands to recombinant human carbonic anhydrase I. *Biochemistry* 44, 3211–3224.

49 Roy, B. C., Hegge, R., Rosendahl, T., Jia, X., Lareau, R., Mallik, S., Srivastava, D. K. **2003**, Conjugation of poor inhibitors with surface binding groups: a strategy to improve inhibition. *Chem. Commun.* 2003, 2328–2329.

50 Banerjee, A. L., Swanson, M., Roy, B. C., Jia, X., Haldar, M. K., Mallik, S., Srivastava, D. K. **2004**, Protein surface-assisted enhancement in the binding affinity of an inhibitor for recombinant human carbonic anhydrase-II. *J. Am. Chem. Soc.* 126, 10875–10883.

51 DeLano, W. L. **2004**, PyMOL, available at: http://pymol.sourceforge.net/.

Part 4: Emerging Technologies in Chemistry

15
Click Chemistry for Drug Discovery
Stefanie Röper and Hartmuth C. Kolb

15.1
Introduction

Nature's secondary metabolites ("natural products") capture the imagination of the synthetic organic chemist, due to their complex structures and carbon frameworks, and their oftentimes interesting biological properties. Despite their promise as lead structures for drug discovery, optimization efforts are usually thwarted by complex and lengthy syntheses, which make analog generation difficult. For this reason, combinatorial chemistry approaches have been developed to rapidly scan the chemical universe for new molecules with interesting biological properties that may serve as lead structures for the development of new therapeutics. Guida et al. estimated the size of the universe of "drug-like" compounds to be on the order of 10^{63} molecules [1]. For this calculation, compounds were defined as being drug-like if they were stable in water, weighed no more than 500 Da and contained only H, C, N, O, P, S, F, Cl and Br. Obviously, only an infinitesimal portion of the potential medicinal chemistry universe has been explored to date; and it seems clear that a chemical toolkit containing only the most reliable and versatile processes is needed if one is to make any kind of impact. Click chemistry, introduced in 2001 by Sharpless, Finn and Kolb, was developed with these ideas in mind. It is a modular approach that employs a set of highly reliable reactions for assembling fragments to facilitate the discovery of new lead structures and their optimization [2]. It focuses exclusively on highly energetic, "spring-loaded" reactants and a pure, kinetically controlled outcome (Fig. 15.1). Click chemistry reactions are wide in scope, stereoselective, insensitive to water and oxygen and they utilize readily available starting materials to produce the desired products in high yields, while giving only inoffensive by-products. The reaction work-up is simple, requiring no purification by chromatography. Click chemistry not only simplifies compound synthesis, but also makes it more reliable, which may result in faster lead discovery and optimization. It does not replace conventional methods for drug discovery but rather complements and extends them.

Fragment-based Approaches in Drug Discovery. Edited by W. Jahnke and D. A. Erlanson
Copyright © 2006 WILEY-VCH Verlag GmbH & Co. KGaA, Weinheim
ISBN: 3-527-31291-9

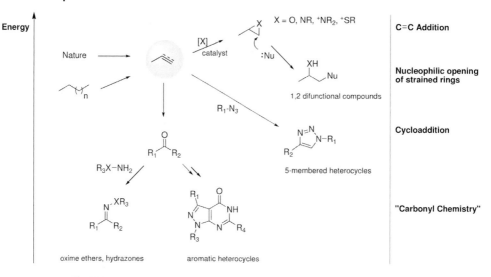

Fig. 15.1
Energetically highly favorable linking reactions, which are part of the click chemistry universe.

15.2
Click Chemistry Reactions

Inspired by how Nature performs combinatorial chemistry, click chemistry focuses on carbon–heteroatom bond forming reactions. Olefins and acetylenes provide the carbon frameworks. Sample reactions include:
- cycloaddition reactions, e.g. hetero-Diels–Alder [3] and 1,3-dipolar cycloaddition [4];
- carbonyl chemistry, e.g. the formation of hydrazones, oxime ethers and heteroaromatic systems;
- addition to carbon–carbon multiple bonds, such as epoxidation [5], aziridination [6], dihydroxylation [7] and sulfenyl [8] and nitrosyl halide additions;
- nucleophilic substitution on strained compounds and intermediates, such as epoxides, aziridines and episulfonium and aziridinium ions [2].

From the beginning, water has played an important role as a reaction medium; and many click chemistry reactions proceed optimally in pure water, particular when the reactants are *insoluble* [9]. Scheme 15.1 illustrates the phenomenon of rate acceleration that occurs under heterogeneous conditions (the reactants are stirred in aqueous suspension). Without water or water/alcohol mixtures, the reactions are slower and less selective, and are more dangerous on a large scale, because of their highly exothermic nature. Water, owing to its large heat capacity, has a great advantage over other solvents on safety issues, but it also facilitates product isolation.

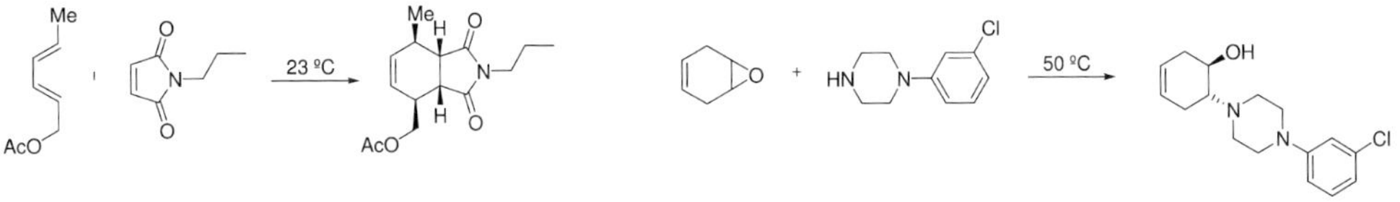

Scheme 15.1
Water-based reactions.

solvent	conc. [M]	time [h]	yield [%]
toluene	1	144	79
MeOH	1	48	82
CH$_3$CN	1	>144	43
neat	3.69	10	82
on H$_2$O	3.88	8	81

solvent	conc. [M]	time [h]	yield [%]
toluene	1	120	<10
neat	3.88	72	76
EtOH	1	60	89
on H$_2$O	3.88	12	88

The triazole forming Huisgen [4] 1,3-dipolar cycloaddition of alkynes and azides is the premier example of a click chemistry reaction. Despite the fact that the reaction is thermodynamically favored by 30–35 kcal mol^{-1}, it possesses a very high activation barrier, which prevents it from taking place at room temperature. In its classic form, the reaction requires elevated temperature or pressure to achieve reasonable rates, yielding mixtures of regioisomers. Despite their high intrinsic energy, azides and alkynes are surprisingly unreactive, which is very advantageous, since it provides for a high degree of stability towards aqueous media and biological molecules. In 2002, the groups of Sharpless/Fokin [10] and Meldal [11] independently discovered a copper(I)-catalyzed, stepwise variant of the 1,3-dipolar cycloaddition between azides and terminal acetylenes, which not only proceeds under mild conditions, but also is regiospecific for the 1,4-disubstituted 1,2,3-triazole (Scheme 15.2).

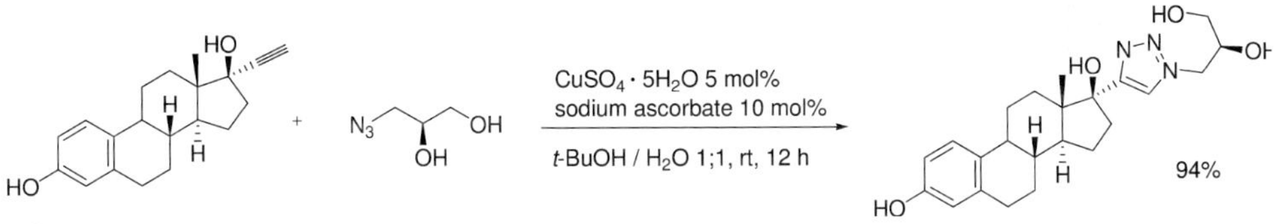

Scheme 15.2
The Cu(I)-catalyzed generation of 1,4-disubstituted 1,2,3-triazoles.

This new method has enjoyed rapid acceptance; and numerous applications for drug discovery and material sciences have been developed, due to its reliability, tolerance to a wide variety of functional groups, regiospecificity and the ready availability of starting materials. The 1,4-disubstituted 1,2,3-triazole unit of the reaction products is an interesting, if underappreciated, pharmacophore that is nearly impossible to oxidize, reduce or cleave. It has been demonstrated to actively engage in interactions with proteins, due to its large dipole moment of ~5 Debye, and the ability of N-3 to function as a hydrogen bond acceptor [12]. These pharmacophoric properties and the rigidity of 1,4-disubstituted triazoles are reminiscent of peptide bonds, suggesting this linking unit to be useful for developing new biologically active molecules.

15.3
Click Chemistry in Drug Discovery

Since the discovery of the regiospecific copper(I)-catalyzed triazole process, numerous chemical biology applications of this linking reaction have been published [13]. Simple experimental procedures, complete conversion, high yields and specificity for 1,4-disubstituted 1,2,3-triazoles make product purification easy and provide excellent opportunities for combinatorial library synthesis and natural product modification. The inertness of azides and alkynes to biological systems makes this ligation process interesting for a wide range of biomolecular applications, ranging from target-templated *in situ* chemistry to DNA and proteomics research, via bioconjugation reactions.

15.3.1
Lead Discovery Libraries

The Kolb laboratory at Coelacanth Corporation [14] developed the click chemistry approach to combinatorial chemistry. Over 170 000 individual compounds across 75 different structural themes were made on a 25–50 mg scale using semi-automated liquid phase synthesis approaches. Some of the "spring-loaded" building blocks that were used for library generation were epoxides and aziridines, which provided 1,2-difunctionalized compounds by nucleophilic ring opening [15], azides and β-ketoesters, which were converted to triazoles via 1,3-dipolar cycloaddition [15], imidoesters for the preparation of 5-membered aromatic heterocycles [17] and 3-aminoazetidines for the synthesis of non-aromatic heterocyclic libraries [18]. Also, targeted amino acid-derived libraries were made, which led to the discovery of potent peroxisome proliferator-activated receptor γ (PPAR-γ) agonists [19].

Recent studies have shown that the copper-catalyzed triazole process is suitable for the solid phase synthesis of combinatorial libraries. The group led by Gmeiner used this process for the preparation of a polystyrene based formyl indol methyl triazole (FIMT) resin for parallel solid phase synthesis of a series of BP-897-type arylcarboxamides (Scheme 15.3) [20]. A library of 42 compounds was generated in a five-step sequence that involved resin loading by reductive amination, followed by amide coupling, deprotection, palladium-catalyzed N-arylation and acidic cleavage. The final products were screened for binding to monoaminergic G protein-coupled receptors without further purification. Five library members showed excellent dopamine D3 receptor affinities; and carboxamide (Scheme 15.3; **1**) displayed a K_i value of 0.28 nM, which exceeds the binding properties of drug candidate BP897.

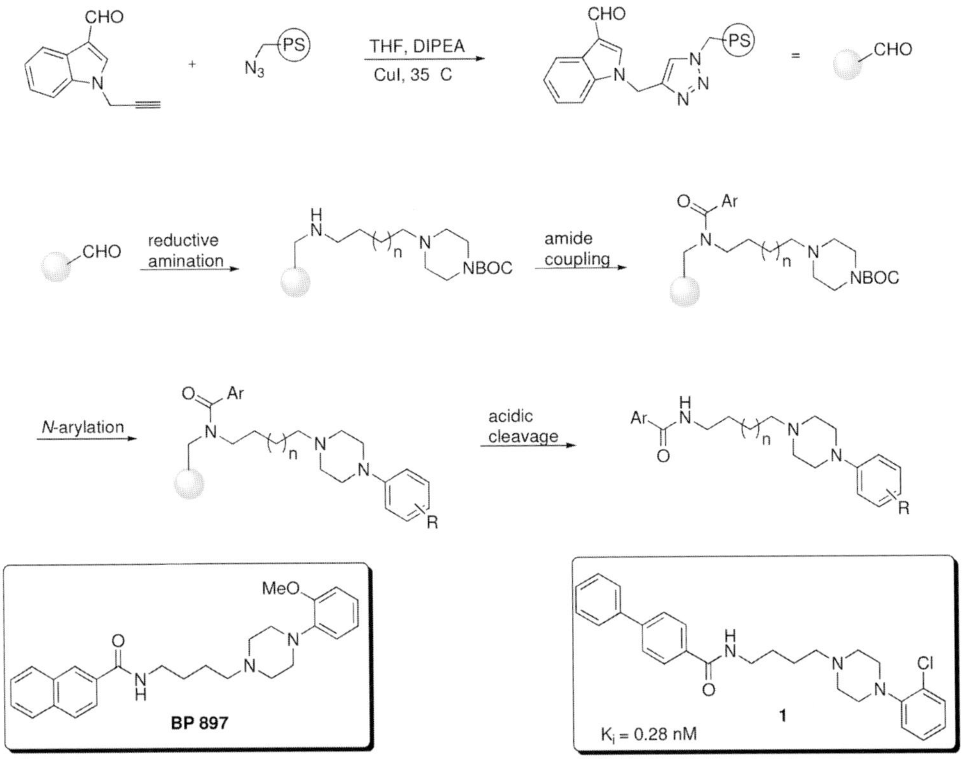

Scheme 15.3
Solid-phase supported synthesis of dopamine D3 receptor ligands.

15.3.2
Natural Products Derivatives and the Search for New Antibiotics

Bacterial resistance to antibiotic drugs is a growing problem in antibacterial therapy; and new classes of anti-infectives are desperately needed [21]. Predictions suggested that by 2005 nearly 40% of all *Streptococcus* strains will be resistant to both penicillin and macrolide antibiotics in the United States [22]. Of special concern is the drug resistance of Gram-positive pathogens such as *Streptococcus pneumoniae* to penicillin, *Staphylococcus aureus* (MRSA) to methicillin and in particular *Enterococcus faecium* (VRE) to vancomycin. The biological activity of natural products is often mediated by their carbohydrate epitopes, which influence the pharmacokinetic properties, targeting and mechanism of action in general. Consequently, diversification of the carbohydrate groups may lead to new therapeutics [23]. With the aid of the Cu(I)-catalyzed process, 15 triazole-linked monoglycosylated derivatives of vancomycin were synthesized (Fig. 15.2) [24]. Antibacterial screens of this library revealed that acid derivative 2 (Fig. 15.2; 2) was twice as active as vancomycin against both *E. faecium* and *S. aureus*.

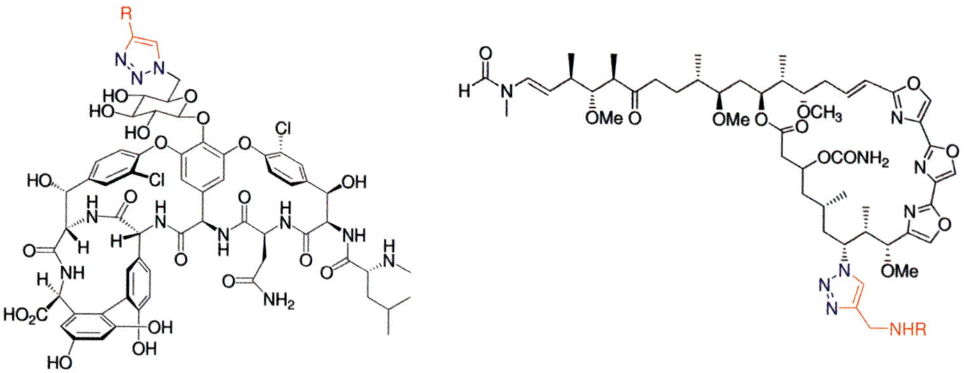

2 R = COOH

Triazole linked monoglycosylated vancomycins

R = H, (*H*-fluoren-9-ylmethoxycarbonyl)

Triazole anlogue of Kabiramide C

Fig. 15.2
Triazole derivatives of natural products.

Another approach for modifying natural products was described by Marriott et al., who synthesized a triazole derivative of Kabiramide C in the presence of a catalytic amount of Cu(I) and NEt$_3$ (Fig. 15.2) [25]. Kabiramide C is a macrolide drug that targets actin, a highly conserved and abundant protein, which plays an essential role in cytokinesis, cell mobility and vesicle transport. Each triazole derivative was shown to tightly bind G-actin with the same stoichiometry as the natural product.

A new class of synthetic antibacterial agents against Gram-positive bacteria, containing the oxazolidinone pharmacophore, targets bacterial protein synthesis. However, growing drug resistance has been observed with *E. faecium* and *S. aureus*. Additionally, oxazolidinone antibiotics can cause severe hypertensive crisis, an undesired side-effect caused by the inhibition of monoamine oxidase (MAO). In an effort to address these issues, researchers at AstraZeneca made analogues of lineozolide [26] by replacing the acetamide functionality with a 1,2,3-triazole and the morpholine ring by a thiapyran sulfone [27]. Both the copper-catalyzed and thermal [3+2]cycloaddition processes were employed to generate triazoles; and vinylsulfones were found to give 1,4-disubstituted triazoles with good selectivities (Scheme 15.4).

SAR studies revealed that 5-substitued triazoles were generally inactive or only marginally active. In contrast, many triazoles with a small substituent in the 4-position showed good antibacterial activities and at the same time displayed significantly reduced inhibition of MAO.

Even though tyrocidine antibiotics are attractive therapeutic agents, they cause lysis of human red blood cells. In an attempt to decrease this toxic side effect, Walsh et al. employed a chemoenzymatic approach to make carbohydrate-modified cyclic peptide tyrocidine (tyc) antibiotics, which are nonribosomal peptides (NRP; Scheme 15.5) [28]. They generated a library of 247 glycopeptides by conducting an enzymatic macrocyclization, followed by Cu(I)-catalyzed triazole for-

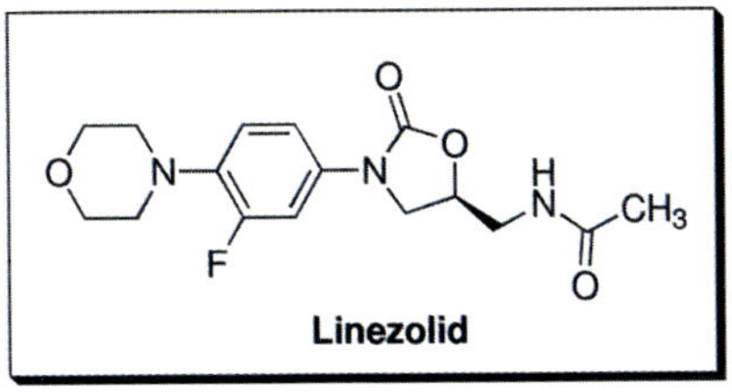

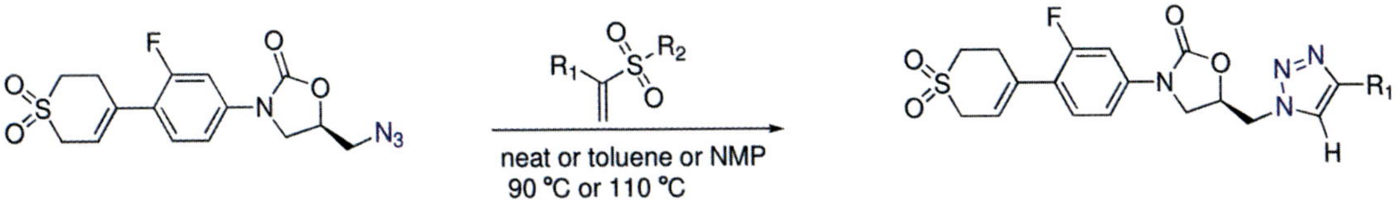

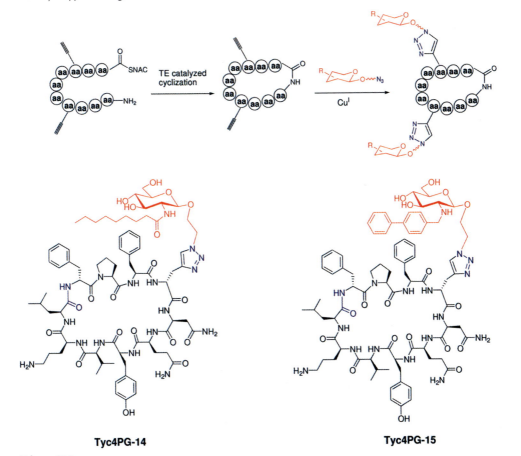

Scheme 15.4
Triazole/thiapyran analogs of the oxazolidinone antibiotic linezolid.

Scheme 15.5
Chemoenzymatic approach to glycopeptides via copper-catalyzed triazole formation.

mation with azidoalkyl glycosides in 96-well plates. The excised thioesterase (TE) domain from the tyrocidine synthetase was used to catalyze the formation of the head-to-tail cyclic peptide derivatives from linear, propargylglycine-containing, peptide N-acetyl cystamine (SNAC) thioesters. The cyclization was followed by copper-catalyzed triazole formation, which proceeded cleanly, allowing the crude reaction mixtures to be screened in antibacterial and hemolytic assays. Screening hits were validated by re-testing the purified compounds, leading to the identification of two glucopeptides that displayed a 6-fold better therapeutic index than wild-type tyrocidine while maintaining its high antibacterial potency.

15.3.3
Synthesis of Neoglycoconjugates

All cell surfaces carry oligosaccharides linked to proteins or lipids. These glycoconjugates modulate a number of cellular recognition processes, for example the inflammatory response; and their importance for the generation of therapeutic agents has been increasingly recognized in recent years. However, there are some inherent disadvantages associated with this class of compounds that complicate drug discovery efforts. The synthesis of N-and O-linked glycopeptides is difficult; and the products are susceptible to chemical and enzymatic hydrolysis [29, 30]. Isosteric and metabolically stable replacements for the glycosidic linkage may solve these issues. Triazoles have been investigated as linkers for stable glycopeptide analogues (Scheme 15.6). They can be made by coupling azide-functionalized glycosides with acetylenic amino acids or acetylenic glycosides with azide-containing amino acids, using the Cu(I)-catalyzed process [31].

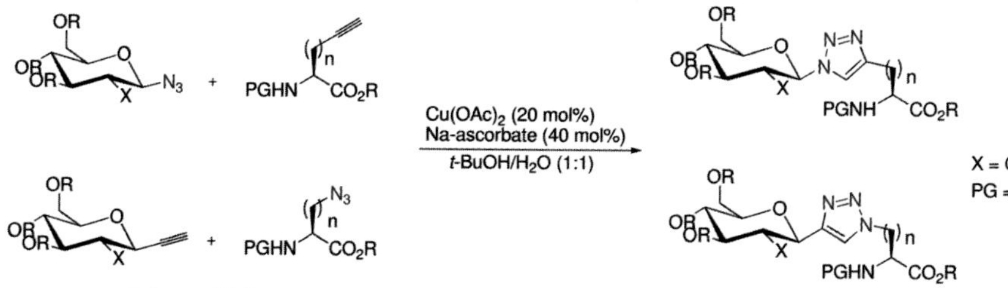

Scheme 15.6
Synthesis of triazole-linked glycopeptides.

Copper-catalyzed triazole formation from carbohydrate-derived acetylenes and azides is considerably faster under microwave irradiation, allowing reaction times to be reduced from several hours at room temperature to a few minutes. This allowed the preparation of a series of multivalent, triazole-linked neoglycoconjugates with complete regiocontrol and in high yields, using organic soluble Cu(I)-complexes as catalysts [32]. Polyvalent mannosylated aromatic and heteroaromatic systems were obtained in this way, in addition to a heptavalent manno-β-cyclodextrin (Scheme 15.7).

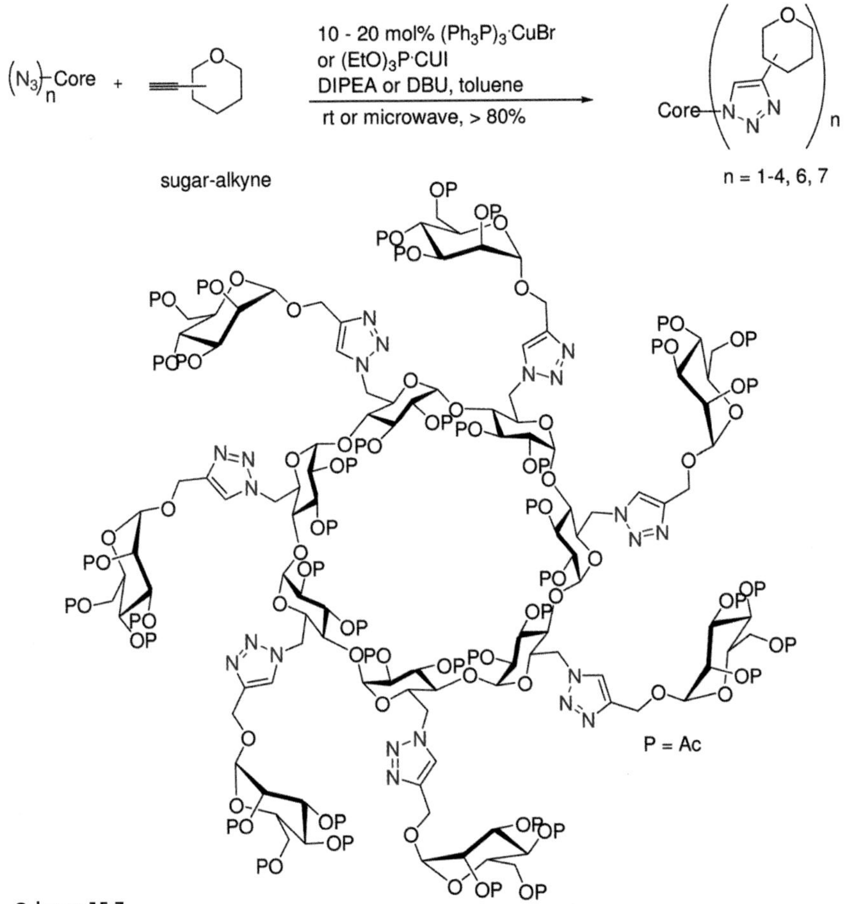

Scheme 15.7
Multivalent neoglycoconjugates.

Norris et al. reported the rapid, regiospecific synthesis of various glucosyl-1,4-disubstituted 1,2,3-triazoles in water, based on the CuSO$_4$/ascorbate system [33]. Sabesan et al. generated a combinatorial library of triazole-linked oligosaccharide analogs [34] by using thermal [3+2]cycloaddition to generate a variety of mimics of natural and unnatural carbohydrates with a wide variety of substituents placed on the carbohydrate groups.

15.3.4
HIV Protease Inhibitors

Every year, particularly in sub-Saharan Africa, human immunodeficiency virus type 1 (HIV-1) causes the deaths of millions of people; and the number of individuals who carry the virus continues to rise. HIV-protease [35] is responsible for the final stage of virus maturation and its inhibitors are used together with reverse transcriptase inhibitors for "highly active anti-retroviral therapy" (HAART). Unfortunately, numerous drug-resistant and cross-resistant mutant HIV-1 proteases have evolved, making the development of new inhibitors necessary [36].

Cu(I)-catalyzed triazole formation was used to prepare a 100-member library based on hydroxyethylamine peptide isosteres (Scheme 15.8; **3–6**) [37]. Two different azide cores (**3** and **4**), inspired by the existing drugs (aprenavir [38], nelfinavir [39], saquinavir [40]), were connected with acetylenes for library production. Direct screening of the aqueous reaction mixtures against HIV-protease and three mutants (G48V, V82F, V82A) led to the identification of four hits with inhibition constants in the 10 nM range, which were all derived from scaffold **3**. In contrast, scaffold **4** failed to provide any hits.

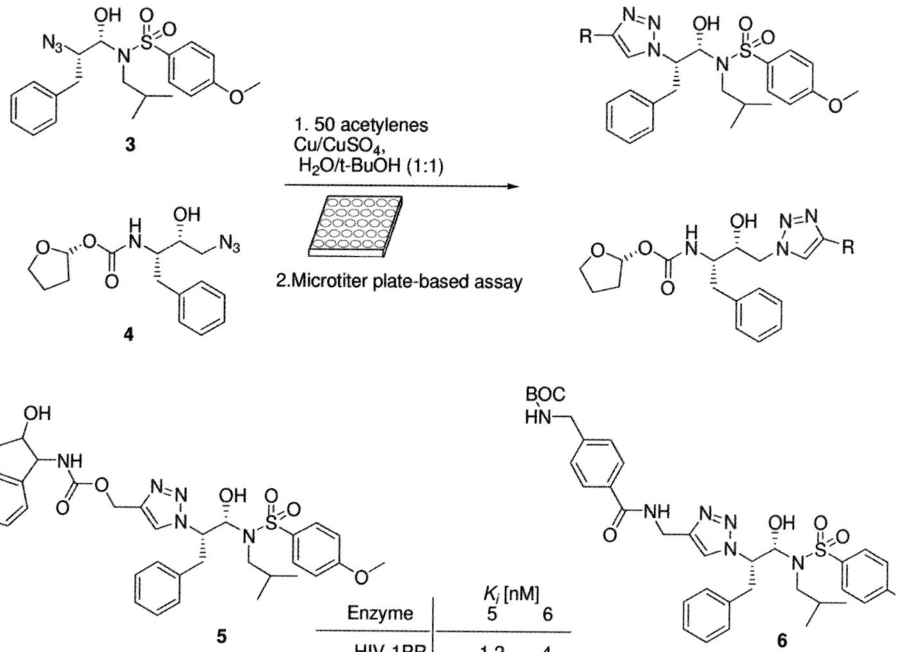

Enzyme	K_i [nM]	
	5	6
HIV-1PR	1.2	4
V82F	10	13
G48V	22	9.7
V82A	27	30

Scheme 15.8
Development of HIV-1 inhibitors through combinatorial triazole formation, combined with high-throughput screening.

15.3.5
Synthesis of Fucosyltranferase Inhibitor

Fucosyltranferases (Fuc-T) catalyze the final glycosylation step in the biosynthesis of the sialyl Lewis x (sLe^x) and sialyl Lewis a (sLe^a) epitopes of cell surface glycoproteins and glycolipids. These fucosylated oligosaccharides mediate a variety of crucial cell–cell recognition processes such as fertilization, embryogenisis, lymphocyte trafficking, immune response and cancer metastasis [41]. The transferases catalyze the attachment of L-fucose to a specific hydroxyl group of sialyl N-acetyl-lactosamine, using guanosine diphosphate β-L-fucose (GDP-fucose) as the fucosyl donor. Selective fucosyltransferase inhibitors may block the generation of these fucosylated products and the pathology they trigger. However, due to low substrate affinity and the absence of enzyme structure data, rational design of inhibitors has been difficult [42].

Wong et al. discovered nanomolar inhibitors from a triazole library prepared by the copper-catalyzed reaction of 85 azides with a GDP-derived alkyne in water [43]. High yields and the absence of protecting groups made it possible to directly screen the crude reaction mixtures (Scheme 15.9). This led to the identification of the biphenyl derivative 7 as a hit, which was purified and tested against a variety of fucosyl and galactosyl tranferases and kinases, to discover that it was not only highly selective for human α-1,3-fucosyltransferase VI but also very potent.

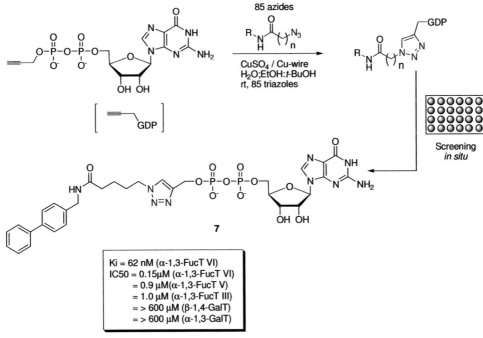

Scheme 15.9
Synthesis and *in situ* screening of fucosyltransferase inhibitors.

15.3.6
Glycoarrays

Carbohydrate arrays (glycoarrays) are used for the high-throughput analysis of sugar–receptor interactions [44, 45] and for studying cell–cell recognition processes mediated by carbohydrates. Wong et al. created a covalent array by reacting azide-bearing saccharides with an alkyne that was bound to microtiter plates via a disulfide linker (Scheme 15.10) [46]. The disulfide bond survives a range of biological assay conditions and can readily be cleaved with reducing agents, allowing characterization and quantitative analysis by mass spectrometry. The utility of this method was demonstrated by performing two binding assays with *Lotus tetragonolobus* lectin (LTL), which recognizes α-l-fuctose [47], and *Erythrina cristagalli* lectin (EC) [48], which recognizes galactose.

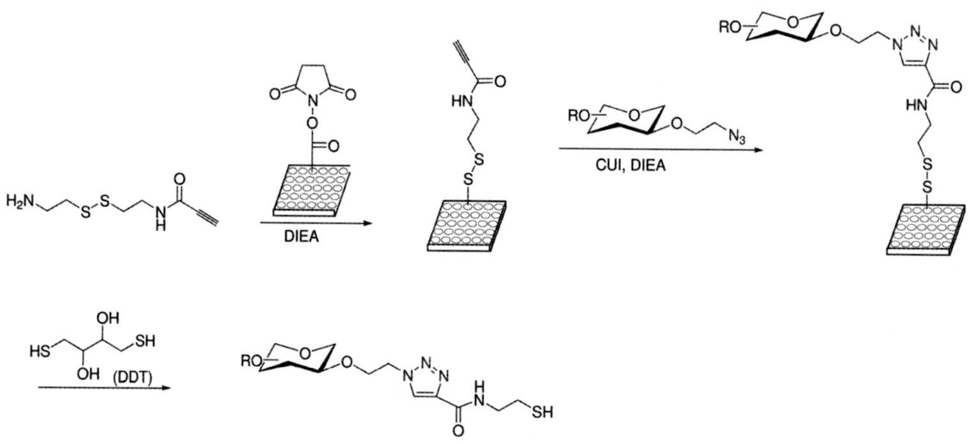

Scheme 15.10
Glycoarray synthesis based on the copper-catalyzed triazole process.

A carbohydrate array was employed for screening a compound library of GDP derivatives for Fuc-T inhibitors (Scheme 15.11) [49]. N-acetyllactosamine (LacNac) was displayed on the surface of a microtiter plate by Cu(I)-catalyzed triazole formation with an immobilized lipid alkyne. The compound library was mixed with Fuc-T and GDP-fucose and then incubated in the LacNAc-containing microtiter plate wells. The transfer of fucose to LacNAc was quantified with a lectin from *Tetragonolobus pupureas* (TP), conjugated to a peroxidase, leading to the identification of four compounds with nanomolar inhibition constants. Three of these hits contained a biphenyl substituent. Remarkably, these hits had not been identified in a preceding fluorescence-based assay [50].

This method can be used for high-throughput screening of large compound libraries at low cost.

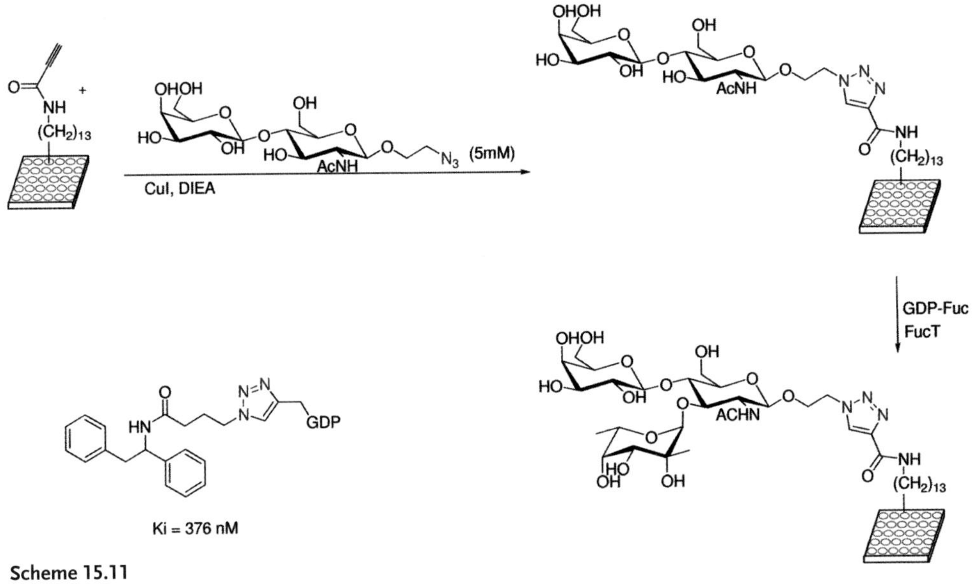

Scheme 15.11
Identification of fucosyl transferase inhibitors using a glycoarray.

15.4
In Situ Click Chemistry

The groups of Sharpless and Kolb developed a new strategy for drug discovery, which relies on the irreversible, target-guided synthesis (TGS) of high infinity inhibitors from small click chemistry fragments (Fig. 15.3) [51–54]. The latter are assembled within the biological target's binding pockets through selective binding and irreversible interconnection to generate potent inhibitors that engage in multiple interactions with the target. Since this target-templated, fragment-based approach requires relatively few reagents (which are sampled by the protein in thousands of different ways), it promises to be more efficient than traditional combinatorial chemistry, which involves the synthesis, purification and screening of thousands of library compounds. Follow-up tests are performed only with the target-generated hits to measure their binding affinities and specificities. These hits are usually very potent, due to the multivalent nature of the interactions with the protein.

15.4.1
Discovery of Highly Potent AChE by *In Situ* Click Chemistry

The target-guided click chemistry approach was first tested with acetylcholine esterase (AChE) (Fig. 15.4). The enzyme plays a key role in neurotransmitter hydrolysis in the central and peripheral nervous system [55, 56]. It has two separate binding sites on either end of a narrow gorge [57]. For fragment assembly by the

Step 1. Identification of Anchor Molecule

Step 2. Biligand Formation

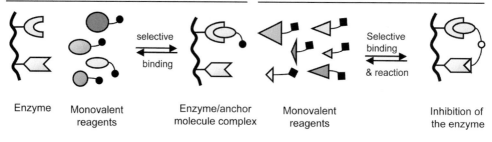

Fig. 15.3
In situ click chemistry.

enzyme, the thermal 1,3-dipolar cycloaddition reaction between azides and acetylenes [4] was employed, since the reaction is extremely slow at room temperature and the reactants do not react with the protein. The fragments were designed to bind to either the peripheral anionic site or the active center. They carried azide and acetylene groups, respectively, via flexible spacers. From a total of 52 possible reagent combinations, the enzyme selected only four pairs, leading to the formation of four reaction products in a highly regioselective fashion. These compounds turned out to be more potent than any other non-covalent organic AChE inhibitor, the most potent one having a dissociation constant K_d of 99 fM in the case of eel AChE (Fig. 15.4, left).

Additional experiments demonstrated that this fragment-based approach is suitable for the discovery of novel AChE inhibitors from reagents that were not previously known to interact with the enzyme's peripheral site. Thus, a total of 24 acetylene reagents were incubated in sets of up to ten compounds with AChE and the active site ligand, tacrine azide (Fig. 15.4, right) [58]. The triazole products, formed by the enzyme, were identified by HPLC-MS analysis of the crude reaction mixtures. The enzyme selected only the two phenyltetrahydroisoquinoline building blocks that were in the reagent library and combined them with the

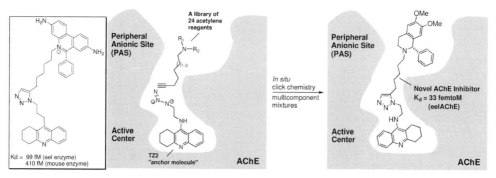

Fig. 15.4
In situ click chemistry with acetylcholine esterase (AChE).

tacrine azide within the active center gorge, to form multivalent inhibitors that simultaneously associate with the active and peripheral binding sites. The winning combination involved reagents that have relatively weak binding affinities for AChE (7.8–34 µM), demonstrating that this target-guided strategy is successful even with micromolar binders. The new inhibitors are not only more "drug-like" than the previous phenylphenanthridinium-derived compounds, due to the lack of permanent positive charge and aniline groups and fewer fused aromatic rings, but they are also three times as potent. With dissociation constants as low as 33 fM, these compounds are the most potent non-covalent AChE inhibitors known.

Recent work in the Kolb and Fokin laboratories has shown that the target-guided lead discovery method also works with targets other than acetylcholine esterase, e.g. other enzymes such as carbonic anhydrase II and HIV protease [59], and non-enzymes, such as transthyretin.

Carbonic anhydrases are Zinc–enzymes that catalyze the interconversion of HCO_3^- and CO_2. They are involved in key biological processes related to respiration and the transport of CO_2/HCO_3^-, bone resorption, calcification and tumorigenicity [60]. Systemically and topically administered carbonic anhydrase inhibitors have long been used to control the elevated intraocular pressure associated with glaucoma [61, 62]. Most inhibitors are aromatic or heteroaromatic systems that carry a sulfonamide functional group [63–65], which coordinates the Zn^{2+} ion in the active site. Based on this information, Kolb et al. designed 4-ethynylbenzenesulfonamide as an anchor molecule for the target-guided formation of carbonic anhydrase inhibitors (Fig. 15.5) [53]. The benzenesulfonamide anchor molecule was incubated with each of 24 azide reagents in the presence of bovine carbonic anhydrase II for 40 h. Analysis of all 24 reaction mixtures by liquid chromatography with mass spectrometry/selected ion mode detection (LC/MS-SIM) revealed

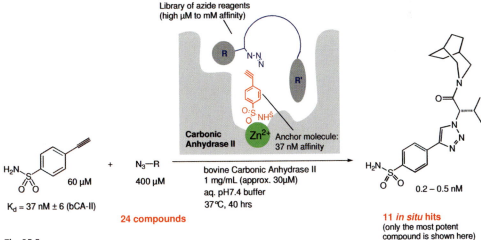

Fig. 15.5
In situ click chemistry with Carbonic anhydrase-II.

that 11 mixtures had formed triazoles inside the enzyme. Control experiments validated these hits: (a) no product was observed in the absence of protein, (b) no product was generated with bovine serum albumin as a result of non-specific protein binding and (c) no product was observed when the reaction was performed in the presence of a low-nanomolar carbonic anhydrase inhibitor. These results demonstrate that carbonic anhydrase is able to assemble triazoles from azide and acetylene building blocks and that the reaction occurred inside or near its active site. The most potent hit compound is 185 times more active ($K_d = 0.2$ nM) than the acetylenic scaffold ($K_d = 37$ nM).

15.5
Bioconjugation Through Click Chemistry

15.5.1
Tagging of Live Organisms and Proteins

Click chemistry was recently employed to selectively tag biomolecules in living cells or organisms to monitor their involvement in biological process and/or detect their biodistribution and metabolism [66, 67].

The azido group is a very useful chemical handle for bioconjugation, since it is largely inert in biological systems and readily undergoes highly selective reactions, such as the Staudinger ligation with phosphines [68] as well as [3+2]cycloaddition with strained [69] and terminal alkynes [70]. The Cu(I)-catalyzed reaction between azides and acetylenes enables the selective tagging of virus particles [71], nucleic acids [72], enzymes [73], cells [74] and proteins from complex tissue lysates [75].

The utility of the Cu(I)-catalyzed triazole formation for this purpose was established by labeling intact cowpea mosaic virus particles (CPMV) [76] with fluorescein (Scheme 15.12). CMPV is composed of 60 identical copies of a two sub-unit protein assembled around a single-stranded RNA. Through traditional bioconjugation techniques, a total of 60 azides were attached per virus particle, setting the stage for Cu(I)-catalyzed tagging with alkynefluorescein. Finn et al. developed spe-

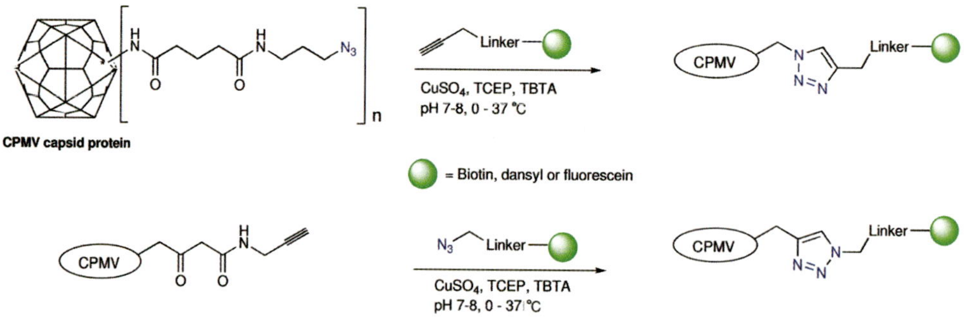

Scheme 15.12
Labeling of cowpea mosaic virus particles with fluorescein.

cial conditions for the triazole formation: CuSO$_4$ was reduced *in situ* with tris(2-carboxyethyl)phosphine (TCEP) and a Cu(I)-stabilizing ligand, tris[(1-benzyl-1H-1,2,3-triazol-4-yl)methyl]amine (TBTA), was used to stabilize Cu(I) in water and protect the virus from Cu-triazole-induced disassembly [71].

Tirrel and Link replaced methione residues in the outer membrane protein C (OmpC) of *Escherichia coli* cells by azidohomoalanine (Fig. 15.6; **8**) and subsequently reacted the azide groups with a biotinylated alkyne reagent, using the TBTA/copper-catalyzed triazole process to tag *E. coli* with biotin (Scheme 15.13) [74].

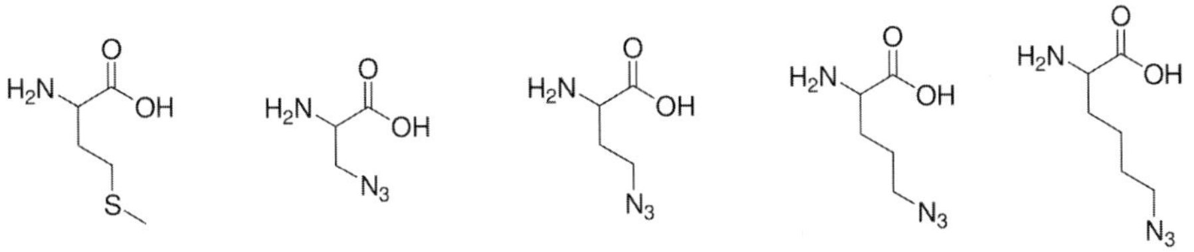

Fig. 15.6
Methionine, azidoalanine, azidohomoalanine azidonorvaline, azidonorleucine.

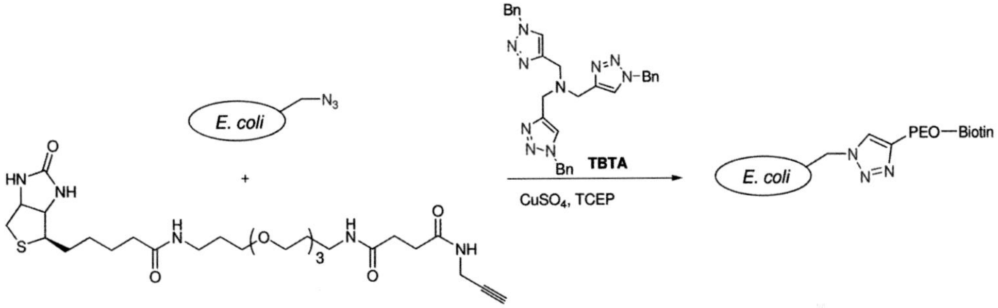

Scheme 15.13
Tagging *E. coli* with biotin.

Further studies revealed that highly pure (99.999%) CuBr allowed cell surface labeling to be enhanced by a factor of 20. This led to the discovery that even azidoalanine (Fig. 15.6; **9**), which was previously not thought to be a good replacement for methionine in protein synthesis, was incorporated in cell surface proteins. Also, azidonorvaline **10** and azidonorleucine **11** were detected [77].

Schultz and co-workers reported a site-specific, fast and irreversible method for protein bioconjugation [73]. The incorporation of genetically encoded azide- or alkyne-containing unnatural amino acids into proteins of *Saccharomyces cerevisiae*, followed by coupling with azide- or alkyne-dyes (Fig. 15.7, **12**), allowed gel-fluorescence scanning of the modified protein. This methodology for genetically encoding unnatural amino acids in prokaryotic and eukaryotic organisms was used in conjunction with phage display to synthesize polypeptide libraries with unnatural amino acid building blocks [78]. For this study, a pIII fusion streptavidin-binding

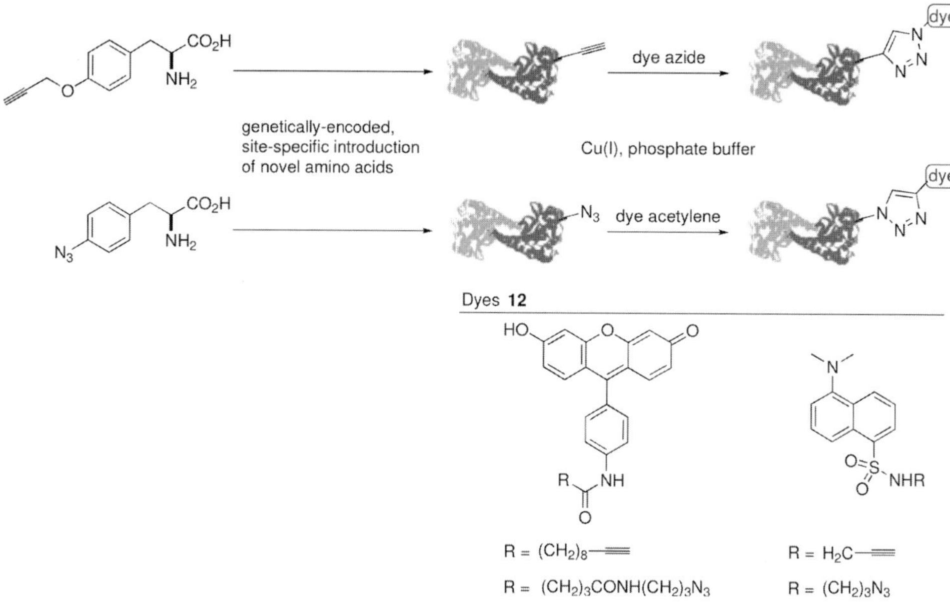

Fig. 15.7
Genetically encoded incorporation of unnatural amino acids for site-specific tagging of proteins.

peptide was expressed in *E. coli* strain TTS (type III secretion) in the presence of unnatural amino acids. The resulting phages displayed the unnatural amino acid-carrying peptides. In order to characterize this phage display system, the resulting phages were tagged with an alkyne dye, using the Tirrell conditions for the triazole-forming process [74]. SDS-PAGE and Western blot analysis showed that the unnatural amino acid was specifically introduced into the pIII coat protein of the unnatural phage. This technology enables the site-specific and irreversible attachment of a single PEG molecule to a protein, making possible the production of well defined and homogenous therapeutic PEGylated proteins.

15.5.2
Activity-based Protein Profiling

The field of proteomics is engaged in the development of methods for analyzing protein expression and function. Conventional methods measure protein variations based on abundance, whereas activity-based protein profiling (ABPP) [75], a chemical proteomics method, utilizes active site-directed probes to determine the functional state of enzymes in complex proteomes (e.g. to distinguish an active enzyme from its inactive precursors and inhibitor-bound forms). The ABPP probes contain two main elements: a reactive group (RG), which binds to and covalently labels the active site, and a reporter tag (e.g. biotin and or a

15.5 Bioconjugation Through Click Chemistry | 331

fluorophore), which enables the rapid detection and isolation of the labeled enzyme [66].

Sorensen et al. synthesized a series of rhodamine biotin-tagged compounds using the Cu(I)-catalyzed triazole process to investigate the FR182877 mechanism of action (Scheme 15.14), [78]. Incubation of the tagged derivatives with mouse proteome *in vitro* led to the identification of carboxylesterase-1 (CE-1) as a specific target of the natural product. The authors examined enzyme activity with a serine hydrolase-directed probe fluorophosphonate rhodamine and determined that Fr182877 is a selective inhibitor of CE-1 with an IC_{50} of 34 nM.

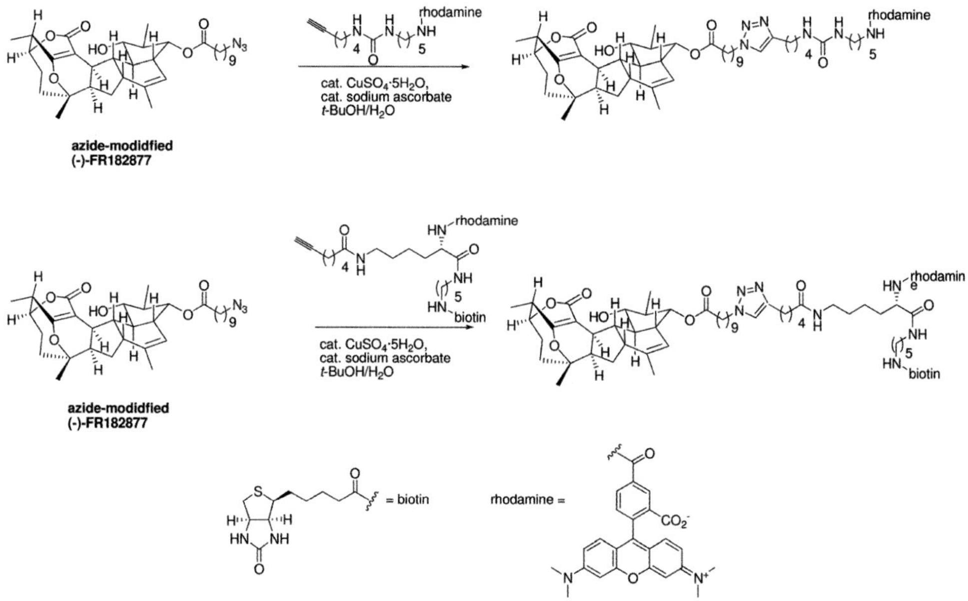

Scheme 15.14
Rhodamine biotin-tagged compounds for ABPP of carboxylesterase-1.

Despite the successful application of ABPP for several enzyme classes, the currently used bulky, cell wall impermeable reporter tags require cell homogenization prior to analysis, making measurement in living cells impossible. Cravatt et al. developed a "tag-free" version of ABPP, in which the reporter group is attached to the activity-based probe after covalent labeling of the protein targets (Scheme 15.15) [75]. Bioorthogonal coupling partners, azides and alkynes, were employed to join the probe and tag with the copper-catalyzed triazole process. With this method, Cravatt et al. detected glutathione *S*-transferase (GSTO 1–1), enoyl COA hydratase-1 (ECH-1) and ALDH-1 at endogenously expressed levels in viable cells. For these studies, an azide-bearing (tag-free) phenylsulfonate ester (PS-N$_3$) probe was incubated with intact cells and then ligated with a rhodamine alkyne reporter group after cell homogenization. The results were consistent with traditional methods.

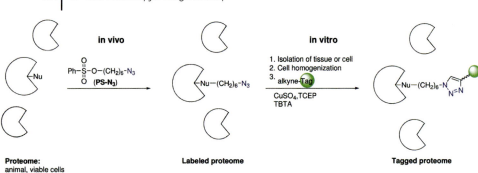

Scheme 15.15
"Tag-free" ABPP.

This click chemistry-based approach to ABPP allowed, for the first time, the determination of protein expression levels in living animals. ECH-1 was observed in the heart tissue of mice by injecting PS-N$_3$ into the living animal and staining the homogenized tissue of the sacrificed animal with rhodamine alkyne.

The drawback of this methodology is a higher degree of background labeling as compared to conventional ABPP, due to low-level non-specific labeling of abundant proteins of the proteome.

Further studies demonstrated that swapping the reaction groups (i.e. PS-alkyne/RhN$_3$ instead of PS-N$_3$/Rh-alkyne) reduced the extent of background labeling and enabled the visualization of lower-abundance enzyme activity [80]. However, the PS-N$_3$/Rh-alkyne combination reacts four times faster, making it the system of choice for fast analysis.

15.5.3
Labeling of DNA

The Ju laboratory has fluorescently labeled single-stranded DNA (ssDNA), using concerted thermal 1,3-dipolar cycloaddition [72]. Thus, DNA was azido-labeled by reacting succinimidyl 5-azido-valerate with an amino linker-modified oligonucleotide (M13–40 universal forward sequencing primer); and its cycloaddition with 6-carboxy-fluorescein (FAM) was carried out thermally in aqueous media (Scheme 15.16a). The resulting FAM-labeled ssDNA was successfully used in the Sanger dideoxy chain termination reaction, using PCR-amplified DNA as a template to produce DNA sequencing products terminated by biotinylated dideoxyadenine triphosphate (ddATP-biotin). These fragments were analyzed by capillary array electrophoresis and the results matched exactly the sequence of the DNA template. An electron-withdrawing group was attached to the alkyne to optimize the cycloaddition reaction for DNA analysis and immobilization [81]. This allowed catalyst-free 1,3-dipolar cycloaddition to proceed smoothly at room temperature in aqueous solution, conditions that are compatible with DNA modification under cell-biological conditions.

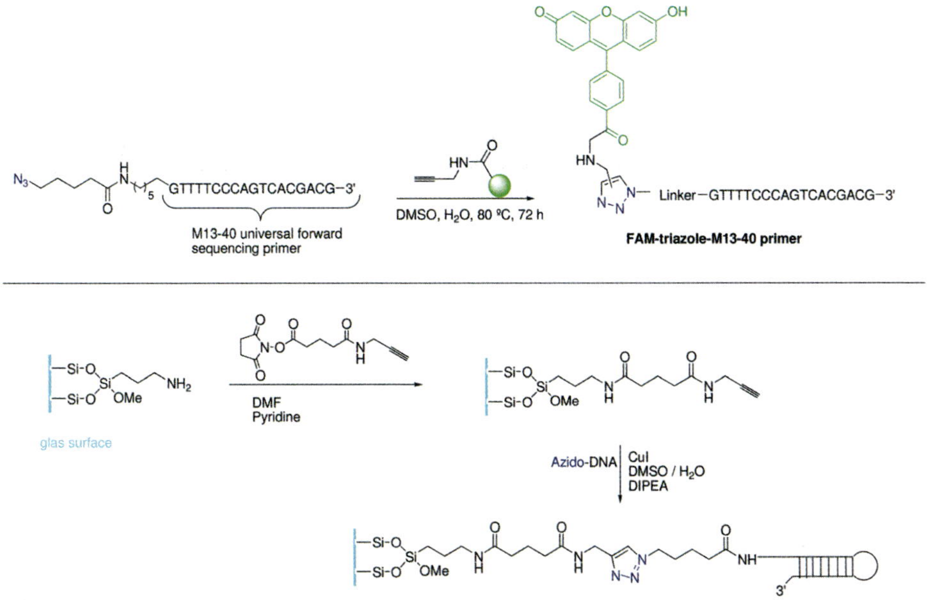

15.5 Bioconjugation Through Click Chemistry | 333

Scheme 15.16
(a) DNA labeling, (b) immobilization of DNA on a glass chip.

The Cu(I) process has been employed to immobilize DNA on a solid support for fast sequencing (Scheme 15.16 b) [82]. Thus, an alkyne-modified glass surface was coupled with azido-labeled DNA, using the very mild and selective Cu(I) triazole process. In order to determine the functionality, stability and accessibility of the immobilized DNA, polymerase extension reactions were successfully carried out with three unique photocleavable (PC) fluorescent nucleotides. In a sequence involving alternate incorporation of PC fluorescent nucleotides, fluorophore detection and photolytic removal of the fluorophore, a seven-nucleotide sequence in the DNA template was correctly identified.

15.5.4
Artificial Receptors

In polar solvents, cyclodextrins are able to bind hydrophobic molecules within their cavities, making them useful as enzyme mimetics [83] and molecular receptors [84]. Due to the large number of hydroxyl groups, selective modification of cyclodextrins is challenging. Most approaches to create functionalized cyclodextrins have very low yields and often lead to complex mixtures of anomeric compounds or to systems with different ring sizes. Gin et al. demonstrated that Cu(I)-catalyzed triazole formation can be used for making functionalized cyclodextrin mimics (Scheme 15.17; **14**, **15**), starting from azido acetylene-bearing trisaccharides (Scheme 15.17; **13**) [85]. In the presence of Cu(I), a β-cyclodextrin analog was obtained in excellent yield, whereas the uncatalyzed reaction led to multiple pro-

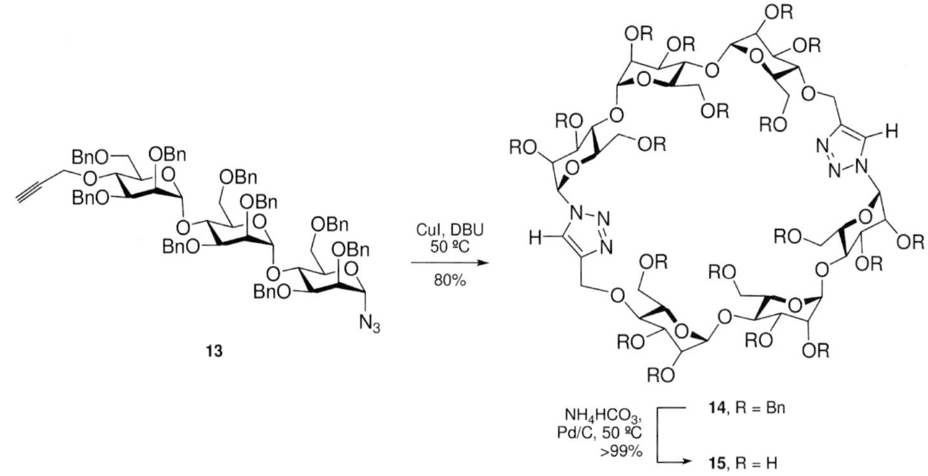

Scheme 15.17
β-Cyclodextrin analogs by cyclodimerization of trisaccharide via [3+2]-cycloaddition.

ducts. The deprotected macrocycle **15** was shown to bind 8-anilino-1-naphthalene-sulfonate with similar affinity as regular β-cyclodextrin. This new strategy allowed the synthesis of functionalized cyclodextrin mimics that would have been difficult to make using traditional means.

15.6
Conclusion

In summary, click chemistry has proven to be a powerful tool in biomedical research, ranging from combinatorial chemistry and target-templated *in situ* chemistry for lead discovery, to bioconjugation strategies for proteomics and DNA research. A number of click chemistry reactions have been developed; and heterogeneous, water-based reactions as well as the formation of triazoles from azides and acetylenes deserve special recognition. Despite their high energy content, azides and acetylenes are stable across a broad range of organic reaction conditions and their irreversible combination to triazoles is highly exothermic, albeit slow. The full potential of this ligation reaction was unleashed with the discovery of Cu(I) catalysis, which allows the reaction to proceed rapidly with near-quantitative yields. Since no protecting groups are used, the crude triazole products can be subjected to biological screens without purification. This triazole-forming process, and click chemistry in general, promises to accelerate both lead finding and lead optimization, due to its great scope, modular design and reliance on extremely short sequences of near-perfect reactions.

References

1 Bohacek, R. S. M., Guida, C., Wayne, C. **1996**, The art and practice of structure-based drug design: a molecular modeling perpective. *Med. Chem. Res.* 16, 3–50.

2 Kolb, H. C., Finn, M. G., Sharpless, K. B. **2001**, Click Chemistry: Diverse Chemical Function from a Few Good Reactions. *Angew. Chem. Int. Ed.* 40, 2004–2021.

3 (a) Tietze, L. F., Kettschau, G. **1997**, Hetero Diels-Alder reactions in organic chemistry. *Topics Curr. Chem.* 189, 1–120; (b) Jorgensen, K. A. **2000**, Catalytic Asymmetric Hetero-Diels-Alder Reactions of Carbonyl Compounds and Imines. *Angew. Chem. Int. Ed.* 39, 3558–3588.

4 (a) Huisgen, R. **1965**, 1,3-Dipolar Cycloaddition introduction survey, mechanism. In: *1,3-Dipolar Cycloadditon Chemistry*, vol. 1, ed. Padwa, A., Wiley, p. 1–176; (b) Huisgen, R. **1963**, 1,3-Dipolar Cycloaddition. Past and Future. *Angew. Chem. Int. Ed.* 2, 565–598.

5 Adolfsson, H., Converso, A., Sharpless, K. B. **1999**, Comparison of amine additives most affective in the new methyltrioxorhenium-catalyzed epoxidation process. *Tetrahedron Lett.* 40, 3991–3994.

6 Gontcharov, A. V., Liu, H., Sharpless, K. B. **1999**, *tert*-Butylsulfonamide. A New Nitrogen Source for Catalytic Aminohydroxylation and Aziridination of Olefins. *Org. Lett.* 1, 783–786.

7 Kolb, H. C., VanNieuwenhze, M. S., Sharpless, K. B. **1994**, Catalytic Asymmetric Dihydroxilation. *Chem. Rev.* 94, 2483–2547.

8 Kühle, E. **1971**, One Hundred Years of Sulfenic Acid Chemistry. IIa. Oxidation, Reduction, and Addition reactions of Sulfenyl Halides. *Synthesis* 11, 563–586.

9 Narayan, S., Muldoon, J., Finn, M. G., Fokin, V. V., Kolb, H. C., Sharpless, K. B. **2005**, "On Water": Unique Reactivity of Organic Compounds in Aqueous Suspension, *Angew. Chem. Int. Ed Engl.* 117, 3339–3343.

10 Rostovtsev, V. V., Green, L. G., Fokin, V. V., Sharpless, K. B. **2002**, A Stepwise Huisgen Cycloaddition Process: Copper(I)-Catalyzed Regioselective Ligation of Azides and Alkynes. *Angew. Chem. Int. Ed Engl.* 114, 2596–2599.

11 Tornoe, C. W., Christensen, C., Meldal, M. **2002**, Peptidotriazoles on Solid Phase: [1,2,3]-Triazoles by Regiospecific Copper(I)-Catalyzed 1,3-Dipolar Cycloadditions of Terminal Alkynes to Azides. *J. Org. Chem.* 67, 3057–3064.

12 Bourne, Y., Kolb, H. C., Radic, Z., Sharpless, K. B., Taylor, P., Marchot, P. **2004**, Freeze-frame inhibitor captures acetylcholinesterase in a unique conformation. *Proc. Natl. Acad. Sci. USA* 101, 1449–1454.

13 Kolb, H. C., Sharpless, K. B. **2003**, The Growing Impact of Click Chemistry on Drug Discovery. *Drug Discov. Today* 8, 1128–1137.

14 Now Lexicon Pharmaceuticals.

15 Kolb, H. C., et al. Large scale synthesis of optically pure aziridines. 2001, US 20033153771.

16 Kolb, H. C., et al. **2003**, Preparation of 1,2,3-triazoles carboxylic acids from azides and β-ketoester in the presence of base. US patent 2003135050.

17 Kolb, H. C., et al. **2003**, Modified safe and efficient process for the environmentally friendly synthesis of imidoester by Pinner addition using hydrogen chloride solution. US patent 20033153728.

18 Chen Z., et al. **1999**, Preparation of 3-aminoazetidines as building blocks for combinatorial libraries. Patent WO9919297.

19 Kolb, H. C. et al. **2000**, Preparation of PPAR-(gamma) agonists as agents for the treatment of type II diabetes. Patent WO2000063161.

20 Bettinetti, L., Löber, S., Hübner, H., Gmeiner, P. **2005**, Parallel Synthesis and Biological Screening of Dopamine Receptor Ligands Taking Advantage of a Click Chemistry Based BAL Llinker. *J. Comb. Chem.* 7, 309–316; see also Löber, S., Rodriguez-Loaiza, P., Gmeiner, P. **2003**, Click Linker: Efficient and High-Yielding Synthesis of a New Family of SPOS Resins by 1,3-Dipolar Cycloaddition. *Org. Lett.* 5, 1753–1755.

21 Woodford, N. **2003**, Novel agents for the treatment of resistant Gram-positive infections. *Expert Opin. Invest. Drugs* 12, 117–137.

22 McCormick, A. W., Whitney, C. G., Farley, M. M., Lynfield, R., Harrison, L. H., Bennett, N. M., Schaffner, W., Reingold A., Hadler, J., Cieslak, P., Samore, M. H., Lipsitch M. **2003**, Geographic diversity and temporal trends of antimicrobial resistance in *Streptococcus pneumoniae* in the United States. *Nat. Med.* 2003, 424–430.

23 Weymouth-Wilson, A. C. **1997**, The role of carbohydrates in biologically active natural products. *Nat. Prod. Rep.* 14, 99–110.

24 Fu, X., Albermann, C., Jiang, J., Liao, J., Zhang, C., Thorson, J. S. **2003**, Antibiotic optimization via *in vitro* glycorandomization. *Nat. Biotechnol.* 21, 1467–1469.

25 Petchprayoon, C., Suwanborirux, K., Miller, R., Sakata, T., Marriott G. **2005**, Synthesis and Characterization of the 7-(4-Aminomethyl-1*H*-1,2,3-triazol-1-yl) Analogue of Kabiramide C. *J. Nat. Prod.* 68, 157–161.

26 Distributed as Zyvox.

27 Reck, F., Zhou, F., Girardot, M., Kern, G., Eyermann, C. J., Hales, N. J., Ramsay, R. R., Gravestock, M. B. **2005**, Identification of 4-Substituted1,2,3-Triazoles as Novel Oxazolidinone Antibacterial Agents with Reduced Activity Against Monoamine Oxidase A. *J. Med. Chem.* 48, 499–506.

28 Lin, H., Walsh, C. T. **2004**, A Chemoenzymatic Approach to Glycopeptide Antibiotics. *J. Am. Chem. Soc.* 126, 13998–14004.

29 (a) Large, D. G., Warren, C. D. (eds) **1997**, *Glycopeptides and Related Compounds: Synthesis, Analysis and Applications.* Marcel Dekker, New York; (b) Kunz, H. **1987**, Synthesis of Glycopeptides, Partial Structures of Biological Recognition Components [New Synthetic Methods]. *Angew. Chem., Int. Ed.* 26, 294; (c) Spiro, R. G. **2000**, Protein glycosylation: Nature, Distribution, Enzymatic Formation, and Disease Implications of Glycopeptide Bonds. *Glycobiology* 12, 43–56; (d) Seitz, R. O. **2000**, Glycopeptide Synthesis and the Effects of Glycosylation on Protein Structure and Activity. *ChemBioChem* 1, 214.

30 Dondoni, A., Marra, A. **2000**, Methods for Anomeric Carbon-Linked and Fused Sugar Amino Acid Synthesis: The Gateway to Artificial Glycopeptides. *Chem. Rev.* 100, 4395–4422.

31 (a) Kuijpers, B. H. M., Groothuys, S., Keereweer, A. R., Quaedflieg, P. J. L. M., Blaauw, R. H., van Delft, F. L., Rutjes, F. P. J. T. **2004**, Expedient Synthesis of Triazole-Linked Glycosyl Amino Acids and Peptides. *Org. Lett.* 6, 3123–3126; (b) Dondoni, A., Giovannini, P. P., Massi, A. **2004**, Assembling Heterocycle-Tethered *C*-Glycosyl and α-Amino Acid Residues via 1,3-Dipolar Cycloaddition Reactions. *Org. Lett* 6, 2929–2932.

32 Perez-Balderas, F. M., Ortega-Munoz, J., Morales-Sanfrutos, F., Hernandez-Mateo, F. G., Calvo-Flores, J. A. Calvo-Asin, J., Isac-Garcia, F., Santoyo-Gonzalez **2003**, Multivalent Neoglycoconjugates by Regiospecific Cycloaddition of Alkynes and Azides using Organic-Soluble Copper Catalysts. *Org. Lett.* 5, 1951–1954.

33 Akula, R. A., Temelkoff, D. P., Artis, N. D., Norris, P. **2004**, Rapid Acess to Glycopyranosyl-1,2,3-Triazoles Via Cu(I)-Catalyzed Reactions In Water. *Heterocycles* 63, 2719–2725.

34 Sabesan, S. **2003**, Preparation of Triazole-linked Oligosaccharides cia Cycloaddition of Alkyne with Azide Sugars. US patent 6664399 (Sabesan, S. **2004**, *Chem. Abstr.* 140, 667).

35 Abdel-Rahman, H. M. et al. **2002**, HIV Protease inhibitors: Peptidomimetic Drugs and Future Perspectives. *Curr. Med. Chem.* 9, 1905–1922.

36 Miller, V. **2001**, Resistance to protease inhibitors. *J AIDS* 26[Suppl.1], 34–50.

37 Brik, A., Muldoon, J., Lin, Y.-C., Elder, J. E., Goodsell, D. S., Olson, A. J., Fokin, V. V., Sharpless, K. B., Wong, C. H. **2003**, Rapid Diversity-Oriented Synthesis in Microtiter Plates for In Situ Screening of HIV Protease Inhibitors. *ChemBioChem* 4, 1246–1248.

38 Kim, E. E., Baker, C. T., Dwyer, M. D., Murcko, M. A., Rao, B. G., Tung, R. D., Navia, M. A. **1995**, Crystal Structure of HIV-1 Protease in Complex with VX-478, a Potent and Orally Bioavailable Inhibitor

of the Enzyme. *J. Am. Chem. Soc.* 117, 1181–1182.

39 Kaldor, S. W., Hammond, M., Dressman, B. A., Fritz, J. E., Crowell, T. A., Hermann, R. A. **1994**, New dipeptide isosteres useful for the inhibition of HIV-1 protease. *Bioorg. Med. Chem. Lett.* 4, 1385–1390.

40 Roberts, N. A., Martin, J. A., Kinchington, D., Broadhurst, A. V., Craig, J. C., Duncan, I. B., Galpin, S. A., Handa, B. K. **1990**, Rational Design of Peptide-Based HIV Protease Inhibitors. *Science* 248, 358–361.

41 (a) Staudacher, E. **1996**, *Trends Glycosci. Glycotechnol.* 8, 391–408; (b) Sears, P., Wong, C.-H. **1998**, Enzyme action in glycoprotein synthesis. *Cell Mol. Life Sci.* 54, 223–252.

42 (a) Compain, P., Martin, O. R. **2001**, *Bioorg. Med. Chem.* 9, 3077–3092; (b) Qian, X., Palcic, M. M. **2000**, in: Ernst, B. (ed.) *Carbohydrates in Chemistry and Biology*, vol. 3, Wiley–VCH, Weinheim, p. 293–312; (c) Mitchell, M. L.; Tian, F.; Lee, L. V.; Wong, C.-H. **2002**, *Angew. Chem. Int. Ed.* 41, 3041–3044.

43 Lee, L. V., Mitchell, M. L., Huang, S.-J., Fokin, V. V., Sharpless, K. B., Wong, C.-H. **2003**, A Potent and Highly Selective Inhibitor of Human α-1,3-Fucosyltransferase via Click Chemistry. *J. Am. Chem. Soc.* 125, 9588–9589.

44 Fazio, F., Bryan, M. C., Lee, H.-K., Chang, A., Wong, C.-H. **2004**, Assembly of sugars on polystyrene plates: a new facile microarray fabrication technique. *Tetrahedron Lett.* 45, 2689–2692.

45 Feizi, T., Fazio; F., Chai; W., Wong, C.-H. **2003**, Carbohydrate microarrays – a new set of technologies at the frontiers of glycomics. *Curr. Opin. Struct. Biol.* 13, 637–645 (and refs. ceited therein).

46 Bryan; M. C., Fazio, F., Lee, H.-K., Huang, C.-Y., Chang, A., Best, M. D., Calarese, D. A., Blixt, O., Paulson, J. C., Burton, D., Wilson, I. A., Wong, C.-H. **2004**, Covalent Display of Oligosaccharide Arrays in Microtiter Plates. *J. Am. Chem. Soc.* 126, 8640–8641.

47 Svensson, C., Teneberg, S., Nilsson, C. L., Kjellberg, A., Schwarz, F. P., Sharon, N., Krengel, U. **2002**, High-resolution Crystal Structures of *Erythrina cristagalli* Lectin in Complex with Lactose and 2′α-L-Fucosyllactose and Correlation with Thermodynamic Binding Data. *J. Mol. Biol.* 321, 69–83.

48 Hasselhorst, T., Weimar, T., Peters, T. **2001**, Molecular Recognition of Sialyl Lewis^x and Related Saccharides by Two Lectins. *J. Am. Chem. Soc.* 123, 10705–10714.

49 Bryan, M. C., Lee, L. V., Wong, C.-H. **2004**, High-throughput identification of fucosyltransferase inhibitors using carbohydrate microarrays *Biorg. Med. Chem Lett.* 14, 3185–3188.

50 Mitchell, M. L., Tian, F., Lee, L. V., Wong C.-H. **2002**, Synthesis and Evaluation of Transition-State Analogue Inhibitors of α-1,3-Fucosyltransferase. *Angew. Chem. Int. Ed.* 41, 3041–3044.

51 Lewis, W. G., Green, L. G., Grynszpan, F., Radic, Z., Carlier, P. R., Taylor, P. Finn, M. G., Sharpless, K. B. **2002**, Click chemistry in situ: Acetylcholinesterase as a reaction vessel for the selective assembly of a femtomolar inhibitor from an array of building blocks. *Angew. Chem. Int. Ed.* 41, 1053–1057.

52 Bourne, Y., Kolb, H. C., Radic, Z., Sharpless, K. B., Taylor, P., Marchot, P. **2004**, Freeze-frame inhibitor captures acetylcholinesterase in a unique conformation. *Proc. Natl. Acad. Sci. USA* 101, 1449–1454.

53 Mocharla, V. P., Colasson, B., Lee, L. V., Röper, S., Sharpless, K. B., Wong, C.-H., Kolb, H. C. **2005**, In situ Click Chemistry: Enzyme-Generated Inhibitors of Carbonic Anhydrase-II. *Angew. Chem. Int. Ed.* 44, 116–120.

54 Manetsch, R., Krasinski, A., Radic, Z., Raushel, J., Taylor, P., Sharpless, K. B., Kolb, H. C. **2004**, In Situ Click Chemistry: Enzyme Inhibitors Made to Their Own Specifications. *J. Am. Chem. Soc.* 126, 12809–12818.

55 Quinn, D. M. **1987**, Acetylcholinesterase: Enzyme Structure, Reaction Dynamics, and Virtual Transition States. *Chem. Rev.* 87, 955–979.

56 Taylor, P., Radic, Z. **1994**, *Annu. Rev. Pharmacol. Toxicol.* 34, 281–320.

57 Sussman, J. L., et al. **1991**, Atomic structure of acetylcholinesterase from *Torpedo*

californica: a prototypic acetylcholine-binding protein. *Science* 253, 872–879.
58 Krasinski, A.; Radic, Z.; Manetsch, R.; Raushel, J.; Taylor, P.; Sharpless, K. B.; Kolb, H. C. **2005**, In situ Selection of Lead Compounds by Click Chemistry: Target-Guided Optimization of Acetylcholinesterase Inhibitors. *J. Am. Chem. Soc.* 127, 6680–6692.
59 Whiting, M., Muldoon, J., Lin, Y.-C., Silvermann, S. M., Lindstrom, W., Olson, H., Kolb, H. C., Finn, M. G., Sharpless, K. B., Elder, J. H., Fokin, V. V. **2005**, Inhibitors of HIV-1 Protease via In Situ Click Chemistry. *Angew. Chem.* (submitted).
60 Dodgson, S. J. **1991**, The carbonic anhydrases. Overview of their importance in cellular physiology and in molecular genetics. In: Dodgson, S. J. (ed.) *Carbonic Anhydrases*. Plenum, New York, p. 3–14.
61 Sugrue, M. F., Smith, R. L. **1985**, Antiglaucoma agents. *Annu. Rep. Med. Chem.* 20, 83–91.
62 Sugrue, M. F. Gautheron, P., Schmitt, C., Viader, M. P., Conquet, P., Smith, R. L., Share, N. N., Stone, C. A. **1985**, On the pharmacology of L-645,151: a topically effective ocular hypotensive carbonic anhydrase inhibitor. *J. Pharmacol. Exp. Ther.* 232, 534–540.
63 Supuran, C. T., Scozzafava, A., Casini, A. **2003**, Carbonic anhydrase inhibitors. *Med. Res. Rev.* 23, 146–189.
64 Gao, J., Cheng, X., Chen, R., Sigal, G. B., Bruce, J. E., Schwartz, B. L., Hofstadler, S.A., Anderson, G. A., Smith, R. D., Whitesides, G. M. **1996**, Screening Derivatized Peptide Libraries for Tight Binding Inhibitors to Carbonic Anhydrase II by Electrospray Ionization-Mass Spectrometry. *J. Med. Chem.* 39, 1949–1955.
65 DuPriest, M. T., Zinke, P. W., Conrow, R. E., Kuzmich, D., Dantanarayana, A. P., Sproull, S. J. **1997**, Enantioselective Synthesis of AL-4414A, a Topically Active Carbonic Anhydrase Inhibitor. *J. Org. Chem.* 62, 9372–9375.
66 van Swieten, P. F., Leeuwenburgh, M. A., Kessler, B. M., Overkleeft, H. S. **2005**, Bioorthogonal organic chemistry in living cells: novel strategies for labeling biomolecules. *Org. Biomol. Chem.* 3, 20–27.

67 Cook, B. N., Bertozzi, C. R. **2002**, Chemical approaches to the investigation of cellular systems. *Bioorg. Med. Chem.* 10, 829–840.
68 Dube; D. H., Bertozzi C. R. **2003**, Metabolic oligosaccharide engineering as a tool for glycobiology. *Curr. Opin. Chem. Biol.* 7, 616–625.
69 Agard, N. J., Prescher, J. A., Bertozzi, C. R. **2004**, A Strain-Promoted [3+2] Azide-Alkyne Cycloaddition for Covalent Modification of Biomelcules in Living Systems. *J. Am. Chem. Soc.* 126, 1504615047.
70 Breinbauer, R., Köhn, M. **2003**, Azide–alkyne coupling: a powerful reaction for bioconjugate chemistry. *ChemBioChem* 4, 1147–1149.
71 Wang, Q., Chan, T. R., Hilgraf, R., Fokin, V. V., Sharpless, K. B., Finn, M. G. **2003**, Bioconjugation by copper(I)-catalyzed azide-alkyne [3+2]cycloaddition. *J. Am. Chem. Soc.* 125, 3192–3193.
72 Seo, T. S., Li, Z., Ruparel, H., Ju, J. **2003**, Click Chemistry to Construct Fluorescent Oligonucleotides for DNA Sequencing. *J. Org. Chem.* 68, 609–612.
73 Deiters, A., Cropp, T. A., Summerer, D., Mukherji, M., Schultz, P. G. **2004**, Site-specific PEGylation of proteins containing unnatural amino acids. *Bioorg. Med.-Chem.* 14, 5743–5745; b) Deiters; A., Cropp, T. A, Mikherji, M., Chin, J. W., Anderson, J. C., Schultz, P. G. **2003**, Adding Amino Acids with Novel Reactivity to the Genetic Code of *Saccharomyces cerevisiae*. *J. Am. Chem. Soc.* 125, 11782–11783.
74 Link, A. J., Tirell, D. A. **2003**, Cell surface labeling of *Escherichia coli* via Copper(I)-catalyzed [3+2]cycloaddition. *J. Am. Chem. Soc.* 125, 11164–11165.
75 (a) Speers, A. E., Adams, G. C., Cravatt, B. F. **2003**, Activity-Based Protein Profiling in Vivo Using a Copper(I)-Catalyzed Azide-Alkyne[3+2] Cycloaddition. *J. Am. Chem. Soc.* 125, 4686–4687; (b) Speers, A. E., Cravatt, B. F. **2004**, Chemical Strategies for Activity-Based Proteomics. *ChemBioChem.* 5, 41–47.
76 Wang, Q., Lin, T., Tang, L., Johnson, J. E., Finn, M. G. **2002**, Icosahedral Virus Particles as Addressable Nanoscale Building Blocks. *Angew. Chem. Int. Ed.* 41, 459–462.

77 Link, A. J., Vink, M. K. S., Tirrell, D. A. **2004**, Presentation and Detection of Azide Functionality in Bacterial Cell Surface Proteins. *J. Am. Chem. Soc.* 126, 10598–15602.

78 Tian, F., Tsao, M.-L., Schultz, P. G. **2004**, A phage display system with unnatural amino acids. *J. Am. Chem. Soc.* 126, 15962–15963.

79 Adam, G. C., Vanderwal, C. D., Sorensen, E. J., Cravatt, B. F. **2003**, (–)-FR182877 is a potent and selective inhibitor of carboxylesterase-1. *Angew. Chem. Int. Ed.* 42, 5480–5484.

80 Speers, A. E., Cravatt, B. F. **2004**, Profiling Enzyme Activities in Vivo Using Click Chemistry Methods. *Chem. Biol.* 11, 535–546.

81 Li, Z., Seo, T. S., Ju, J. **2004**, 1,3-dipolar cycloaddition of azides with electron-deficient alkynes under mild condition in water. *Tetrahedron. Lett.* 45, 3143–3146.

82 Seo, T. S., Bai, X., Ruparel, H., Li, Z., Turro, N. J., Ju, J. **2004**, Photocleavable fluorescent nucleotides for DNA sequencing on a chip constructed by site-specific coupling chemistry. *Proc. Natl. Acad. Sci.* 101, 5488–5493.

83 (a) Breslow, R., Dong, S. D. **1998**, Biomimetic Reactions Catalyzed by Cyclodextrins and Their Derivatives. *Chem. Rev.* 98, 1997–2012; (b) Rizarelli, E., Vecchio, G. **1999**, Metal complexes of functionalized cyclodextrins as enzyme models and chiral receptors. *Coord. Chem. Rev.* 188, 343.

84 Takahashi K. **1998**, Organic Reactions Mediated by Cyclodextrins. *Chem. Rev.* 98, 2013–2034.

85 Bodine, K. D., Gin, D. Y., Gin, M. S. **2004**, Synthesis of Readily Modifiable Cyclodextrin Analogues via Cyclodimerization of an Alkynyl-Azido Trisaccharide. *J. Am. Chem. Soc.* 126, 1638–1639.

16
Dynamic Combinatorial Diversity in Drug Discovery
Matthias Hochgürtel and Jean-Marie Lehn

16.1
Introduction

Combinatorial chemistry evolved as a key technology for the rapid generation of large populations of structurally distinct molecules that can be screened efficiently *en masse* for desirable properties. This approach, in combination with high-throughput screening, evolved into a powerful technique capable of significantly accelerating the drug discovery process [1]. Initially developed to produce peptide libraries for screening against antibodies or receptors, combinatorial synthesis and the screening of combinatorial arrays became an integral part in lead generation and optimization in drug discovery [2–5]. Furthermore, combinatorial methodologies also demonstrated their potential in other areas, such as the development and discovery of catalysts, and material science [6, 7].

Modern combinatorial libraries are characterized by the generation of numerous different, but structurally related compounds that exist discretely as static entities, under similar conditions, in a systematic manner. Currently, there are several distinct methodologies for the generation of combinatorial libraries [8, 9]. These combinatorial arrays can be produced in a parallel fashion, either as a library of individual compounds or as pooled mixtures. In applying a parallel format, the compounds are handled individually in separated compartments. This removes difficulties associated with compound mixtures; and it allows for every individual compound a straightforward evaluation of chemical integrity and structure–activity relationship after biological testing. The library size handled with this methodology is more limited than in a pooling strategy. Pooled mixture formation is straightforward and less time-consuming than the synthesis of individual compounds. More difficult is the analysis and screening of mixtures of components. By the split and mix technology, very large libraries can be formed progressively over multiple synthetic steps. Also, several technological advances in solid phase automated synthesis, analysis, robotics and miniaturization allow the rapid parallel synthesis and characterization of very large arrays of individual compounds. The substances generated are then subsequently evaluated for biological activity by high-throughput screening, applying fully automated systems towards a chosen target protein.

Fragment-based Approaches in Drug Discovery. Edited by W. Jahnke and D. A. Erlanson
Copyright © 2006 WILEY-VCH Verlag GmbH & Co. KGaA, Weinheim
ISBN: 3-527-31291-9

16.2
Dynamic Combinatorial Chemistry – The Principle

Dynamic combinatorial chemistry (DCC) is a recent and rapidly developing approach for ligand or receptor identification, based on the implementation of dynamic assembly and recognition processes [10–20]. It offers a possible alternative to the static approaches of traditional combinatorial chemistry. The method is based on the generation of a dynamic combinatorial library (DCL) of interchangeable constituents. Such a DCL consists of continually interchanging library members generated by reversible reaction between a set of building blocks (components). The set of all potentially accessible self-assembly combinations of the components represents a virtual combinatorial library (VCL). The interaction of the library constituents with a molecular target drives the equilibrium towards the formation of those individual library members that bind to the target and thus are thermodynamically stabilized. Such a self-screening process, by which the active species are preferentially generated at the expense of the non-active ones and retrieved from the dynamic mixture, drastically simplifies the identification of effective ligands in the pool. The DCC approach is target-driven and combines two basic supramolecular themes [21]: self-assembly in generating a DCL and molecular recognition in the interaction of all possible entities with a target species.

The DCC concept is represented schematically in Fig. 16.1 using the "lock and key" image of Emil Fischer. The process can be divided into three simple steps: (a) selecting the initial building blocks, capable of connecting reversibly with each other, (b) establishing the library generation conditions, where the building blocks are allowed to form interchanging, individual molecular entities, and (c) subjecting the library to selection, in this case, binding affinity to a target species. If the target molecule acts as a trap for a good binder, the ensemble of candidates can be

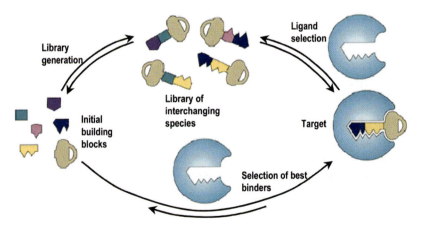

Fig. 16.1
Schematic representation of the concept underlying dynamic combinatorial chemistry and virtual combinatorial chemistry based on the "lock and key" image of Emil Fischer.

forced to rearrange to favor the formation of this species. The process described where ligands are screened towards a given receptor is termed "substrate casting"; and the corresponding process where a synthetic receptor is selected by the addition of a certain ligand can is termed "receptor molding" [22].

16.3
Generation of Diversity: DCC Reactions and Building Blocks

The generation of DCLs can be accomplished by using any type of molecular reaction or supramolecular interaction. A key feature is the reversible interconnection between the library constituents. Besides reversibility, several other criteria have to be fulfilled by a chemical reaction for the efficient generation of a DCL. The reversible exchange should be sufficiently fast to reach equilibrium within a short time. The generation of the dynamic library should proceed under thermodynamic control. The reaction should be compatible with a broad range of functional groups; and suitable building blocks should be accessible in order to increase the diversity of the libraries. The reactivity of linker functionalities should be similar in order to form unbiased, iso-energetic libraries. This will guarantee the formation of all possible combinations in comparable amounts.

Applying DCC to a biological target requires additional features. Working with biological systems means the reaction needs to be compatible with physiological aqueous conditions. In buffered aqueous media, there can be problems with the solubility of the starting materials, which are used in part in excess, together with biological molecules. The reaction conditions must be optimized for keeping the biological targets intact in their active form. An additional desirable characteristic of the reaction used is that, once the system has reached equilibrium, it should be possible to freeze or lock-in the dynamic exchange process [22]. By this step, the dynamic library is transformed into a static library which can be more easily analyzed. This can be done either by changing the reaction conditions (pH, temperature, solvent composition) or by adding a chemical quenching reagent (oxidation/reduction reagent).

Several types of reversible reactions are known (Table 16.1). Each of them has its own characteristics, advantages and drawbacks. For the construction of DCLs in the presence of a biological target imine, acylhydrazone formation [13, 22–29], disulfide exchange [30–33], transamidation [34], transthioesterification [35], enzyme-catalyzed aldol reaction [36, 37], cross-metathesis [38, 39] and boronate transesterification [40] have been reported. The C=N bond formation processes, by condensation of an amino group with carbonyl functionalities, are (besides disulfide exchange) the most important class of DCC reactions and have proven to be efficient when biological systems are targeted. The library formation and exchange kinetics can be optimized by careful selection of nucleophiles and carbonyl compounds in respect to suitable conditions for the biological target. In the case of aliphatic amines as nucleophiles, imines are rapidly formed and exchanged with carbonyl compounds [41]. At neutral pH, the equilibrium favors the starting materi-

Table 16.1 Selection of dynamic processes for potential use in DCC. All processes represented are "self-contained" [45], except imine formation, disulfide formation and boronic ester formation.

Reversible covalent bond formation

Process	Reactants	Products
Imine formation	>C=O + H₂N–R	>C=N–R
Hemiketal formation	>C=O + HO–R	>C(OH)(O–R)
Transacylation	R–C(=O)–XR₁ + Y–R₂	R–C(=O)–YR₂ + X–R₁
Aldol formation	R–CHO + R–C(=O)–R	R–CH(OH)–CH₂–C(=O)–R
Michael reaction	CH₂=CH–C(=O)– + H–X	X–CH₂–CH₂–C(=O)–
Disulphide formation	R–SH + HS–R	R–S–S–R
Diels Alder reaction	diene + dienophile	cyclohexene
Metathesis reaction	R¹–CH=CH–R¹ + R²–CH=CH–R²	R¹–CH=CH–R² + R¹–CH=CH–R²
Boronic ester formation	R–B(OH)₂ + HO–CHR¹–CHR²–OH	cyclic boronic ester

Reversible interactions

Process	Reactants	Products
Metal coordination	M^{m+} + nL	[ML_n]^{m+}
Electrostatic interaction	R–COO⁻ + H₃N⁺–R'	R–COO⁻····⁺H₃N–R'
Hydrogen bonding	>C=O + HN<	>C=O····H–N<
Donor-acceptor interaction	D + A	[DA]

Table 16.1 (continued)

Reversible intramolecular processes	
Configurational	
Cis-trans isomerization	(structure)
Conformational	
Internal rotation	(structures)
Ring inversion	(structures)
Structural	
Tautomerism	(structure)

als. Therefore, detection of imines may be difficult and analysis of a dynamic imine library requires an additional quenching step by reductive amination. In the case of acyl hydrazones or hydroxylamines, the imines are stabilized, the equilibrium is on the product side and the exchange is much slower [25, 42], so that efficient exchange rates require more acidic conditions. By using selected acylhydrazones having electron-withdrawing groups, the formation and exchange can be tuned for nearly neutral pH [43].

An important challenge in the development of DCC is still the search for new or the rediscovery of known suitable reversible reactions for the generation of dynamic libraries. Nucleophilic addition to conjugated carbonyl compounds is known to be a reversible reaction. Recently, the Michael addition of a thiol to ethacrynic acid derivates was studied in a model system to evaluate their potential in the formation of DCLs [44]. Under mild, nearly physiological conditions, the library was rapidly equilibrated. The scrambling process was shown to be selectively switched on and off by lowering the pH. To ensure the generation of near iso-energetic libraries, enone building blocks were functionalized relatively far from the linker Michael acceptor.

Cycloaddition represents another class of reactions which are attractive for the application in DCC. First, a demonstration of its potential use in DCC was reported recently [45]. By the reaction of dienes with a set of dienophile Diels–Alder building blocks under mild, non-aqueous conditions, a dynamic library was rapidly formed; and component exchange in adduct formation was demonstrated. The Diels–Alder reaction allows a straightforward generation of structural and functional diversity; and, starting from two planar constituents, a three-dimensional structure is generated. As in case of the Michael and aldol reaction, the

Diels–Alder reaction belongs to the class of self-contained reversible reactions, where all atoms present in the starting materials are also present in the products.

Much higher diversity may be achieved by the generation of dynamic libraries through implementation of more than one reversible linker chemistry [13, 46]. By using a similar number of building blocks and two or more reactions, highly complex multi-dimensional dynamic libraries can be realized. Preferably, the applied chemistry should act in an orthogonal fashion that allows independent control of the exchange processes, for example the combination of metal coordination and imine chemistry [46]. Within such multi-dimensional DCC experiments, the diversity increases rapidly, so that a smaller number of building elements may be used, in a compact and more easily controllable experimental setup.

16.4
DCC Methodologies

Different approaches for the generation and screening of DCLs have been developed, including the adaptive approach, the pre-equilibrated approach [25, 30] and the iterative approach [47, 48]. These approaches differ mainly in the screening and selection processes. The deletion approach [49–51] is only indirectly related to DCC, because it utilizes so-called pseudo-dynamic libraries which include irreversible library generation.

The adaptive approach represents the ideal case in applying DCC. The generation and screening take place simultaneously in one compartment in the presence of the biological target. The system, because of its fully dynamic nature, is able to respond to an external pressure, so as to adapt and amplify the best binding library constituents. If one constituent in the DCL interacts better than the others with a certain target species, then this constituent will be withdrawn from the equilibrating pool and all of the components that make up this constituent will also be masked by the binding. Because of the equilibrium situation, the system has to rearrange so as to produce more of this constituent at the expense of the other species in the library. On re-equilibration, the most active constituent therefore experiences a certain degree of amplification, in comparison with the situation in which no target molecule is added.

The pre-equilibrated approach includes the formation of a dynamic library under reversible conditions and the screening for active species after freezing the equilibration under static conditions. This approach is especially useful when sensitive and delicate biological targets that are unavailable in large quantities are used or the linker chemistry requires equilibration conditions which are not compatible with the biological target. The library is generated under reversible conditions, before screening of the reversible exchange is stopped by changing the reaction condition to yield a static library. Biological testing of the static library is then accomplished under standard assay conditions.

Active compounds can be identified in the static library by using different deconvolution protocols. In the iterative approach, the generation of the DCL proceeds in one compartment under reversible conditions. In an additional step, the library members can interact with the biological target either in the same or in an additional reaction chamber. By using an immobilized target, the bound species are separated from the unbound species, which are then redirected to the reaction chamber, re-scrambled and subsequent again interact with the biological target. After several scrambling and selection circles, the accumulated bound species on the immobilized target is analyzed.

The deletion approach is based on a irreversible formation and destruction process of library constituents in the presence of a biological target. These types of libraries are so-called pseudo-dynamic libraries (pDCLs), which involve irreversible instead of reversible transformations, as is the case for dynamic libraries. After irreversible library generation, the constituents are interacted with the biological target; and the destruction process removes library constituents from the mixture continuously, removing the unbound ligands faster than the bound ligands. By tightly binding to the target, the library members with highest affinity are protected from destruction. Generation, destruction and selection proceed in three different compartments separated by a dialysis membrane. Periodically, the library is regenerated by the addition of starting material; and both the selection and the destruction process start again. After several cycles, the tightest binders are accumulated and only small amounts of the non-binders are left.

16.5
Application of DCC to Biological Systems

The applications of dynamic combinatorial chemistry, as described so far in numerous reports, can be divided into two categories. One set is primarily focused on the development of reversible chemistries, to establish optimized conditions for a controllable rapid exchange of library constituents without subjecting these libraries to any target-directed screening. The conditions applied are often not suitable for biological systems; and the studies aim at understanding the basic features of DCL formation, their characteristics and their analysis. [24, 32, 40, 52–57]. The other set addresses both the generation and the screening phases in the presence of a target; and various protocols have been developed. In particular, several applications that target different classes of biological macromolecules, including lectins, enzymes and polynucleotides have been reported (Table 16.2). Biological molecules are the most interesting and the most challenging target molecules. The dynamic libraries have been generated using a range of different building blocks. Most of these have been based on non-natural construction elements, but attempts have also been made with amino acids, nucleotides, and carbohydrates.

An early related case made use of a monoclonal antibody (against β-endorphin) as a molecular target capable of driving the reversible synthesis/hydrolysis of a peptide mixture towards the formation of the specific antibody binder [34] (Fig. 16.2). A protease, thermolysin, exhibiting a broad specificity, was used to equilibrate a library of oligopeptides from two initial peptides. It was shown that

Table 16.2 Application of dynamic combinatorial chemistry to biological systems.

Target [refs.]	Reversible chemistry	Library size	Hit(s)
Trypsin [40]	Alcohol–boronate exchange	n.d.	Tripeptidyl-boronate
Carbonic anhydrase [22]	Transimination	12	Sulfamoylbenzaldimine
Carbonic anhydrase [49, 50]	Transamidation; pseudo-dynamic combinatorial	8	Sulfamoyldipeptides
Acetylcholinesterase [25]	Acyl hydrazone exchange	66	Bis-pyridinium
Acetylcholinesterase [35]	Transthioesterfication	10	Acetylthiocholine
HPr kinase [59]	Acyl hydrazone exchange	440	bis-benzimidazole
β-Galactosidase [28, 29]	Transimination	8	N-Alkyl-piperidine
Neuraminidase [26, 27]	Transimination	>40 000	Tamiflu analogs
Hen egg white lysozyme HEWL [58]	Transimination	12	D-glcNAc derivate
Cyclin-dependent kinase 2 (CDK2) [60]	Hydrazone exchange	30	Oxindol-sulfamoylbenz-hydrazone
Cysteine aspartyl protease-3 (caspase 3) [31]	Disulfide exchange "extended tethering"	7000	Salicylic acid fragment
GalNAc-specific lectins [63, 64]	Metal coordination	4	Tris-GalNAc
Concanavalin A [30]	Disulfide exchange	21	Bis-mannoside
Concanavalin A [65]	Acyl hydrazone exchange	474	Tris mannoside
Weat germ agglutinin [36, 37]	Enzyme-catalyzed aldol reaction	4	Sialic acid
Interleukin 2 (IL-2) [61]	Disulfide exchange "extended tethering"	7000	Small aromatic acid fragments
Anti-β-endorphin [34]	Transamidation	n.d.	Peptide (YGG-FL)
Fibrinogen [34]	Transamidation	n.d.	Peptides
DNA [23, 68]	Transimination + metal coordination	36	Bis-salicylaldimine-Zn(II) complex
DNA; RNA [71, 72]	Transimination	4	Naldixic acid
DNA [73]	Disulfide exchange	9	Peptidic hetero- and homodisulfides
RNA [67]	Metal coordination	>27	Salicylamide Cu(II)-complex
Ac$_2$-L-Lys-D-Ala-D-Ala [39]	Metathesis/disulfide exchange	36	Bis-vancomycin
S. aureus [33]	Disulfide exchange	3828	Psammaplin A analogues

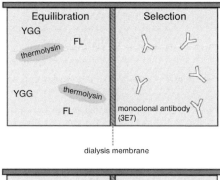

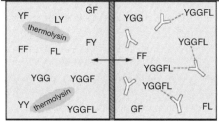

Fig. 16.2
Generation and screening of β-endorphine peptides by protease-catalyzed transamidation.

the peptide exchange by simultaneous peptide synthesis and hydrolysis catalyzed by the protease can take place under mild conditions. To prevent digestion of the target molecule by thermolysin, the library generation and selection process were performed in two compartments separated by a dialysis membrane, permeable only for the small equilibrating peptides. Analysis of the peptide mixture by HPLC and sequencing showed the formation of a pentapeptide known to bind to the antibody target with low nanomolar affinity. By indirect determination, it could be demonstrated that, by applying the protease-catalyzed reaction, the presence of the antibody leads to an amplification of a specific peptide sequence, known as an antibody binder.

16.5.1
Enzymes as Targets

One of the most attractive reaction for dynamic library generation is the imine formation from amines and carbonyl compounds, in particular aldehydes. The reaction is compatible with aqueous media and is characterized by the rapid formation of a stable equilibrium and fast exchange rates. The proof of concept was first performed with the enzyme carbonic anhydrase, subjected to a DCL generated by means of the transimination reaction [22] (Fig. 16.3). This enzyme has an active site comprising a Zn(II)-containing pocket and a neighboring lipophilic site; and known potent inhibitors contain a sulfonamide coupled to a hydrophobic moiety. The building blocks were designed to present structural features close to those of the known efficient inhibitors. By the reaction of three different aromatic aldehydes and four different primary amines, a dynamic library of twelve different compounds was generated. An excess of amine was applied to overcome the effect from potential amino groups at the surface of the enzyme. Because imines possess low stability in aqueous solution, cyanoborohydride was added to freeze out the formed imines by selective reduction to the corresponding secondary amines and thus simplify analysis of the library. Two sets of experiments, wherein the dynamic library was formed in the presence and the absence of the enzyme followed by reduction of library constituents, showed that the presence of target modified the abundance of certain constituents of the library. The formation of one of the imines resulting from a sulfonamide aldehyde and benzylamine was markedly amplified. The structure correlated well with the structural features of one of the strongest known inhibitors and corresponded to SAR data on the carbonic anhydrase active site.

Generation of dynamic combinatorial libraries by transimination has also been successfully applied to identify inhibitors of other enzymes, α-galactosidase [28],

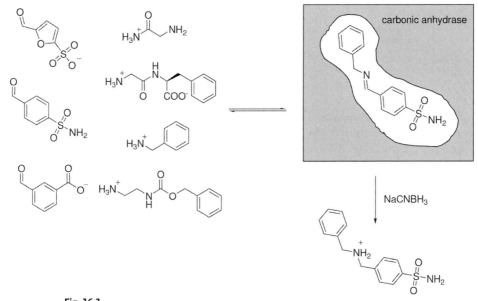

Fig. 16.3
Dynamic combinatorial library of imines interacting with carbonic anhydrase. One amine/aldehyde combination was selected and amplified. The corresponding amine was analyzed by HPLC after reduction with sodium cyanoborhydride.

neuraminidase [26, 27] and hen egg white lysozyme (HEWL) [58]. In the former case, a dynamic imine library was generated from eight different aldehydes and 4-hydroxypiperidine, and resulted in the preliminary identification of two amines that showed inhibitory activity after the reductive step. Neuraminidase, a key enzyme responsible for influenza virus propagation, has been used as a template for selective synthesis of small subsets of its own inhibitors from theoretically highly diverse dynamic combinatorial libraries. A diamine scaffold which was structurally similar to known inhibitors was used, together with various aldehyde or ketones building blocks, for the generation of dynamic libraries (Fig. 16.4). The experiment was performed in the presence and absence of the target. The imines were frozen by reduction with sodium cyanoborhydride to secondary amines. The distribution of formed products was analyzed by HPLC/MS. Addition of the enzyme target led to a dramatic amplification of selected amine peaks.

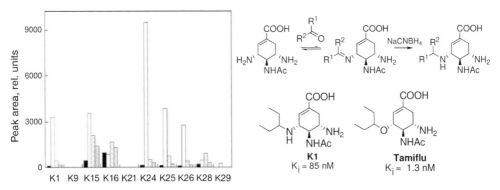

Fig. 16.4
Dynamic combinatorial library generated from a diamino scaffold and ten ketones, showing the relative amounts of library member in the absence of NA (black bars), in the presence of NA (white bars), in the presence of BSA (stippled bars) and in the presence of NA and Zanamivir (gray bars).

Routinely, two control experiments were performed to verify the amplification of hits and to exclude any non-specific effects of the protein: Library generation in the presence of bovine serum albumin (BSA) and in the presence of Neurmanidase together with a known potent inhibitor. Zanamivir inhibited the hit amplification successfully in all cases except one ketone. This building block was also amplified by BSA; and a non-specific amplification through interaction with the protein surface can be assumed.

The neuraminidase example showed that the size of a dynamic library can easily be made very large and that the potential diversity level of such virtual libraries can be very high. The set of 20 aldehydes together with an amine scaffold containing two linker units potentially yields over 40 000 constituents, as shown in Fig. 16.5. In a recent example, the transimination was used to identify weak

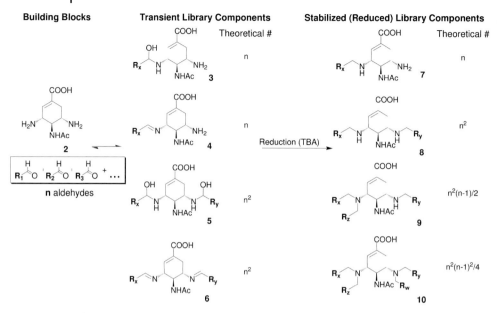

Fig. 16.5
Potential diversity generated from one scaffold with two linker sites and *n* aldehyde components.

binders of HEWL. This protein belongs to the class of carbohydrate-binding proteins (termed Group II proteins). Their shallow carbohydrate binding site results in weak carbohydrate protein interactions, recorded by dissociation constants in the millimolar range. The dynamic library was designed from two amine scaffolds with relatively poor binding affinities and six aromatic aldehydes (Fig. 16.6). After reductive freezing of the library, the resulting amines were analyzed by HPLC. In the presence of the target, the distribution of amines was detectably changed. Two amines derived from N-acetyl-glucosamine out of 12 possible were enriched. By competition experiments with chitotriose, a good HEWL inhibitor, the amplification could be completely suppressed. Resynthsized hits showed a 100-fold improved affinity in respect to the starting scaffold N-acetyl-D-glucosamine.

Because of the imine instability, direct analysis and isolation of a DCL of imines is difficult. Thus the exchange process needs to be frozen by the addition of a reducing agent that converts labile imines into stabile amines. Other chemical reactions have been studied to circumvent this problem. One of these is acylhydrazone formation from hydrazides and aldehydes. The reaction proceeds reversibly under mild acid catalysis and turns out to be slow enough at higher pH for analysis of the DCLs generated. Hydrazone exchange was used to form and screen DCLs towards the inhibition of acetylcholinesterase [25]. This enzyme has two binding sites, the esterasic site at the bottom of a deep gorge and a so-called "peripheral" site near the rim of this gorge. Both sites are selective for positively charged func-

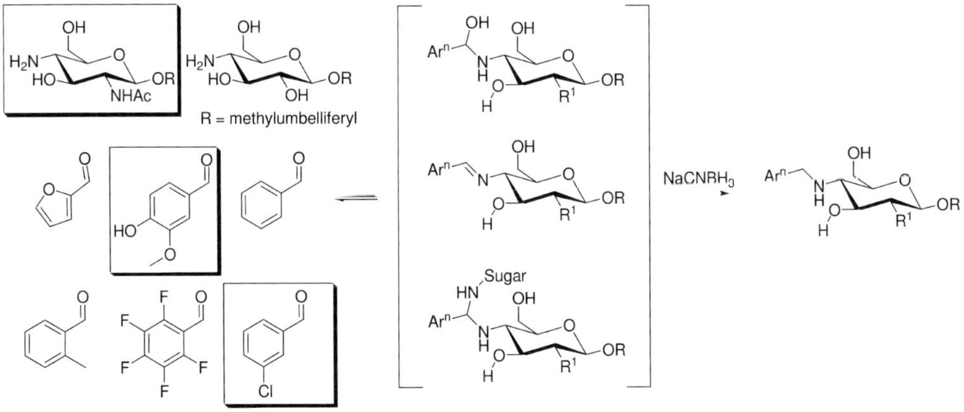

Fig. 16.6
Dynamic combinatorial library of imines interacting with hen egg white lysozyme. Highlighted amine/aldehyde combinations were selected and amplified.

tionalities, such as quaternary ammonium groups; and dicationic compounds bridging the two sites can act as very efficient inhibitors. Based on this structural information, a set of 13 components, hydrazides and aldehydes, was used to form a dynamic library of 66 acyl hydrazones (Fig. 16.7). A dynamic deconvolution procedure based on the sequential removal of starting building blocks revealed a very potent inhibitor of acetylcholinesterase that contained two terminal cationic (pyridinium) recognition groups separated by a spacer of appropriate length.

In a similar example, ditopic dynamic combinatorial libraries were generated and screened toward inhibition of the bifunctional enzyme HPr kinase/phosphatase from *Bacillus subtilis* [59]. The libraries were composed of all possible combinations resulting from the dynamic exchange of 16 hydrazides and five monoaldehyde or dialdehyde building blocks, resulting in libraries containing up to 440 different constituents. By using different dynamic deconvolution procedures, active compounds could be rapidly identified. Potent inhibitors consist of two terminal cationic heterocyclic recognition groups separated by a spacer of appropriate structure and length.

A new approach, termed dynamical combinatorial X-ray crystallography (DCX), was used to identify rapidly potent inhibitors of cyclin-dependent kinase CDK2 [60]. A dynamic library was formed by the reaction of six hydrazines and five oxindoles in the presence of CDK2 protein crystals. The best binding ligands were directly identified by X-ray crystallography by interpreting electron-density maps from crystals exposed to a DCL. With this approach, the direct identification of amplified ligands from the DCL as complex with the target is possible and simultaneous information about the ligand binding mode is accessible.

The approach of pseudo-dynamic chemistry was first explored by targeting carbonic anhydrase with a library of dipeptides. Two amino acids reacted irreversibly

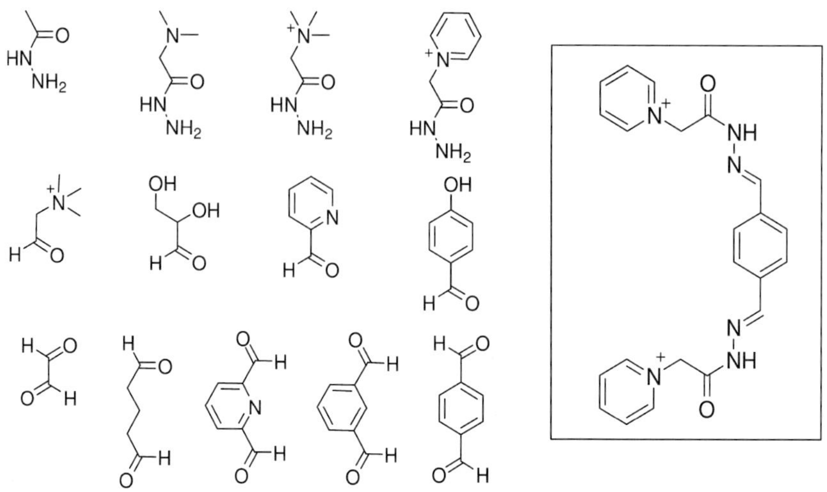

Fig. 16.7
Targeting of acetylcholinesterase by applying prequilibrated libraries of acyl hydrazones. Dynamic deconvolution protocols – sequential removal of each building block and biological testing – identified the optimum combination and so the best binder (highlighted).

with four different polymer-supported active esters to generate a library of eight dipeptides (Fig. 16.8). One of the amino acid contained a sulfonamide group that was expected to bind to the Zn(II) center in the active side of carbonic anhydrase. Non-binding dipeptides were hydrolyzed by a non-selective pronase. Library generation, ligand selection and destruction were conducted in three different compartments separated by a dialysis membrane to prevent degradation or inactivation of carbonic anhydrase by the pronase or the polymer-supported active ester. By action of pronase, hydrolyzed amino acids regenerated the eight possible dipeptides by reaction with periodically added new active ester, to form a pseudo-dynamic library. The process of library generation, selection and destruction was repeated with optimized conditions up to seven times. Finally, only the most potent dipeptide derived from proline and a sulfonamide containing amino acid was detected in an excess of 100:1 over the second strongest binder, which was only 2.3-fold less active than the tightest binder.

A recent variation of the DCL concept, "extended tethering", uses the known disulfide exchange reaction for library generation [31, 61] (chapter 14). This strategy combines the technology of covalent tethering and dynamic combinatiorial chemistry. By protein engineering, cysteins are added specifically near to the enzyme active side. The thiol of cystein reacts with thiol, building blocks to a dynamic set of disulfides; and the most favored combinations are stabilized by binding to the active side of the protein. Under thiol-exchange conditions, 7000 disulfides in pools of 9–12 compounds were screened in the presence of cysteine aspartyl protease-3,

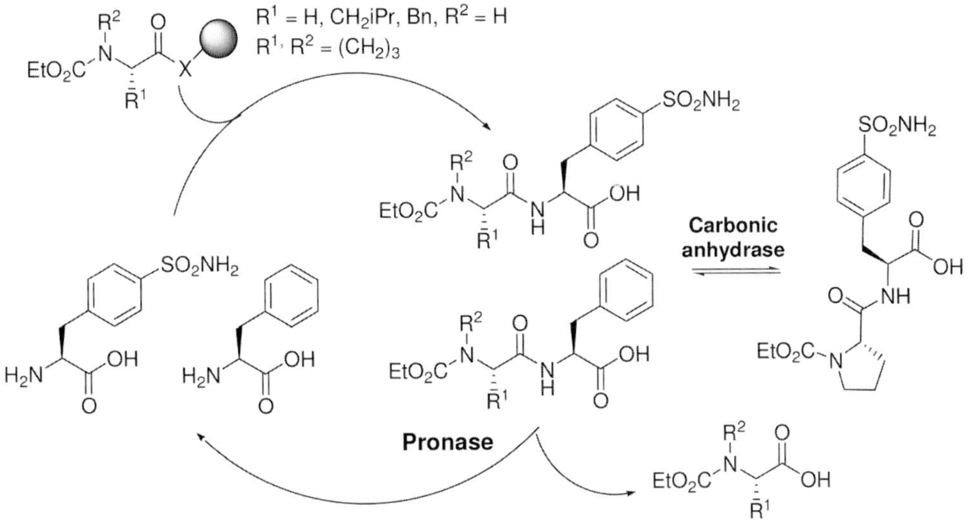

Fig. 16.8
Pseudo dynamic combinatorial library of eight dipeptides generated by irreversible reaction between four active esters and two amino acids. By the action of pronase, amino acids formed by hydrolysis are recoupled by periodical addition of new active ester. The best binding ligand is protected from destruction by complexation with carbonic anhydrase.

an anti-inflammatory target. By MS analysis of the set of enzyme–disulfides, the weak binding fragments which selectively bind to a subsite of caspase 3 were discovered. By linking these fragments to aspartyl aldehyde, a bivalent inhibitor in the low micromolar range was obtained. The same methodology was used for the identification of structural features required for binding to interleukin-2 [61], a target in immune-disorder therapy. Traditional medicinal chemistry may make use of this information for the design of inhibitors of IL-2 with increased potency.

16.5.2
Receptor Proteins as Targets

Molecular recognition of carbohydrates represents a research area with strong potential bearing on drug discovery, as well as various biotechnology applications [62]. Carbohydrates play central roles in many biological processes, such as cell–cell interactions and cell communication; and numerous enzymes are involved in various carbohydrate-mediated processes associated with cell proliferation and cell death, for example. An early example in which DCC has been applied in glycoscience described the formation of a prototype library of four different interchanging stereoisomers from the Fe^{2+}-assisted assembly of a carbohydrate-decorated bipyridine unit [63, 64]. On interaction with a range of GalNAc-selective lectins, the distribution of these isomers was adjusted depending on the lectin.

More recently, the disulfide interchange was applied to produce ditopic carbohydrate structures with binding affinity to concanavalin A [30]. Disulfides can be scrambled under mild conditions by controlling the redox properties of the system. The library was generated at neutral or slightly basic pH from six different thiol-derivatized carbohydrate head groups (Fig. 16.9). Under these conditions, the exchange is fast. The dynamic process can be efficiently locked by lowering the pH from pH 7 to pH 2. In the absence of the target, after equilibration, all 21 different disulfide species were generated, as determinated by HPLC. In the presence of immobilized concanavalin A, a shift in equilibration caused an amplification of a bis-mannoside unit and, to a lesser extent, of mannose heterodimers, at the expense of other library constituents.

Concanavalin A was also targeted by using the acylhydrazone exchange reaction [65]. Dynamic carbohydrate libraries were generated from six carbohydrate alde-

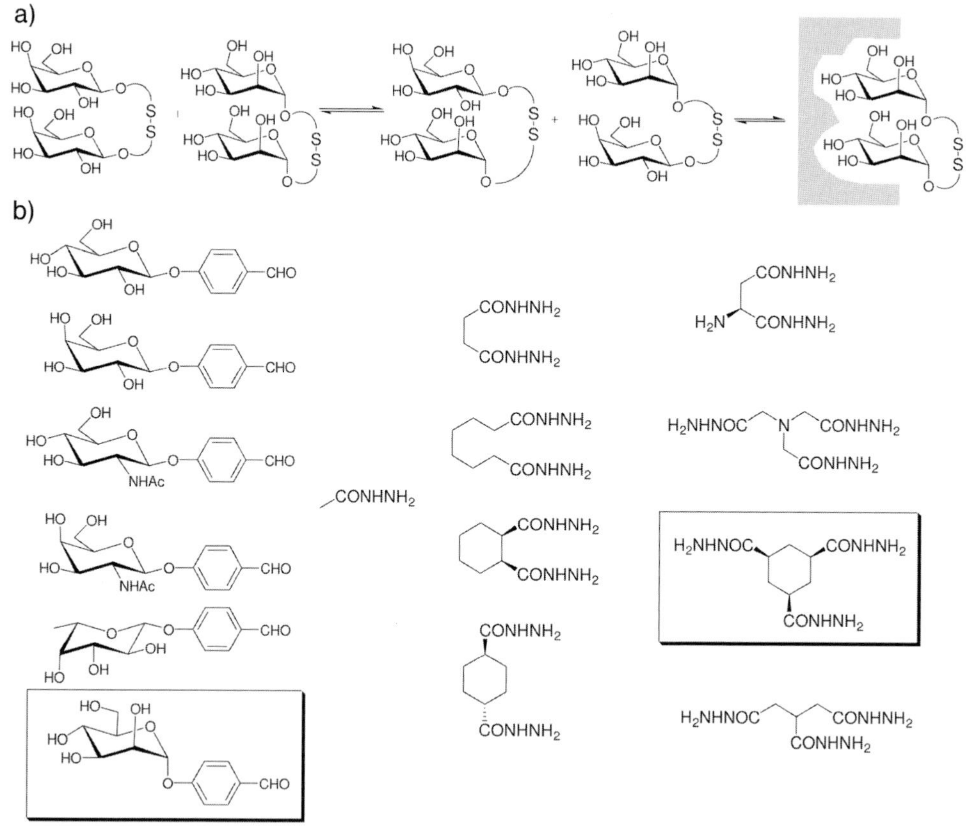

Fig. 16.9
Carbohydrate libraries targeting concanavaline A, showing (A) a dynamic combinatorial library of disulfide containing carbohydrates and (B) a dynamic library of carbohydrates containing hydrazone spacer units.

hydes and eight hydrazide scaffolds containing one, two, or three attachment groups through reversible hydrazone exchange, resulting in a library with up to 474 members (Fig. 16.9). As in the example with acetylcholinesterase, dynamic deconvolution strategies were successfully utilized to identify structural features crucial for binding to concanavalin A. As a result, a tritopic mannoside was selected from the dynamic system and demonstrated to be a strong binder of concanavalin A, showing an IC_{50} of 22 µM.

Another plant lectine, wheat germ aglutidinin (WGA), was chosen as a molecular target to explore enzyme-catalyzed aldol reactions for the generation of DCLs [36, 37]. Three different carbohydrate building blocks were mixed with an excess pyruvate in the presents of NANA aldolase (Fig. 16.10). The known weak WGA binder, sialic acid, was significantly amplified in the presence of the target while others where suppressed or only slightly amplified. By applying the enzyme-catalyzed reaction, the DCL was straightforwardly generated by stereoselective carbon–carbon bond formation. This example also demonstrates that weak binding can be sufficient to shift the equilibrium distribution.

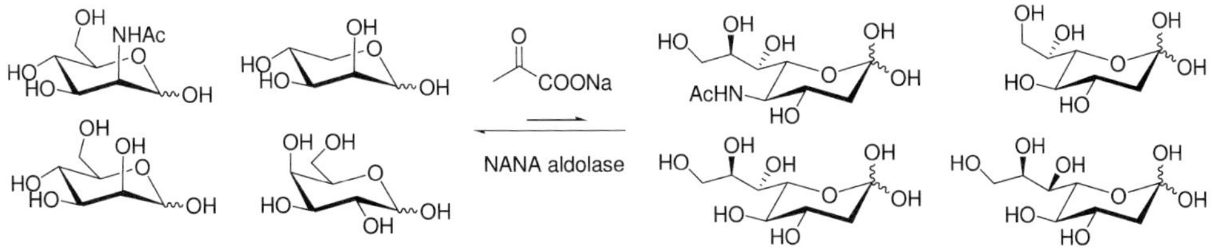

Fig. 16.10
Dynamic carbohydrate libraries generated by enzyme-catalyzed aldol reaction.

16.5.3
Nucleotides as Targets

Several approaches to apply nucleotides as targets to dynamic combinatorial chemistry are reported [23, 66–68]. Oligonucleotides have been targeted with non-nucleotide building blocks and selected metal–ligand complexes that bind DNA with high affinity and selectivity [23, 68]. In this example, Zn^{2+} was used in conjunction with a library of non-nucleotide imines and probed against binding to immobilized duplex DNA:oligo(dA) bound to solid-phase poly(dT). A pool of up to 36 different bidentate Zn(II) complexes with potential affinity to the DNA double helix were generated under physiological conditions (Fig. 16.11). The immobilized target was used to select zinc complexes from the equilibrating library. Two of the salicylaldimines were active, one of which showed a higher binding to the double-stranded polynucleotide than all of the other library constituents.

Modified oligonucleotides play an important rule in understanding the molecular recognition between oligonucleotide ligands and nucleic acid targets in molecular biology and biotechnology. Natural oligonucleotides suffer from several drawbacks – like degradation by nuclease and poor bioavailability – that prevent

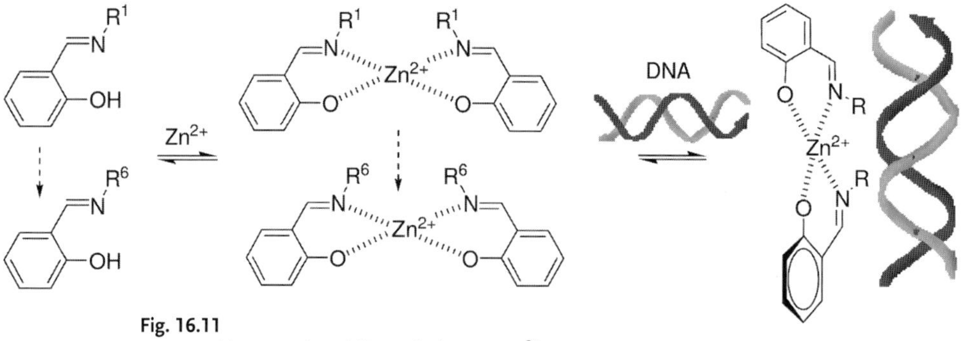

Fig. 16.11
Dynamic libraries of 36 different bidentate Zn^{2+} complexes generated from six starting elements (salicylaldimine) interacting with duplex DNA.

their development toward therapeutically useful drugs. Therefore, modified oligonucleotides with high target affinity and enhanced bioavailability have been the goal of many investigations. However, the generation and screening of large arrays of oligonucleotides for the identification of new ligands is not a trivial task. Combinatorial approaches have been tried to find new ligands [69, 70]. Recent reports described the use of dynamic combinatorial chemistry for the identification of modified oligonucleotides that stabilize nucleic acid complexes [71, 72]. A small dynamic library was generated by the reaction of an oligonucleotide ligand bearing a reactive amino group with a set of three aldehydes under physiological conditions (Fig. 16.12). The addition of a nucleic acid target shifted the equilibrium towards the formation of the strongest binder. The equilibrated library was subsequently selectively reduced with sodium cyanoborohydride to form stable secondary amines, thus allowing isolation and analysis of the library. A DNA duplex and a tertiary-structured RNA–RNA complex were successfully targeted. In both cases, a chemically stable conjugated oligonucleotide ligand with increased affinity for the target was identified. For verification of the amplification effect and to exclude non-specific effects, nucleic acid targets were replaced by modified nucleic acids (non-self-complementary DNA or a mutated target MiniTar 3A, respectively). No significant amplification of products was observed in these control experiments.

In another example, a four-stranded G-quadruplex formed by DNA sequences with stretches of G-nucleotides was used as molecular target [73]. For the screening of quadruplex stabilizing ligands, a dynamic library of six disulfides was generated out of three thiol-containing fragments. The thiol exchange operates under mild conditions with high exchange rates. After reaching the equilibrium, the exchange was stopped by lowering the pH from pH 7.5 to pH 2.0. The quadruplex target with bound ligands was isolated and subsequently heat-denatured to release all bound ligands, which then were analyzed and quantified by HPLC. Results were compared with the experiment in the absence of target. One hetero-disulfide and one homodisulfide containing a small peptide fragment were specifi-

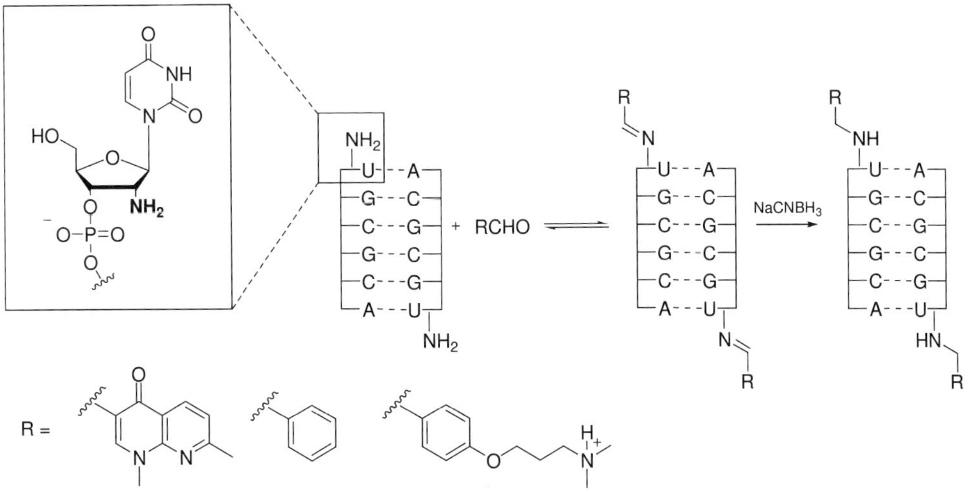

Fig. 16.12
Dynamic library of conjugated duplexes generated from self-complementary oligonucleotides containing reactive amino functionalities and three different aldehydes.

cally selected from the dynamic library with a good quadruplex binding affinity in the low micromolar range.

16.6
Summary and Outlook

Compared to traditional combinatorial chemistry, dynamic combinatorial chemistry (DCC) is a young but fast-developing technology featuring simultaneous generation of molecular diversity and self-screening of the continuously equilibrating library by direct interaction with a target. As a tool for drug discovery, DCC enables the rapid generation of adaptive libraries, capable of responding to external selection pressure by evolving all latent, virtual library constituents. The methodology has been established and implemented with different classes of biological molecules using various linker chemistries, targeting for example enzymes, receptor proteins and polynucleotides.

The potential of a novel methodology is best demonstrated when novel features are revealed before independent confirmation. Such is indeed the case for instance for the acetylcholine inhibitor shown in Fig. 16.7 [25]. Indeed, recent work has uncovered the presence of a binding pocket in the middle of the gorge of the enzyme, a fact in line with the strong activity of the bis-pyridinium inhibitor containing a disubstituted para-phenylene spacer [74]. The field of DCC has been the subject of great interest and activity since the formulation of the concept, as shown by the development of new approaches, e.g. dynamic combinatorial X-ray crystallography [60] or pseudo-dynamic chemistry [49], application of new reversible reactions for DCL generation [44, 45] and application to materials science, as for instance in the case of dynamic polymers–dynamers [75–77] and the formation of G-quartet-based hydrogels [78].

However, a number of challenges remain to be addressed to advance the technique and make DCC a competitive drug discovery technology. Of particular value would be new and controllable reversible chemistries, compatible with physiological conditions and inert to sensitive biological molecules. A toolbox of ready to use linker chemistries with diverse building blocks is desired, so as to cover a large diversity space in order to target different kinds of biological molecules. Actually, most reversible reactions generate extended structures due to the size of the linker group itself. Thus, receptors with extended binding sides are appropriate targets for DCLs. Particularly attractive and challenging in this respect are the protein–protein interactions in biological communication and regulation processes. For targeting biological molecules comprising active or receptor sites of limited size, the available connections capable of forming dynamic libraries with suitably compact ligands are limited.

In addition, in most DCL protocols, the target protein is used in stoichiometric amounts. The need for relative large amounts of sensitive proteins may be overcome by miniaturization of the screening process and by implementation of new analytical techniques. These are in fact needed for the rapid analysis and precise quantification of amplified constituents within extended libraries, which are essential in order to be competitive with traditional combinatorial chemistry and high-throughput screening. In contrast, the combination of DCC with robotics provides powerful means for the efficient exploration of dynamic diversity.

The concept of DCC offers new perspectives for ligand identification in drug discovery. However, it is by no means limited to drug discovery processes. Dynamic combinatorial chemistry may be and has been extended to the field of material science for the generation of dynamic materials like dynamic combinatorial polymers [75–77, 79] and condensed phases [78]. In fact, it represents just the reversible covalent domain of the global field of *constitutional dynamic chemistry* (CDC) [79] that encompasses both supramolecular entities, constitutionally dynamic by nature due to the lability of non-covalent interactions, and molecular entities, made constitutionally dynamic by design through the introduction of reversible covalent connections. CDC introduces a paradigm change with respect to constitutionally static chemistry and enables the emergence of adaptive chemistry [80, 81].

References

1 Terrett, N. K. **1998**, *Combinatorial Chemistry*, Oxford University Press, Oxford.
2 Geysen, H. M., Meloen, R. H., Barteling, S. J. **1984**, Use of peptide synthesis to probe viral antigens for epitopes to a resolution of a single amino acid. *Proc. Natl. Acad. Sci. USA* 81, 3998–4002.
3 Furka, A., Sebestyen, F., Asgedom, M., Dibo, G. **1991**, General method for rapid synthesis of multicomponent peptide mixtures. *Int. J. Pept. Protein Res.* 37, 487–493.
4 Houghten, R. A., Pinilla, C., Blondelle, S. E., Appel, J. R., Dooley, C. T., Cuervo, J. H. **1991**, Generation and use of synthetic peptide combinatorial libraries for basic research and drug discovery. *Nature* 354, 84–86.
5 Lam, K. S., Salmon, S. E., Hersh, E. M., Hruby, V. J., Kazmierski, W. M., Knapp, R. J. **1991**, A new type of synthetic peptide library for identifying ligand-binding activity. *Nature* 354, 82–84.
6 Xiang, X.-D. **1999**, Combinatorial materials synthesis and screening: an integrated materials chip approach to discovery and optimization of functional materials. *Annu. Rev. Mater. Sci.* 29, 149.
7 Brocchini, S. **2001**, Combinatorial chemistry and biomedical polymer development. *Adv. Drug. Deliv. Rev.* 53, 123–130.
8 Jung G. **1999**, *Combinatorial Chemistry*, VCH-Verlagsgesellschaft, Weinheim.
9 Bunin B. **1998**, *The Combinatorial Index*, Academic Press, San Diego.
10 Ganesan, A. **1998**, Strategies for the dynamic integration of combinatorial synthesis and screening. *Angew. Chem. Int. Ed.* 37, 2828–2831.
11 Lehn, J. M. **1999**, Dynamic combinatorial chemistry and virtual combinatorial libraries. *Chem. Eur. J.* 5, 2455–2463.
12 Timmerman P., Reinhoudt, D. N. A. **1999**, combinatorial approach to synthetic receptors. *Adv. Mater.* 11, 71–74.
13 Klekota, B., Miller, B. L. **1999**, Dynamic diversity and small-molecule evolution: a new paradigm for ligand identification. *Trends Biotechnol.* 17, 205–209.
14 Cousins, G. R. L., Poulsen, S. A., Sanders, J. K. M. **2000**, Molecular evolution: dynamic combinatorial libraries, autocatalytic networks and the quest for molecular function. *Curr. Opin. Chem. Biol.* 4, 270–279.
15 Lehn, J. M. **2001**, Dynamic combinatorial chemistry and virtual combinatorial libraries. *Essays Contemp. Chem.* 2001, 307–326.
16 Lehn, J. M., Eliseev, A. V. **2001**, Dynamic combinatorial chemistry. *Science* 291, 2331–2332.
17 Huc, I., Nguyen, R. **2001**, Dynamic combinatorial chemistry. *Comb. Chem. High Throughput Screen.* 4, 53–74.
18 Ramström, O., Lehn, J.-M. **2002**, Drug discovery by dynamic combinatorial libraries. *Nat. Rev. Drug Discov.* 1, 26–36.
19 Ramström, O., Bunyapaiboonsri, T., Lohmann, S., Lehn, J.-M. **2002**, Chemical biology of dynamic combinatorial libraries. *Biochim. Biophys. Acta* 1572, 178–186.
20 Otto, S., Furlan, R. L. E., Sanders, J. K. M. **2002**, Dynamic combinatorial chemistry. *Drug Discov. Today* 7, 117–125.
21 Lehn, J.-M. **1995**, *Supramolecular Chemistry – Concept and perspectives*, VCH, Weinheim.
22 Huc, I., Lehn, J.-M. **1997**, Virtual combinatorial libraries: dynamic generation of molecular and supramolecular diversity by self-assembly. *Proc. Natl. Acad. Sci. USA* 94, 2106–2110.
23 Klekota, B., Hammond, M. H., Miller, B. **1997**, Generation of novel DNA-binding compounds by selection and amplification from self-assembled combinatorial libraries. *Tetrahedron Lett.* 38, 8639–8642.
24 Cousins, G. R. L., Poulsen, S. A., Sanders, J. K. M. **1999**, Dynamic combinatorial libraries of pseudo-peptide hydrazone macrocycles. *Chem. Commun.* 1999, 1575–1576.
25 Bunyapaiboonsri, T., Ramstrom, O., Lohmann, S., Lehn, J. M., Peng, L., Goeldner, M. **2001**, Dynamic deconvolution of a pre-equilibrated dynamic combinatorial library of acetylcholinesterase inhibitors. *ChemBioChem* 2, 438–444.

26 Hochgürtel, M., Kroth, H., Piecha, D., Hofmann, M. W., Nicolau, C., Krause, S., Schaaf, O., Sonnenmoser, G., Eliseev, A. V. **2002**, Target-induced formation of neuraminidase inhibitors from in vitro virtual combinatorial libraries. *Proc. Natl. Acad. Sci. USA* 99, 3382–3387.

27 Hochgürtel, M., Biesinger, R., Kroth, H., Piecha, D., Hofmann, M. W., Krause, S., Schaaf, O., Nicolau, C., Eliseev, A. V. **2003**, Ketones as Building Blocks for Dynamic Combinatorial Libraries: Highly Active Neuraminidase Inhibitors Generated via Selection Pressure of the Biological Target. *J. Med. Chem.* 46, 356–358.

28 Thomas, N. R., Quéléver, G., Walsh, A. **2001**, *unpublished data*.

29 Borman, S. **2001**, Combinatorial Chemistry. *Chem. Eng. News* 79, 49–63.

30 Ramström, O., Lehn, J.-M. **2000**, In situ generation and screening of a dynamic combinatorial carbohydrate library against concanavalin A. *Chem. Bio. Chem*, 1, 41–48.

31 Erlanson, D. A., Lam, J. W., Wiesmann, C., Luong, T. N., Simmons, R. L., DeLano, W. L., Choong, I. C., Burdett, M. T., Flanagan, W. M., Lee, D., Gordon, E. M., O'Brien, T. **2003**, In situ assembly of enzyme inhibitors using extended tethering. *Nat. Biotechmol.* 21, 308–314.

32 Otto, S., Furlan, R. L. E., Sanders, J. K. M. **2000**, Dynamic Combinatorial Libraries of Macrocyclic Disulfides in Water. *J. Am. Chem. Soc.* 122, 12063–12064.

33 Nicolaou, K. C., Hughes, R., Pfefferkorn, J. A., Barluenga, S. **2001**, Optimization and mechanistic studies of psammaplin A type antibacterial agents active against methicillin-resistant *Staphylococcus aureus* (MRSA). *Chem. Eur. J.* 7, 4296–4310.

34 Swann, P. G., Casanova, R. A., Desai, A., Frauenhoff, M. M., Urbancic, M., Slomczynska, U., Hopfinger, A. J., Le Breton, G. C., Venton, D. L. **1996**, Nonspecific protease-catalyzed hydrolysis/synthesis of a mixture of peptides: product diversity and ligand amplification by a molecular trap. *Biopolymers* 40, 617–625.

35 Larsson, R., Pei, Z., Ramström, O. **2004**, Catalytic self-screening of cholinesterase substrates from a dynamic combinatorial thioester library. *Angew. Chem. Int. Ed.* 43, 3716–3718.

36 Lins, R. J., Flitsch, S. L., Turner, N. J., Irving, E., Brown, S. A. **2004**, Generation of a dynamic combinatorial library using sialic acid aldolase and in situ screening against wheat germ agglutinin. *Tetrahedron* 60, 771–780.

37 Lins, R. J., Flitsch, S. L., Turner, N. J., Irving, E., Brown, S. A. **2002**, Enzymatic generation and in situ screening of a dynamic combinatorial library of sialic acid analogues. *Angew. Chem. Int. Ed.* 41, 3405–3407.

38 Nicolaou, K. C., Hughes, R., Cho, S. Y., Winssinger, N., Smethurst, C., Labischinski, H., Endermann, R. **2000**, Target-Accelerated Combinatorial Synthesis and Discovery of Highly Potent Antibiotics Effective Against Vancomycin-Resistant Bacteria. *Angew. Chem. Int. Ed.* 39, 3823–3828.

39 Nicolaou, K. C., Cho, S. Y., Hughes, R., Winssinger, N., Smethurst, C., Labischinski, H., Endermann, R. **2001**, Solid- and solution-phase synthesis of vancomycin and vancomycin analogues with activity against vancomycin-resistant bacteria. *Chem. Eur. J.* 7, 3798–3823.

40 Katz, B. A., Finer-Moore, J., Mortezaei, R., Rich, D. H., Stroud, R. M. **1995**, Episelection: novel Ki approximately nanomolar inhibitors of serine proteases selected by binding or chemistry on an enzyme surface. *Biochemistry* 34, 8264–8280.

41 Godoy-Alcántar, C., Yatsimirsky, A. K., Lehn, J.-M. **2005**, Structure–stability correlations for imine formation in aqueous solution. *J. Phys. Org. Chem.* 18 (in press).

42 Polyakov, V. A., Nelen, M. I., Nazarpack-Kandlousy, N., Ryabov, A. D., Eliseev, A. V. **1999**, Imine exchange in O-aryl and O-alkyl oximes as a base reaction for aqueous "dynamic" combinatorial libraries. A kinetic and thermodynamic study. *J. Phys. Org. Chem.* 12, 357–363.

43 Nguyen, R., Huc, I. **2003**, Optimizing the reversibility of hydrazone formation for dynamic combinatorial chemistry. *Chem. Commun.* 2003, 942–943.

44 Shi, B., Greaney, M. F. **2005**, Reversible Michael addition of thiols as a new tool for dynamic combinatorial chemistry. *Chem. Commun.* 2005, 886–888.

45 Boul, P. J., Reutenauer, P., Lehn, J. M. **2005**, Reversible Diels-Alder Reactions for the Generation of Dynamic Combinatorial Libraries. *Org. Lett.* 7, 15–18.

46 Goral, V., Nelen, M. I., Eliseev, A. V., Lehn, J. M. **2001**, Double-level "orthogonal" dynamic combinatorial libraries on transition metal template. *Proc. Natl. Acad. Sci. USA* 98, 1347–1352.

47 Eliseev, A., Nelen, M. **1997**, Use of molecular recognition to drive chemical evolution. 1. Controlling the composition of an equilibrating mixture of simple arginine receptors. *J. Am. Chem. Soc.* 1997, 1147–1148.

48 Eliseev, A., Nelen, M. **1998**, Use of molecular recognition to drive chemical evolution: mechanism of an auomated genetic algorithm implementation. *Chem. Eur. J.* 4, 825–834.

49 Cheeseman, J. D., Corbett, A. D., Shu, R., Croteau, J., Gleason, J. L., Kazlauskas, R. J. **2002**, Amplification of Screening Sensitivity through Selective Destruction: Theory and Screening of a Library of Carbonic Anhydrase Inhibitors. *J. Am. Chem. Soc.* 124, 5692–5701.

50 Cheeseman, J. D., Corbett, A. D., Gleason, J. L., Kazlauskas, R. J. **2005**, Receptor-assisted combinatorial chemistry: thermodynamics and kinetics in drug discovery. *Chem. Eur. J.* 11, 1708–1716.

51 Reymond, J. L. **2004**, Outrunning the bear. *Angew. Chem. Int. Ed.* 43, 5577–5579.

52 Ro, S., Rowan, S. J., Pease, A. R., Cram, D. J., Stoddart, J. F. **2000**, Dynamic hemicarcerands and hemicarceplexes. *Org. Lett.* 2, 2411–2414.

53 Calama, M. C., et al. **1998**, Libraries of noncovalent hydrogenbonded assemblies – combinatorial synthesis of supramolecular systems. *Chem. Commun.* 1998, 1021–1022.

54 Cardullo, F., Calama, M. C., Snellink-Ruel, B. H. M., Weidmann, J.-L., Bielejewska, A., Timmerman, P., Reinhoudt, D. N., Fokkens, R., Nibbering, N. M. M. **2000**, Covalent capture of dynamic hydrogen-bonded assemblies. *Chem. Commun.* 2000, 367–368.

55 Star, A., Goldberg, I., Fuchs, B. **2000**, New supramolecular host systems. Part 12. Dioxadiazadecalin/salen tautomeric macrocycles and complexes: prototypical dynamic combinatorial virtual libraries. *Angew. Chem. Int. Ed.* 39, 2685–2689.

56 Brady, P. A., Sanders, J. K. M. **1997**, Thermodynamically-controlled cyclisation and interconversion of oligocholates: metal ion templated ‚living' macrolactonization. *J. Chem. Soc. Perkin Trans.* 1997, 3237–3254.

57 Rowan, S. J., Lukeman, P. S., Reynolds, D. J., Sanders, J. K. M. **1998**, Engineering diversity into dynamic combinatorial libraries by use of a small flexible building block. *New J. Chem.* 22, 1015–1018.

58 Zameo, S., Vauzeilles, B., Beau, J. M. **2005**, Dynamic combinatorial chemistry: Lysozyme selects an aromatic motif that mimics a carbohydrate residue. *Angew. Chem. Int. Ed.* 44, 965–969, S965–1.

59 Bunyapaiboonsri, T., Ramström, O., Ramstroem, O., Haiech, J., Lehn, J.-M. **2003**, Generation of Bis-Cationic Heterocyclic Inhibitors of Bacillus subtilis HPr Kinase/Phosphatase from a Ditopic Dynamic Combinatorial Library. *J. Med. Chem.* 46, 5803–5811.

60 Congreve, M. S., Davis, D. J., Devine, L., Granata, C., O'Reilly, M., Wyatt, P. G., Jhoti, H. **2003**, Detection of ligands from a dynamic combinatorial library by X-ray crystallography. *Angew. Chem. Int. Ed.* 42, 4479–4482.

61 Raimundo, B. C., Oslob, J. D., Braisted, A. C., Hyde, J., McDowell, R. S., Randal, M., Waal, N. D., Wilkinson, J., Yu, C. H., Arkin, M. R. **2004**, Integrating fragment assembly and biophysical methods in the chemical advancement of small-molecule antagonists of IL-2: an approach for inhibiting protein-protein interactions. *J. Med. Chem.* 47, 3111–3130.

62 Gabius, H. J., Andre, S., Kaltner, H., Siebert, H. C. **2002**, The sugar code: functional lectinomics. *Biochim. Biophys. Acta* 1572, 165–177.

63 Sakai, S., Shigemasa, Y., Sasaki, T. **1997**, A self-adjusting carbohydrate ligand for

GalNAc specific lectins. *Tetrahedron Lett.* 38, 8145–8148.
64 Sakai, S., Shigemasa, Y., Sasaki, T. **1999**, Iron(II)-assisted assembly of trivalent GalNAc clusters and their interactions with GalNAc-specific lectins. *Bull. Chem. Soc. Jpn* 1999, 72, 1313–1319.
65 Ramström, O., Lohmann, S., Bunyapaiboonsri, T., Lehn, J.-M. **2004**, Dynamic combinatorial carbohydrate libraries: probing the binding site of the Concanavalin A lectin. *Chem. Eur. J.* 10, 1711–1715.
66 Karan, C., Miller, B. L. **2000**, Dynamic diversity in drug discovery: putting small-molecule evolution to work. *Drug Discov. Today* 5, 67–75.
67 Karan, C., Miller, B. L. **2001**, RNA-Selective Coordination Complexes Identified via Dynamic Combinatorial Chemistry. *J. Am. Chem. Soc.* 123, 7455–7456.
68 Klekota, B., Miller, B. L. **1999**, Selection of DNA-binding compounds via multistage molecular evolution. *Tetrahedron* 55, 11687–11697.
69 Alam, M. R., Maeda, M., Sasaki, S. **2000**, DNA-binding peptides searched from the solid-phase combinatorial library with the use of the magnetic beads attaching the target duplex DNA. *Bioorg. Med. Chem.* 8, 465–473.
70 Guelev, V. M., Harting, M. T., Lokey, R. S., Iverson, B. L. **2000**, Altered sequence specificity identified from a library of DNA-binding small molecules. *Chem. Biol.* 7, 1–8.
71 Bugaut, A., Bathany, K., Schmitter, J. M., Rayner, B. **2005**, Target-induced selection of ligands from a dynamic combinatorial library of mono- and bi-conjugated oligonucleotides. *Tetrahedron Lett.* 46, 687–690.
72 Bugaut, A., Toulime, J. J., Rayner, B. **2004**, Use of dynamic combinatorial chemistry for the identification of covalently appended residues that stabilize oligonucleotide complexes. *Angew. Chem. Int. Ed.* 43, 3144–3147.
73 Whitney, A. M., Ladame, S., Balasubramanian, S. **2004**, Templated ligand assembly by using G-quadruplex DNA and dynamic covalent chemistry. *Angew. Chem. Int. Ed.* 43, 1143–1146.
74 Campiani, G., Fattorusso, C., Butini, S., Gaeta, A., Agnusdei, M., Gemma, S., Persico, M., Catalanotti, B., Savini, L., Nacci, V., Novellino, E., Holloway, H. W., Greig, N. H., Belinskaya, T., Fedorko, J. M., Saxena, A. **2005**, Development of molecular probes for the identification of extra interaction sites in the mid-gorge and peripheral sites of butyrylcholinesterase (BuChE). Rational design of novel, selective, and highly potent BuChE inhibitors. *J. Med. Chem.* 48, 1919–1929.
75 Skene, W. G., Lehn, J.-M. **2004**, Dynamers: polyacylhydrazone reversible covalent polymers, component exchange, and constitutional diversity. *Proc. Natl. Acad. Sci. USA* 101, 8270–8275.
76 Giuseppone, N., Lehn, J.-M. **2004**, Constitutional dynamic self-sensing in a zinc(II)/polyiminofluorenes system. *J. Am. Chem. Soc.* 126, 11448–11449.
77 Ono, T., Nobori, T., Lehn, J.-M. **2005**, Dynamic polymer blends – component recombination between neat dynamic covalent polymers at room temperature. *Chem. Commun.* 2005, 1522–1524.
78 Sreenivasachary, N., Lehn, J.-M. **2005**, Gelation-driven component selection in the generation of constitutional dynamic hydrogels based on guanine-quartet formation. *Proc. Natl. Acad. Sci. USA* 102, 5938–5943.
79 Lehn, J. M. **2000**, *Supramolecular Polymer Chemistry – Scope And Perspectives*, Marcel Dekker, New York, pp 615–641.
80 Lehn, J. M. **1999**, *Supramolecular Science: Where it is and Where it is Going*, Kluwer Academic, Dordrecht, pp 273–286.
81 Lehn, J. M. **2002**, Toward complex matter: supramolecular chemistry and self-organization. *Proc. Natl. Acad. Sci. USA* 99, 4763–4768.

Index

a
acetylcholine esterase   39, 325
adenosine kinase   184
adenylosuccinate synthetase   60
adipocyte lipid-binding protein   169
affinity   13, 20
– definition   13 f.
alanine scanning   69
allosteric binding site   162 f., 300
– caspase-3   300
ancillary binding sites   162 f.
antibody   40 ff.
avidin   60
avidity   12 f., 20, 22 f.
– definition   13 f.
– effective concentration   22 f.

b
Barrier, rigid body   55 ff., 61 f.
BCR-Abl   222
binding affinity, intrinsic   56 ff.
binding modes, multiple   243
binding site   67, 78, 83
– location of   67
– mapping   67 ff., 83
– – computational methods   83
– water   78
binding to p38α   207
biochemical screening   231
bisubstrate inhibitor   39

c
C5 a receptor   303
carbonic anhydrase   39, 275, 309, 327, 350, 353
– pseudo-dynamic libraries   353
caspase-1   299
caspase-3   296 ff., 300
– allosteric site   300
CDK2   205

chelate effect   18, 21, 29, 150
– cooperativity   29
chemical graphs   92
chemical space   92
CHO interactions   205
chymotrypsin   72
classification   158
click chemistry   313 ff., 328 f., 325 ff., 328 ff.
– bioconjugation   328, 338 f.
– – 1,3-dipolar cycloaddition   332
– – DNA   332
– – for proteomics   330
– – tagging in living cells   328
– in drug discovery   316 ff., 320
– HIV protease inhibitors   321
– – library generation   316
– – natural products derivatives   317
– – synthesis of neoglycoconjugates   320
– *in situ*   325, 327
– – acetylcholine esterase   325
– – carbonic anhydrase   327
– – reactions   324 ff.
– – 1,3-dipolar cycloaddition   315 ff., 326
– – water-based   315, 326, 332
clinical candidates   219
– for approval   219
– probability   219
combinatorial chemistry, see also dynamic combinatorial chemistry   341
combinatorial libraries   341
– generation   341
– – split and mix technology   341
complexity, molecular   5
cooperativity   24 ff.

d
dendrimers   36, 40
dihydrodipicolinate reductase   171

disulfide bond   286, 290, 324
– conjugation energy   286
– for glycoarrays   324
– reducing agent   290
– thermodynamics   286
diversity, chemical   4 f.
dock   200
docking   125 ff., 200, 242
– for fragment optimization   242
– for library design   200
druggability   163
drug-like properties   95 f.
drugs   113 ff., 197
– fragments present in   113
– marketed injectable   115
– marketed oral   115
dynamic combinatorial chemistry   342 ff., 346 f., 350 ff., 356 f.
– application to biological systems   347 ff., 350, 356
– – acetylcholinesterase   352
– – carbonic anhydrase   350
– – concanvalin A   356
– – lysozyme   352
– – neuraminidase   351
– – oligonucleotides   357
– – wheat germ aglutidinin   357
– dynamical combinatorial X-ray crystallography (DCX)   353
– – CDK2   353
– methodologies   346
dynamic combinatorial library   342 ff., 347, 354
– building blocks   343 ff.
– extended tethering   354
– pseudo-dynamic libraries   347
– reactions   343 ff.
– receptor molding   343
– self-screening process   342
– substrate casting   343

e

elastase   72 ff., 78 ff.
– binding pockets   74
– crystal structures of   72 f.
– – dielectric constant   73
– inhibitors   75
– internal water   79
– plasticity   75
– surface waters   80
– water molecule distribution   78
enthalpy   19, 21 f.
enthalpy/entropy compensation   21, 26, 32 f.

entropic barrier to binding   55
entropy   19, 21, 55 ff.
– consequences of Linking Ligands   55
– rotatable bond   58
entropy, conformational   25
entropy, rigid body   151
Erm-AM   188
ESI-MS   267 ff., 271, 275
– affinity chromatography   271 ff.
– as a screening technique   257 ff., 275
– basic parameters   275 f.
– buffer conditions   275
– control experiments   271
– detected interactions   271
– ionization conditions   275
– signal abundances   275

f

factor VIIa   228 ff.
– exosite   229 f.
– fragment screen   229
FK506 binding protein, FKBP   78, 184
FKBP   163
fragment   219 ff.
fragment assembly, see fragment linking   167, 171
– NMR-guided   167
– NMR SOLVE   171
fragment elaboration   181, 187 f., 190
– energetic analysis   188
– iterative library design   188, 190
– potency gains   190
fragment engineering, see fragment elaboration   240
fragment library   89, 91, 96 ff., 99, 113, 152 ff., 156, 158, 160, 195 ff., 199, 201, 217 ff., 222
– ADMET properties   153
– bromine atoms   217, 219, 222
– characterization   152
– cheminformatics   91
– cofactors   160
– design   89 ff., 152
– drug fragment library   197
– filtering of   153
– for X-ray crystallography   195, 217 f.
– fragments of marketed drugs   113 ff.
– intellectual property   92
– kinase-library   199
– lead-like   98
– MCSS   125
– phosphatase library   201
– practical example   99
– privileged fragment library   197

- purity 152
- quality control 201
- REOS 153
- rule-of-five 94
- SAR by NMR 156, 158
- SHAPES 154
- solubility 156
- synthetic handles 217
- targeted libraries 197
- WOMBAT 99
fragment library, linking 149
fragment linking 57, 59, 150, 168, 170, 173, 181 ff., 187, 240, 257 ff., 277, 293, 297
- combinatorial advantage 178
- energetic analysis 183
- expected potency 187
- experimental examples 59
- interligand NOE 170
- NMR ACE 173
- Potency Gains 181
- SAR by MS 277
- SAR by NMR 168, 180
- spin label 170
- Tethering 293, 297
- – with Extenders 297
- theoretic treatment 57, 150 f.
fragment optimization 241, 243 f.
- activity increase 244
- combinatorial libraries 243
- increase in activity 244
- linear libraries 241
fucosyltranferase 323
functional genomics 158 ff.

g
glycogen phosphorylase 61
GOLD 200

h
HCV protease 184
hemagglutination 28, 40
high-throughput screening 3, 64, 215, 341
- low-quality interactions 64
hit rate 5, 107, 219, 221
hot spots 69, 193, 293
human kinome 224

i
interleukin-2 293 ff.
- dynamic suffrage 294
- Tethering 295
- X-ray structures of 295
intrinsic binding affinity 56
Isothermal Titration Calorimetry (ITC) 26 f.

j
Jencks, William 6, 149 f.
Jnk3 169

k
kinome, human 224

l
lead-like properties 94, 215, 218
LFA 184
library 219 ff.
- diversity 220
ligand efficiency 6, 62 f., 206
- intrinsic 63
ligand, bivalent 13
ligand, oligovalent 12 f., 33
- scaffolds 33 f.
ligand, polyvalent 15, 35, 37
- scaffolds 35 f.
- synthesis 37 f.
linear library 241
linker 25, 36 f., 288, 291
- binding mode 291
- conformational entropy 25, 36
- influence on 291
- Tethering 288, 291
- – entropic costs 288
linking efficiency 243
lysozyme 78

m
Mass Spectrometry, see ESI-MS 250
Mass Spectrometry, see Tethering 285
MCSS, see multiple copy simultaneous search 168
MMP3 61
multi-copy simultaneous search 78, 125 ff., 132
- comparison with experiment 137 f.
- comparison with GRID 135 ff.
- functional groups 128
- functionally maps 132 ff.
- interaction energies 127
- ligand design 138 ff.
- method 137
- multiple solvent crystal structure 137 f.
- NMR 138
- protein flexibility 141
- topology file 127
Multiple Solvent Crystal Structure Method 67 ff.
multivalency 11 ff., 15 ff., 19, 24 f., 26, 32, 39 f.
- additivity 19

- antibodies 40
- cooperativity 24
- definitions 12
- design rules 32
- enhancement 15
- experiment studies 16
- experimental studies 26 ff.
- hetero-bivalent systems 39
- hetero-multivalency 11
- homo-multivalency 11
- lead discovery 39
- mechanism for dissociation 18
- thermodynamics 19 ff.

MurF 261
- open conformation 261
- closed conformation 261

n

NMR 159 ff., 167, 249 ff., 261
- and X-ray crystallography 249 ff.
- hit triaging by 251, 261
- in the drug discovery process 250
- SAR by NMR 149, 167

NMR screening 208
- ligand-based 208
- - STD 208
- - Water-LOGSY 208

NMR-based screening 163 ff., 167, 169 f., 253
- ligand-based 163 ff., 167
- measurmat by NMR 164 f.
- - K_d values 164
- - linewidths 164
- - STD 164
- - transferred NOE 164
- - WATERLOGSY 164
- paramagnetic probes 169
- protein-based 165, 253
- - TINS 169
- PTP 1B 253
- SLAPSTIC 170
- spin labels 170

non-classic hydrogen bonds 205
Nuclear Magnetic Resonance, see NMR 7

o

oral drugs 113
- structural fragments 113

p

p38 MAP kinase 174, 207
p38α 174
Pak4 235 ff.
paramagnetic probes 169

peptide deformylase 78
probe molecules 67 ff., 71, 74
- binding sites 71
- clustering 74

protein classification 158 ff.
protein tyrosin phosphoton 18, PTP-18 184
protein tyrosine phosphatase 1B 251 ff., 257 ff., 292 f.
- core replacement 259
- crystallography 251 ff.
- from 259
- HTS 252
- inhibitors 257 ff.
- monoacid replacements 258
- NMR 251 ff.
- open conformation 293
- open form 293
- polar surface area 258
- potency and selectivity 257
- Tethering 292

PTP-1B, see protein tyrosine phosphatase 1B 253, 256
- inhibitors 253
- noncatalytic 256
- open form 253
- site 256
- - NMR screening 256
- states of oxidation 253
- - NMR screening 256

r

RECAP 113 f., 153
recognition pocket 64
ribonuclease A 71
rigid body 55 ff.
rigid body barrier 61
rigid body barrier to binding 55
rigid body entropy 55
RNA 276 ff.
rotatable bonds 58
- entropic penalty 58
rule of 3 96 ff., 153, 196, 219
- commercially available compounds 153
rule of 5 153, 196, 218

s

SAR 63
- hypersensitive 63
SAR by MS 277
SAR by NMR 7, 149, 151, 156, 167, 181 ff.
Scaffold for 33
self-assembled monolayer 18
SHAPES 154, 169

SMILES 93
solvent mapping 68, 77
solvent, organic 68
solvent, organic probe molecule 73
– binding sites 73
stromelysin 60, 184 f.
structure-activity relationship 63
structure-based drug design 4
subtilisin 72
Surface Plasmon Resonance Spectroscopy 27
Syk 234 f.
– Pak4 as surrogate 235
synchrotron 203, 217

t
Tar RNA 276
telavancin 39
Tethering 285 ff., 289, 292 f., 297, 303, 305 ff., 354
– advantages and disadvantages 289
– general process 285 f.
– GPCRs 303
– protein tyrosine phosphatase 1B 292
– protein-protein interactions 293
– related approaches 306
– thymidylate synthase 289 ff.
– using functional screening 305
– – disulfide formation 306
– – imine formation 307
– – metal-mediated 307
– with extenders 289, 292, 297, 354
– – dynamic combinatorial libraries 354
– – for fragment linking 297
thermolysin 78
– crystal structures 78
– solvent interaction sites 78
thrombin 207

thymidylate synthase 289 ff.
toxins, AB_5 18, 29
tryptase 60

v
vancomycin 17, 28, 60, 272
– conformational entropy 28

w
water binding sites 78 ff., 81 f.
– conservation of 81
– general properties of 82
WDI 95
WOMBAT 96
World Drug Index 93, 95

x
X-ray crystallography 194 f., 201 ff., 207, 215 ff., 222, 226, 233 f., 249 ff., 262
– and NMR 207 ff., 249 ff.
– as a screening technique 194 ff., 216 ff., 222, 226, 233 f.
– – concentrations 226
– – fragment optimization 234
– – selection of hits 233
– co-crystallization 262
– Data Collection 202
– Data Processing 203
– flow chart 222 ff.
– fragment screening 201 f., 217, 262
– – co-crystallization 201
– – soaking 201, 217
– in the drug discovery process 250
– NMR pre-screening 262
– structural genomics 249
– structure solution 204
– tpdel
– – flow-chart 204